James Dyson was born in Norfolk in 1947 and studied at the Royal College [illegible] Art in [illegible]
the S[illegible]uck, a high-speed flatbottomed boat, with Jeremy [illegible] known for his revolutionary cyclonic vacuum cleaner, his products are disruptive market leaders around the globe. He believes that it is engineers who will solve the world's problems and he helps them to do so through the Dyson Institute of Engineering and Technology, The James Dyson Foundation and the annual James Dyson Award.

Praise for *Invention*

'I left [this book] enlightened about engineering and design, and impressed by Dyson's dogged, trailblazing spirit.' **Oliver Shah, *Sunday Times***

'The ferociously creative entrepreneur who emerges in these pages revels in practical experiment and learns from failure... Dyson's struggles against the advice of experts and in the teeth of endless commercial and technical challenges haven't lost their power to fascinate... Those who persist with *Invention* despite its flaws will be rewarded.' **Marc Sidwell, *Sunday Telegraph***

'Stories of consumer products and gadgets drive Mr Dyson's narrative forward, but parallel... [he] tells a story of the struggles of entrepreneurship. A final thread... is his passionate case for more and better engineering and science education in British schools.' **Henry Petroski, *Wall Street Journal***

INVENTION

A Life Through Failure

JAMES DYSON

SIMON &
SCHUSTER

London · New York · Sydney · Toronto · New Delhi

First published in Great Britain by Simon & Schuster UK Ltd, 2021
This edition published in Great Britain by Simon & Schuster UK Ltd, 2022

1 3 5 7 9 10 8 6 4 2

Simon & Schuster UK Ltd
1st Floor
222 Gray's Inn Road
London WC1X 8HB

www.simonandschuster.co.uk
www.simonandschuster.com.au
www.simonandschuster.co.in

Simon & Schuster Australia, Sydney
Simon & Schuster India, New Delhi

A CIP catalogue record for this book is available from the British Library

Paperback ISBN: 978-1-4711-9877-9
eBook ISBN: 978-1-4711-9876-2

Typeset in Minion 3 and Dyson Futura by Anna Green at Siulen Design

Printed in the UK by CPI Group (UK) Ltd, Croydon, CR0 4YY

This book is dedicated to Deirdre, without whose love, encouragement, advice, tolerance and forbearance, none of this would ever have happened. Also, to Emily, Jake and Sam and our wonderful grandchildren, who make the closest of families and whose enthusiasm for creating things flourishes so strongly.

Contents

INTRODUCTION		9
CHAPTER ONE	Growing Up	13
CHAPTER TWO	Art School	28
CHAPTER THREE	Sea Truck	47
CHAPTER FOUR	The Ballbarrow	74
CHAPTER FIVE	The Coach House	91
CHAPTER SIX	DC01	117
CHAPTER SEVEN	Core Technologies	144
CHAPTER EIGHT	Going Global	194
CHAPTER NINE	The Car	223
CHAPTER TEN	Farming	256
CHAPTER ELEVEN	Education	274
CHAPTER TWELVE	Making the Future	307
POSTSCRIPT	*Jake Dyson*	319
POSTSCRIPT	*Deirdre Dyson*	325
AFTERWORD		331
ACKNOWLEDGEMENTS		347
IMAGE CREDITS		349
INDEX		350

Introduction

In 1983, after four years of building and testing 5,127 hand-made prototypes of my cyclonic vacuum, I finally cracked it. Perhaps I should have punched the air, whooped loudly and run down the road from my workshop shrieking 'Eureka!' at the top of my voice. Instead, far from feeling elated, which surely after 5,126 failures I should have been, I felt strangely deflated.

How could this have been? The answer lies in failure. Day after day, with the wolf at the door, I had been pursuing the development of an ever more efficient cyclone for collecting and separating dust from a flow of air. I built several cyclones each day, conducting tests on each one to evaluate its effectiveness in collecting dust as fine as 0.5 microns – the width of a human hair is between 50 and 100 microns – while using as little energy as possible.

This might sound boring and tedious to the outsider. I get that. But when you have set yourself an objective that, if reached, might pioneer a better solution to existing technologies and products, you become engaged, hooked and even one-track-minded.

Folklore depicts invention as a flash of brilliance. That eureka moment! But it rarely is, I'm afraid. It is more about failure than ultimate success. I even thought of calling this book *James Dyson: Failure*, but was talked out of it because it might give the wrong impression.

People want to read about success. Funnily enough, engineers who are good at inventing things are never satisfied with their latest creation. They tend to look at it quizzically and say, 'I now know how to make it better', which is a wonderful opportunity! This is the start of their re-invention, marking another leap in performance.

And yet, if would-be young inventors could see that pioneering a cure for Alzheimer's, for example, is not about eureka flashes of Archimedean brilliance but more about an intelligent pursuit of diligent research, they might be less discouraged by the notion that brilliance is a prerequisite for their research. Research is about conducting experiments, accepting and even enjoying failures, but going on and on, following a theory garnered from observing the science. Invention is often more about endurance and patient observation than brainwaves.

I wanted to write my story now, just as the first cohort of Dyson students graduate at their convocation from the Dyson University. It caused me to reflect on what I felt on the same occasion fifty-two years ago when I graduated from the Royal College of Art, as well as what has happened to me since. It is a story told through a life of creating and developing things, as well as expressing a call to arms for young people to become engineers, creating solutions to our current and future problems.

My tale is one of not being brilliant. I wasn't even trained as an engineer or scientist. I did, however, have the bloody-mindedness not to follow convention, to challenge experts and to ignore Doubting Thomases. I am also someone who is prepared to slog through prototype after prototype searching for the breakthrough. If a slow starter like me could succeed, surely this might encourage others. I remember setting out with my degree, excited that I would be making a product that I had helped design but not having a clue about volume manufacturing or marketing. I was naïve and eager to learn all that as I progressed. It was a steep learning curve, though, and still is. Everything changes all the time, so experience is of little use. I didn't know that at the time and assumed that it would become easier with experience. It must be encouraging for those just graduating to know that is not the case. Fifty-two years later I can assure my graduates that I am no better for my experience.

The big change for me is that I now have around me an extraordinarily talented team, hell-bent on creating new technology and delivering it to our millions of customers throughout the world. We share the same belief in pioneering our own way, with the same determination to overcome the difficulties. More than anything their loyalty and dedication is what has built Dyson into a global technology company. Working with them is as inspirational as it gets.

The vacuum story began forty-two years ago. Day after day I crossed my yard at home to a small coach house to continue my quest to develop a cyclonic system for separating dust without a clogging bag. The early days were spent, as with most research and development, testing prototypes with different parameters in order to discover certain truths about dimensions and shapes. This is the essential groundwork of learning the art, before starting the experimental work to make the quantum leap. I looked forward to going to work, hoping that maybe today I would discover something new and inch my way forwards.

The failures began to excite me. 'Wait a minute, that should have worked, now why didn't it?' I was scratching my head, mystified, but then had another idea for an experiment that might lead to solving the problem. I was usually covered in dust, getting deeper and deeper into debt, yet happy and absorbed. Fortunately, my wife, Deirdre, allowed me to put our house and home life at risk, while the bank was kind enough to lend us money. She and our children never expressed doubt about what I was doing every day. They offered encouragement, love and understanding. Without that I would have given up. The same is true of every one of our friends. They must have thought I was mad and wasting my time, leading my family into penury. They never said so. Instead, they supported us and gave the unstinting encouragement without which I also doubt I could have lasted the course. They are true and close friends.

The reason I think invention is so very important today is that young people are passionate about saving the planet, improving the environment and finding cures for life-threatening diseases. I happen to believe that all these problems can be solved by the diligent application of research and development. I would love to see more school children and university students motivated to becoming engineers and scientists to make the breakthroughs that they so want to happen. We should be encouraging the young to become doers, rather than virtue signallers, to help them strive to solve the problems of their age while looking forward to a better future.

During my career, I have tried to seek out those young people who can make the world a better place. I have seen what miracles they can achieve. This book is aimed at encouraging them. Some may well become heirs to my heroes – inventors, engineers and designers – who make their appearance in these pages. Like them, they will not find it easy and they will need oodles of determination and stamina along the way. They will have to run and run hard, which is how my life story began . . .

CHAPTER ONE

Growing Up

There are moments – long moments – when the sea, sky and sands of north Norfolk beaches coalesce to form a seemingly infinite horizon. And when the early tide washes in and the ground beneath your feet reflects the big skies like a giant mirror, it can feel as if you are running through some ethereal space free of visible boundaries or restraint.

The first thing I knew I was good at, and something that I had taught myself as a teenage schoolboy, was long-distance running. Once through the pain barrier, I found I had the determination, or sheer bloody mindedness, to keep on running. Running, early in the morning or late at night, through that hauntingly beautiful landscape proved to be more than a ritual challenge. It was an escape from school, allowing me to think that anything and everything was possible.

Not that my thoughts were particularly focused. When I left school at eighteen, my headmaster, Logie Bruce-Lockhart, wrote to my mother saying, 'We shall be sorry to part with James. I cannot believe that he is not really quite intelligent, and I expect it will be brought out somehow somewhere.' And to me he wrote, 'The academic side, although we have to pretend it is important, matters comparatively little. You will do all the better for not having masses of tiresome degrees full of booklearning [sic] hanging around your neck. Good luck at the Art School.' At the time I enjoyed the double negative in the letter to my mother and hoped that something would be brought out but, like him, I had no clue where that might be. Thinking about it later, I thought how refreshing it was that a headmaster was quick to say that life isn't just about academic achievement.

Bruce-Lockhart, a kindly, lively and witty man, was a champion of the individual. A true countryman who loved music, birds and painting in watercolours, he had won five rugger caps for Scotland and served with the Household Cavalry during the thick of the action as his armoured car pushed into Germany in the fiercely fought ember days of the Second World War. He was to become a lifelong friend. I last saw Logie shortly before he died in 2020.

School was Gresham's in Holt, a handsome Norfolk market town, remote and largely car-free then, where my father was Head of Classics. Founded in the reign of 'Bloody' Mary, it had educated any number of highly individual and unorthodox young men – notably between the two world wars – who went on to both fame and infamy. Among them were the poets W. H. Auden and Stephen Spender, composer Benjamin Britten, artist Ben Nicholson and Christopher, his talented architect brother, killed in a gliding accident in 1948, and the notorious spy Donald Maclean. Lord Reith, who founded the BBC, had been at Gresham's, too.

There have been several celebrated engineers and inventors, among them Sir Martin Wood, who developed the whole-body superconducting magnets that led to MRI (magnetic resonance imaging), a key instrument in hospitals and laboratories worldwide today, and Sir Christopher Cockerell of hovercraft fame. Then there was Leslie Baynes, the aeronautical engineer who came up with the world's least powerful powered aircraft, the lightweight Carden-Baynes Auxiliary of 1935.

As a teenager in the early '60s, I had, though, taken my foot off the academic pedal. Not that I was lazy. Quite the opposite. I threw myself into almost any non-academic activity. All kinds of sport. Music, too. I chose to play the bassoon at the age of nine because I hadn't heard of it, because it was different and because it promised another challenge.

And then there was drama as both actor and set designer, although my design of programmes for our house play, Sheridan's *The Critic*, was not so well received. I'd produced these in the guise of scrolls rather than folded paper in spirit, or so I thought, with the era of the late eighteenth-century play. 'Your programmes are a disgrace, Dyson', said my housemaster. 'Programmes should be flat.' And that was that. My last school play was *The Tempest*, playing Trinculo alongside Tim Ewart's – the future ITN *News at Ten* anchor – Caliban.

Art was not a subject much celebrated at school. Sitting down in my sixth form with a careers officer – an ex-RAF chap with a handlebar moustache – it was suggested, given my love of the great outdoors, that I might want to think of becoming an estate agent. Or possibly – my contribution – a surgeon. I did go to see an estate agent, in Cambridge, who told me I ought to be an artist. I also went for an interview at St George's Hospital, in those days, before its move to Tooting, on Hyde Park Corner, where it was suggested I might be happier taking up . . . art.

Though the idea of being a surgeon held a certain if fleeting appeal, my one great love aside from long-distance running was art. I had been painting seriously, as I thought, from the age of eight or nine. What I really wanted to do was to go to art school. Ever since my encounter with that RAF careers chap at Gresham's, I have been wary of either accepting or giving advice. It may be well meant, yet it is often wrong. Encouragement is something else. My view is that if the advice chimes with one's instincts, then the advice may be good. It should be more of a reaffirmation.

Art, however, was not the post-school career or way of life my teachers had imagined for me. Given the fact that my father, Alec, was a Classics master, my elder brother Tom an Open Classics scholar to Cambridge and my elder sister Shanie equally brilliant academically, it had felt as if my scholastic career, at least, was preordained. And I was good at Latin and really enjoyed Greek and Ancient History. I was, though, a third child and like many third children felt an almost pathological longing from early on to prove myself by going my own way.

I went to Gresham's when I was just eight. We had rented part of a draughty Victorian house next to the school, icy in winter, since 1946. My father, a Cambridge Exhibitioner who had taught in Kenya and fought in Abyssinia, had come back from serving in Burma with the Northants Regiment in Bill Slim's 14th, or 'Forgotten', Army, in 1946 minus teeth and hair. The war in Burma had been extremely tough and Dad – twice Mentioned in Dispatches – had been in the thick of the action at Imphal and in sniper-infested jungles. He had fought alongside Indian, African, Chinese and American troops and with militias of Shan, Chin, Kachin, Karen and Naga tribes, too. The British themselves, with allies and troops drawn from many parts of the world far beyond Europe, had been a minority of those who defeated the Japanese in Burma and the Naga Hills.

I remember my father as an ever-cheerful polymath. He ran the school cadet force as a major, coached hockey and rugger and taught me to sail dinghies on the Norfolk Broads. He used to wake me up early in the morning to catch the spring tide down at Morston. He did so following a stormy night in 1954 when the sea floods had swept across the mudflats, marshes and valleys of north Norfolk. This wasn't a matter of simply jumping in the car and speeding out to find the boat. Our car, an old Standard 12, powered, I recall, by a Jaguar engine, demanded a hand-crank start with a violent kickback and suffered frequent breakdowns. It was certainly an adventure.

My father played tenor recorder in a group, produced school plays – I still have his margin notes in miniature volumes of Shakespeare – and was happy pouring molten lead to cast miniature soldiers and working wood in his workshop. He wrote a children's book about India that he illustrated with charming watercolours. It was called *The Prince and the Magic Carpet.* Happily, my grandchildren love me reading it to them and chime in with the magic words 'Dhurry dhurry ooper jow' to set the carpet flying. He was able to spout spontaneous and suggestive limericks. In his obituary a member of staff said how much they 'enjoyed his humour which often veered into the Rabelaisian'. He was an amateur photographer who developed his own prints, sticking them into precious albums. He was always up to something that involved us, whether it was feeding the chickens, allowing us to stand leaning out from the running boards of the car or putting stage make-up on the Gresham's School actors.

When my father met my mother, Mary, she was only seventeen, the daughter of the slightly remote vicar of Fowlmere in Cambridgeshire and a very artistic mother who painted glorious watercolours. My paternal grandfather was a distinguished retired headmaster who lived with my grandmother at nearby Thriplow. My mother and father had met at a local social event and arranged a quick wartime marriage in 1941. Their honeymoon was brief. My father, of course, was in the army. Missing out on university, having attended the Perse School in Cambridge – I don't know how her parents could have afforded it – my mother had volunteered for the WAAF (Women's Auxiliary Air Force). She pushed aircraft positions across a vast map of Europe at RAF Tangmere in West Sussex. An important strategic airfield from the Battle of Britain onwards, Tangmere was much popularised in war films with Winston Churchill looking down from a balcony to the map below.

My sister Alexandra (Shanie) was born in 1942 and my elder brother Tom in 1944. They were wartime children. I was born in 1947. At home in peacetime Norfolk, there was no television, never enough heating, no new toys, and few, if any, consumer goods. We had just enough money to get by. This was the Age of Austerity and, until I was seven, that of ration books, too. We grew our own vegetables and collected eggs from our hens. Sometimes, we walked to the Regal in Holt to watch films. As children, though, we had something even more entertaining and truly priceless: the free run of Gresham's grounds, playing fields, tennis courts and swimming pool throughout the holidays. Gresham's, it is said, has more acres than pupils. The vast and often empty Norfolk beaches were close by.

The large Victorian house was divided between three families. We children were a tribe of sorts, all very *Famous Five*, *Secret Seven* and

Swallows and Amazons, and all children of teachers at Gresham's. I was the youngest and so, I suppose, had something to prove. I was also the smallest of my tribe, and in class at school. I shot up, quite suddenly, when I was fifteen. As children we did all the things running around Gresham's that would be looked down on or probably banned today for being far too dangerous. We built hazardous tunnels, climbed challenging trees and were as often as not grubby, grazed and out of breath. The tunnels were built by each digging our own den and then connecting them by trenches. Logs or timbers were spread across the trenches and rusty corrugated iron placed on top. It was an interesting lesson in structures and amazingly nobody was buried alive. Those days were truly idyllic.

In 1955, when I was eight, we were motoring back from a beach holiday in Polzeath in Cornwall, which I would remember until that moment mostly for uncomfortable boils on my bottom. We stopped for a picnic on Dartmoor. I set off alone along a track exploring the high bracken. Round a corner I discovered my father being violently sick. Before I could say anything, he said, 'Don't tell Mummy'. It was typical of him not to want to cause alarm. I felt immense love and compassion for him as we made our way back to our family.

Dad died in 1956. I was nine. He was forty. He had been thirty when he came back from Burma. Three years later he was diagnosed with cancer. Throat and lungs. He took school lessons speaking through a loudhailer. Jim Wilson, a former pupil, recalled in the 2016 *Old Greshamian Magazine*, 'looking back, one can readily recognise his courage continuing to teach using a microphone and speakers to amplify his voice. At the time, did we really appreciate the bravery and determination this must have taken?'

His last days were spent in Westminster Hospital. He had said goodbye holding a small leather suitcase as we waved from the back door. He set off to Holt station and caught the train that steamed him to London. That was the last time I saw him. His brave cheerfulness chokes me every time I recall the scene. It is impossible to imagine my father's emotions as he waved goodbye knowing that he might be on his way to London to die. All the more tragic as the years he had spent fighting in Burma had been a very long sea voyage away from his young bride and family.

Sixty years have not softened these memories, nor the sadness that he missed enjoying his three children growing up and marrying wonderful people. How he would have relished playing with his grandchildren, of whom there are seven. This was all the more poignant when I observed one of my own grandchildren, Mick, at the age I was when my father died. Mick is loving, bright as a button and self-possessed, yet still at that age took his ruffled, soft toy puppy to bed with him. He was far too vulnerable to lose his father. I realise how much I missed mine as I watch Mick playing ping-pong with his creative and loving father, Ian.

My elder brother, Tom, my mother and I had been drinking asparagus soup when the phone rang at home in Holt that day in 1956. As my mother answered the call, I had a naïve premonition of the news. This was surprising as I was unaware that cancer was an inevitable killer. We feared for our talented elder sister, Shanie, who was away at boarding school. How would she be able to take the news on her own?

I had only just started at Gresham's and there I was, days later, in the school chapel in short trousers and knobbly knees going to my own father's memorial service. For reasons I cannot understand, rather than sitting with my family I was in a row of seats with all the other boys – my peers – who didn't really know why they had been dragged to a service that as far as they were concerned was a waste of time. I found this traumatic. I still find it uncomfortable to think about. They hadn't meant to be disrespectful, but this was my father.

I felt the devastating loss of my dad, his love, his humour and the things he taught me. I feared for a future without him. Having recently become a boarder at the school, away from my family, I was suddenly alone. It didn't do to cry or show emotion, just a stiff upper lip. Ever since, a part of me has been making up for that painfully unjust separation from my father and for the years he lost. Perhaps I had to learn quickly to make decisions for myself, to be self-reliant and be willing to take risks. Little could be worse than my father dying when he did.

Logie Bruce-Lockhart, the school's generous headmaster, and his kind wife Jo had arranged for Tom and me to become boarders for a nominal fee as this would allow my mother to go out to work. She took up dressmaking before training as a teacher. Later on, she went stoically up to Cambridge as a mature student to take a degree in English. It was my mother who brought me up after my father's death and influenced my childhood learning. My parents were married for fifteen years, but only really at home together for three, before my father was diagnosed with cancer. This might explain my mother's ability to survive on her own while raising three children and studying for two higher-education qualifications.

At 5ft 11, my mother was tall. I don't remember her having any difficulty with discipline. To me, though, she was mild, loving and indulgent, although not with money. There was none to spare. She was very encouraging. She read voraciously, holding her own with academics at Gresham's, and spoke French flawlessly even though she had never been to France. When she did finally go, she took us in her Morris Minor. Camping in a cheap ridge tent, she showed me, among many other treasures, the flying buttresses of Chartres Cathedral, the pantiled roofs of Vézelay and the beautifully austere Cistercian cloisters of the Abbaye du Thoronet. We pitched our tent by the Dordogne and swam in the river long before this patch of France became a British colony.

Determined to live well on little money, at home we picked samphire on Stiffkey Marshes and dug cockles from the sand. We went to first performances of Benjamin Britten operas and works conducted by Britten himself, who lived in Suffolk. My mother played her Kathleen Ferrier and Peter Pears LPs. We read, played charades and made things. Lead soldiers, model gliders and diesel-powered aeroplanes were my thing. I didn't play with the soldiers or collect them. What I enjoyed was making them, using my father's equipment to melt lead in a crucible and to pour the dangerous molten lead into moulds.

In 1957, when my mother decided on her vocation, she went to Norwich Teacher Training College for a two-year course, presumably on a student grant. She taught at Sheringham Secondary Modern before Runton Hill, a rather good local girls' public school, offered her a job teaching and as housemistress of a new boarding house. Having to visit a girls' boarding school suited me, too.

In 1968, three years after I had left home to become a student, my mother decided to study for a degree at New Hall, Cambridge. As a wartime bride, she must have regretted not finishing her secondary education, nor having had the chance to go up to Cambridge as had both my dad and my brother. Even so, it must have been depressing for her to have had to survive on yet another student grant and to live in basement digs, as I was doing in London. Although she was ill and hospitalised up to and during her finals, she got a 2:1. She then taught English at Fakenham Grammar School for five happy years, where she also thoroughly enjoyed producing plays. Through a bitter twist of fate, in 1978 she was diagnosed with and very quickly died of liver cancer.

My wife now claims that I have inherited my mother's determination and warrior spirit. My mother did have high expectations. She was also very broad-minded and had a catholic choice of friends of all ages. Enjoying conversation on any subject, hers was a modern outlook. Ahead of her time, she was tolerant of others of all walks of life. She was happy to discuss anything. This may seem odd for a religious daughter of a vicar, yet perhaps any Edwardian attitudes she may have had were changed by the hardships and levelling of society caused by the war.

She coaxed me into seeing and understanding a broad culture. She encouraged my acting in plays, my playing the bassoon and painting, all things I had chosen to do myself. Occasionally she came along to watch me playing a sport I loved. Perhaps, instinctively, she understood the lessons that sport can teach children. She was never too disappointed with my academic achievements. A keen amateur artist herself, as was my father, she was also secretly pleased that one of her children might be an artist. Later on, she was intrigued when I branched out into manufacturing as well as design.

She and my headmaster Logie shared a vision of education. While academic achievement is its primary purpose, there are other educational lessons schools can impart. I took part in academic life, and quite enjoyed it, but I didn't feel competitive about it. I left that to sport and the creative side of life. At thirteen I had to choose between the sciences and the arts. My father's and brother's influence led me to choose Classics and, after my O levels, sat when I was fifteen, I specialised in Latin, Greek and Ancient History. Other subjects appealed to me more than these and, in hindsight, it is easy to say that I should have pursued maths and science. I enjoyed these and I was good at maths. Nobody at the time, though, would have expected me to have made that choice and nor did I imagine these subjects were for me. In the event, I was a frustration to my teachers, and they were disappointed in me. I do not advocate children following my selective and irreverent example. Later on, I did embrace academia. I worked hard and competitively at art school and the Royal College of Art. Today, I am an avid reader of history books, while mathematics, engineering and writing are a part of my everyday life.

It was playing games, however, that taught me the need to train hard and to understand teamwork and tactics. The planning of surprise tactics, and the ability to adapt to circumstance, are vital life lessons. These virtues are unlikely to be learned from academic life and certainly not from learning by rote. Acting in plays, which I very much enjoyed, taught me about character, learning to express thoughts and to emphasise dramatically in speech. Long-distance running allowed me the freedom to roam the wilds of Norfolk while depending on no one but myself. Running also taught me to overcome the pain barrier: when everyone else feels exhausted, that is the opportunity to accelerate, whatever the pain, and win the race. Stamina and determination along with creativity are needed in overcoming seemingly impossible difficulties in research and other challenges in life.

It makes me sad and concerned that schools are failing to teach creativity. Yet life today demands it more and more. We need to create fresh solutions to seemingly intractable problems, to devise new software, to create something different in order to compete in the global economy. These abilities are a prerequisite today. No longer can we lead life by repeating what we have learned and what has worked in the past. The world, thankfully, is becoming better educated, and competition has never been so fierce. The advantage we in the West have relied upon for so long is being diminished. In order to stay ahead we need to focus increasingly on our creativity.

At Dyson, we've always aspired to what we called lean engineering. Happily, this is now celebrated as sustainability and using ever fewer resources, to which can be added increased performance, as desirable characteristics of a new design. But the team needed to achieve these breakthroughs has ballooned. Take engineering. Thirty years ago, we could survive with just mechanical engineers. Now we need electronic engineers, software engineers, robotics engineers and AI scientists. The list goes on.

The world has become more complex and yet more integrated. Almost every country develops technology and exports it around the world. This means that alongside producing the best engineers or scientists, we need to apply the advanced technology we develop at an ever-faster pace, otherwise aggressive competitors will do it first.

Home life has much to teach us, too. It certainly did for me. From the age of eight, I grew up in a single-parent home, where we had to share the many chores. In our rented part of a crumbling Victorian house in the 1950s, we had no machines for gardening or cultivating. We had a push mower for what was quite a big garden, and a spade to dig the vegetable garden. The washing machine was a static boiler that merely soaked the washing, which was then rinsed in a large butler sink and fed into a hard-to-turn mangle. The one motorised machine we did have was an old upright vacuum cleaner with a cloth bag hanging from its handle. We had no wall power sockets, so we had to stand on a stool in each room and plug it into the central light socket and not allow the vacuum to pull too hard on the cord. It was smelly, dusty and ineffective. It haunted me for many years!

I have good reason to be grateful to my mother for introducing me to all these home chores. She taught me to sew, knit, make rugs and cook. My father had taught me to sail. I watched him do carpentry. I taught myself to make model aeroplanes, to start their engines, fly them and to repair my bicycle. Doing things with my hands, often as an autodidact and with an almost total absence of fear, became second nature. Learning by making things was as important as learning by the academic route. Visceral experience is a powerful teacher. Perhaps we should pay more attention to this form of learning. Not everyone learns in the same way.

I think I am someone who likes to learn on his own, by experiencing failure and discovering his own way to make things work. I could put that down to not having had a father after the age of eight to show me how things are done. Yet I have noticed the same trait in both my sons. I witnessed Jake using a lathe before I had the chance to show him how. Sam is a self-taught musician. Emily, on the other hand, had skiing lessons and became stylishly proficient, while Jake, Sam and I eschewed them. We need the visceral experience of trying something out to understand and to be convinced that we are doing this the right way for us. Learning by trial and error, or experimentation, can be exciting, the lessons learned deeply engrained. Learning by failure is a remarkably good way of gaining knowledge. Failure is to be welcomed rather than avoided. It is a part of learning. It should not be feared by the engineer or scientist or indeed by anyone else.

I did, though, miss my father. Many years later, I was intrigued to learn in a book by Virginia Ironside that 85 per cent of all British prime ministers, from Robert Walpole to John Major, and twelve US presidents, from George Washington to Barack Obama, lost their fathers as children. It would be wrong to say the loss of a father is some sort of macabre ticket to success. Perhaps early loss can sometimes inspire people to great achievements?

Even so, my own adventure, in manufacturing and technology, has been quite different from that of my much brighter siblings. My brother became a schoolteacher, my sister a nurse. My own inner demons meant I had much further to run, or sail. As a child I was intrigued by the wording of the telegram sent to Mrs Walker in Arthur Ransome's *Swallows and Amazons* by her absent naval-officer husband responding to the Walker children's request to sail the *Swallow* to a lake island and camp out on their own. 'Better drowned than duffers, if not duffers won't drown.' I had no intention of being a duffer.

After my father's death, I had continued to pursue the *Swallows and Amazons* life in school holidays with my tribe. I helped with housework and made balsa-wood planes, some with small diesel engines. Because I had become a boarder, I came home only on holidays. There was no such thing as half-term then and, although physically near, home could seem a long way away. At school in those days boys were not allowed to have feelings. Any feelings caused by injustice, bullying or compassion, I suppressed. Teachers there, as anywhere else at that time, could be cruelly sarcastic and wholly insensitive to the emotional lives of their young charges. For fourteen weeks at a stretch there was no escape, no parent to explain or tell us not to worry. I looked forward very much to the holidays.

Whatever the ups and downs of my schooldays, I was aware of a much wider life around me. To be British in the 1950s was nothing to sneeze at. Roger Bannister ran the first mile in under four minutes. Edmund Hillary and Tenzing Norgay planted the Union Jack at the top of Everest. Peter Twiss, at the controls of the supersonic Fairey Delta 2, was the first person to fly at over 1,000mph. D-Type Jaguars won Le Mans three times in row. Crick and Watson deciphered DNA. There was low unemployment, the emergence from Austerity and Prime Minister Harold Macmillan riding on the slogan 'You've never had it so good'. The Commonwealth, meanwhile, seemed a noble replacement for an empire that had coloured a quarter of the land in our school atlases pink.

Each week, *Eagle*, a particularly well-illustrated boys' comic boasting a huge circulation, featured a centre-spread colour cutaway drawing by the technical illustrator Leslie Ashwell Wood of some new jet aircraft, turbine locomotive or nuclear power station and any number of inventions conjured in British factories, workshops and laboratories, all too many of them long since bulldozed and replaced by dreary new housing estates or anodyne supermarkets. Funnily enough, when I was nine, I won the 1957 *Eagle* painting competition with an oil painting of a Norfolk seascape, Blakeney Point. This was a timely fillip for me coming shortly after the death of my father. I had been recognised for my painting and nationally, too. I've since learned that David Hockney and Gerald Scarfe, both at the Royal College of Art just before I was there in the mid-'60s, made their debuts with paintings in *Eagle* in which art, engineering and derring-do adventure on land, under the sea and into space, shared imaginative pages.

As schoolboys, we did believe British to be best. After all, we had just won wars against the Germans and the Japanese and, clearly, we were capable of peacetime records, inventions and victories even if our families couldn't afford new clothes, much less washing machines or refrigerators, while feeble coal-burning stoves provided just a few precious inches of hot bath water.

We championed our British cars, too, and as far as I was concerned we went on to own two of the best. These were the Morris Minor and the Mini. They were both designed by one of my engineering heroes, Alec Issigonis, not that I knew his name at the time, and were, I'm sure, superior in terms of steering, roadholding, suspension and all-round visibility to the Volkswagen Beetle, their only real foreign rival at the time. In a nod to tradition, both our cars sported decorative timber trim as if, despite their modern engineering, they were miniature farmhouses on wheels.

We once squeezed thirteen schoolboys into Mum's Morris Traveller. This must have been some sort of record. I knew ours were interesting cars from an engineering point of view, but the word 'design' meant nothing to me at the time. This might seem odd from the perspective of the second decade of the twenty-first century, but not through the lens of north Norfolk or, for that matter, pretty much anywhere in England more than half a century ago. In those days, however, north Norfolk was quite cut off and it can still be a remote, and enchanting, corner of the world.

As for school, I did get by fairly well despite not applying myself as I might have to academic studies. I passed O levels – the equivalent of GCSEs today – in Art, Maths, Latin, Greek, French, English Language, English Literature and History, and A levels in Art, Ancient History and General Studies. In the holidays I took on any local job going, from loading sacks of cold and wet potatoes into lorries, topping and tailing icy Brussels sprouts, picking blackcurrants at two bob (10p) a bucket-load and picking parsley to drive to what was then the ultra-modern Campbell's soup factory at King's Lynn, the largest of its kind this side of the Atlantic.

Schooldays were also made more exciting by our teaming up for lessons at Gresham's with the sixth-form girls at Runton Hill, where my mother taught, and now long gone. It certainly didn't lack academic ambition. My girlfriend there, Caroline Rickaby, went on to get a First at Cambridge and, later, a PhD at Durham University on King John's relationship with his French advisers and, as a consequence, how history may have misrepresented the English king ever since. I have never lost my interest in history, yet my talent, such as it was, lay elsewhere.

CHAPTER TWO

Art School

While my friends were going off on gap years before university, I still wasn't sure what I wanted to do. My gap year proved to be a year spent at art school to see how it suited me. In the autumn of 1965, I rode down to Kensington on my Honda 50 to join the Foundation Year course at the Byam Shaw School of Drawing and Painting. It was all very exciting, not least because I was now self-reliant. I had no one to turn to in London. I needed to find my own home and make my own way. Equally, I was free of restrictions, with a bedsit in Herne Hill in Victorian south-east London suburbia and a motorcycle to get around on.

Perhaps I should have thought a little more than I did at the time about my little Honda. Launched in 1958, the Honda 50, or Super Cub as the Japanese know it, is the most produced motor vehicle ever. In 2017, Honda assembled the one hundred millionth Super Cub in one of its fifteen international production plants. The invention of the engineer Soichiro Honda – who I admire greatly for his addiction to the continuous improvement of products – and salesman Takeo Fujisawa, the Super Cub was simplicity itself to own and ride. Clean, too, with its enclosed drive chain, plastic bodywork and leg guards, and nippy with its tiny 49cc four-stroke engine producing 4½ bhp at a fast-spinning 9,500rpm.

This was an early example of an inventive manufacturer taking an existing product – in this case the low-cost motorcycle – and transforming it into a much better and much more attractive proposition than anything available at the time. Soichiro Honda and Takeo Fujisawa's genius was to think against the grain while focusing on continuous improvement. They also subcontracted basic components while focusing their efforts on inventing and manufacturing inhouse those that were not available elsewhere. Honda made, for example, its own more efficient gearbox. The company continues to invest a sizeable chunk of its income into research and development, aiming for constant improvement and innovation.

I was lucky to be taught at Byam Shaw by the Op Art artists Bridget Riley and Peter Sedgeley, who taught colour and the relationship between colours. While learning to innovate, we were taught to draw well. For me, drawing remains a fundamental and essential skill. What was equally important for me, though, was that I was fortunate enough to be taught under the studious eye of the school's principal, Maurice de Sausmarez. Maurice would have been in his early fifties then. A fine painter in his own right, he was an intellectual with the ability to discuss complex ideas in clear, straightforward language. He never spoke in art-world or any other jargon. He published a number of highly perceptive interviews with artists – Peter Sedgeley, Naum Gabo, Bridget Riley, Ben Nicholson and Henry Moore – encouraging these talented artists to think clearly as they spoke with him.

Painting aside, his passion was education, and he was very good at it. His special genius was an uncanny and unerring ability to understand what a particular student was good at it even if they themselves had no idea. He encouraged us to believe that innate talent, when backed by natural enthusiasm, will shine through if only the right doors are opened. This is something I've certainly never forgotten.

When I reached the end of my year at the Byam Shaw, Maurice sat me down and suggested that I might be interested in doing design. 'What's that?' I asked. The Byam Shaw had only offered drawing and painting. This was 1965 and 'design' was not covered by magazines or newspapers and was certainly not available in shops. Perhaps the immediate post-war public felt grateful to buy anything and design was not a consideration. The Design Centre in Haymarket was the only exhibition space showing good design, although its efforts had not yet reached me in Kensington, engrossed as I was in fine art.

Maurice explained, 'There is fashion design, industrial design, furniture design ...' I stopped him there. I at least knew what chairs were. 'You can shape and make your own chairs,' Maurice suggested. Furniture design, then? I had to think about this. I had wanted to be a painter and I thought that is what artists did. But I also knew that I liked making things and because I had sat on chairs that I had broken and then mended, then maybe I could design them, too. A better chair, perhaps?

Maurice, who had studied there himself before the war, knew that the Royal College of Art, a wholly postgraduate college, was conducting an experiment by taking three students without first degrees to see how they compared. I applied, sat the exam, was interviewed, although possessing a scant knowledge of furniture woods, and was thrilled to be accepted, along with someone who would also switch disciplines, the sculptor Richard Wentworth, and Charles Dillon. Charles embarked on some original designs such as the Kite Light, a cloth kite hanging from ceilings with a lamp suspended inside, a project he worked on with his wife Jane, a fellow RCA graduate, before his tragic and early death.

There was a condition to bypassing a first degree: I would have to spend four years of study instead of the normal three, although this was not exactly a hard condition to accept. Thrilling though this opportunity was, it was dwarfed by a *coup de foudre*: I had fallen in love and have remained so for more than fifty years. Deirdre was the most naturally beautiful of girls in her '60s garb, often from Biba. She was modest yet incredibly talented, a far better painter at the Byam Shaw than I. She is quite the warmest human I have ever met, with an undying curiosity and enthusiasm for life. Her instinctive understanding of the emotions of others is uncanny and utterly endearing. But I'm jumping ahead of myself. How did all this happen?

At the Byam Shaw we made field trips to draw at various interesting places around central London – dinosaurs at the Natural History Museum, people walking in London parks, dancers at the Ballet Rambert and ice skaters at the Queens ice rink in Bayswater. Have you ever tried drawing spinning skaters? I managed to tag along with Deirdre among our group on some of these outings, chatting ceaselessly when not working. Luck was on my side. We both travelled in from south-east London and sometimes met on London Underground's Circle Line, not the most romantic of places perhaps, yet it provided another chance to talk. Things came to a spontaneous head outside the monkey cage on a drawing visit to the London Zoo. I dared to hold Deirdre's hand. Unlike the monkey whose hand I had just grabbed, she didn't snatch it away though she claims that she was truly shocked.

Our first date had me driving the replacement for the Honda 50, an ancient starter-less Morris Oxford, all the way to the charming Ark restaurant in Notting Hill. It is still there and still looks like an ark, although it is now called The Shed. The evening seemed to be a success and many more followed. Deirdre was her understanding self to put up with someone who was little more than an immature public schoolboy. My mother was shocked that I had fallen in love so young, but quickly came to adore Deirdre.

Although we were both living on grants, in 1967 we decided to marry. This may now seem somewhat premature, but five years at university is a long time and we couldn't wait! In the '60s everyone was less worried about job security – neither of us had a job anyway – and we had no thoughts about buying a house. We were far too poor and, as married life is tough on a grant, we both took part-time jobs to buy food and pay the rent. We ran up a huge overdraft that got ever bigger through our married life until it had risen to the astronomic level of £10,000, the equivalent of £50,000 today. It was only paid off when I was forty-eight, by which time it had reached £650,000. I like to think that prodigious borrowing is putting money to good use.

Deirdre has had the most creative of lives while nurturing three loving and talented children, Emily, Jake and Sam. Fortunately they have inherited her extraordinary emotional intelligence and humanity. All three have pursued artistic careers: Emily as a fashion designer, Jake as a designer and Sam as a musician (a talent he has inherited from Deirdre). Early on Deirdre used to do illustrations for *Vogue* of clothes or goods selected for its 'Shophound' page and she has continuously painted and exhibited throughout our married life. She has a distinctive style that is hers alone. Her paintings are of many subjects on a large scale, beautifully drawn and with exquisite use of colour. I only wish Deirdre had kept more of the paintings she has sold.

Her second career as a bespoke rug designer and supplier, from her King's Road gallery, with a showroom in Saint-Germain-des-Prés, illustrates her mastery of colour, as you can see in her book *Walking on Art*. *World of Interiors* credits her as Britain's leading carpet designer. To live with such a creative partner, to walk in bare feet on her artistic silk rugs, to gaze at her paintings on the walls, to be surrounded by her subtle wall colours, makes me the most fortunate of men. I haven't mentioned her opera singing.

Until quite recently, Deirdre has had to put up with a perpetual lack of funds. She made clothes for herself and sometimes for our children. To save money, we grew our own vegetables before it was 'green' to do so. Craziest of all, during the first thirty years of our marriage, she agreed unselfishly, and so typically of her, to keep putting her signature to endless bank guarantee forms in front of solicitors, signing away all our possessions. If we had defaulted on the bank loans, we would have been evicted from our home. She has a mild countenance and manner, but is not one to give in or give up.

Deirdre's background was rather different from mine. She was born and grew up in Bell Green, Lower Sydenham, a south London suburb, she says, of neat semis and orderly commuters. All those green Southern Region electric trains growling and flashing sparks from live third rails up to Catford Bridge and Charing Cross. Deirdre's mother was a legal secretary and her father had driven tanks with the Eighth Army in the Allied push up through Italy between 1943 and 1945. He had also played trombone with George Formby, a big and highly paid star at the time.

She passed her 11-plus, but her small school was folded into Sydenham School for girls, one of Britain's first comprehensives, a type of school open to all pupils regardless of previous academic achievement. It says much about the country, and attitudes towards women at the time, that despite its nod to inclusiveness and equal opportunities, it nevertheless prepared pupils first and foremost for domestic and secretarial servitude. She learned sewing, shorthand and domestic science. The school was proud to show visitors its rows of Singer sewing machines and gleaming electric typewriters.

What Deirdre wanted to be, though, was an artist. Her kindly headmistress explained the perils of painting. If only Deirdre pursued typing and shorthand, she would leave school ready to take notes, type, sew and cook, ideal qualifications for a young woman from Sydenham shortly before the '60s threw so many conventions, ways of life and art, too, high as a kite into the air. Deirdre did win one concession. She was allowed to take O level Art, alongside her eight RSA exams, which she was told were equivalent to O levels. There was no RSA exam for art.

Serendipitously, perhaps, it was Deirdre's shorthand and typing as much as her prized O level that opened the door for her to art school. Her second job after leaving school was as a secretary in the studio of the highly regarded architects Chamberlin, Powell and Bon, busy at the time with the design and construction of the City of London's vast and heroic Barbican estate. This was a highly civilised and inventive office. One of the key design architects on the Barbican projects was Leopold Rubinstein, who had trained with Le Corbusier in Paris. The practice also did a wonderful job of designing New Hall, Cambridge, my mother's university college.

Deirdre was an inveterate sketcher. Her employers noticed her drawings. They suggested she should go to art school. While this was music to Deirdre's ear, she was horrified to learn that colleges did not recognise RSA exams. The only school running its own diploma programme that might accept her was the Byam Shaw. Even then, money was a problem.

Deirdre had saved hard but had nothing like the funds she needed for a year at Byam Shaw. Maurice de Sausmarez came to her rescue. His secretary, he said, did not take shorthand. Now, if Deirdre were to come by their office every afternoon at four o'clock to take shorthand, her fees would be waived. As long, that is, as she attended night school and sat those missing O levels. These would be essential for her to move on after Byam Shaw. Although she had taken a full set of RSA exams, it was necessary to retake these as O levels to enter a degree course at Wimbledon School of Art.

I was as impressed with Deirdre's resolve and determination as Maurice was. And these were characteristics we shared. And ones we look for, although not forensically so, in the many talented young people we employ at Dyson from whatever background or walk of life.

The Royal College of Art was, in so many ways, a real eye-opener. I signed up to study furniture design, but soon found that it was possible to change courses and to steer a line between and through them, meeting fascinating people and learning much along the way. The RCA's thinking was radical. At the time it went against the grain for an engineer to be a designer as well. You didn't switch professions. Designers were mostly thought of as consultants or those, in my imagination, who didn't get their hands dirty and who were concerned with looks rather than function. They were wholly remote from engineers in white lab coats who gave structure to things and made them work. I loved my time at the RCA not least because of its lively and inventive cross-disciplinary approach. Here, as I progressed, I realised that art and science, inventing and making, thinking and doing could be one and the same thing. I dared to dream that I could be engineer, designer and manufacturer at one and the same time.

In 1959, the scientist and novelist C. P. Snow gave a famous lecture, 'The Two Cultures', on the ever-growing and unhealthy divide he saw between science and the humanities. Snow went on to make his point tellingly. 'A good many times,' he wrote, 'I have been present at gatherings of people who, by the standards of the traditional culture, are thought highly educated and who have with considerable gusto been expressing their incredulity at the illiteracy of scientists. Once or twice, I have been provoked and have asked the company how many of them could describe the Second Law of Thermodynamics. The response was cold: it was also negative. Yet I was asking something which is the scientific equivalent of 'Have you read a work of Shakespeare's?'

Snow found the British system of education guilty. Since the Victorian era, science had been overshadowed in schools by humanities and especially by the teaching of Greek and Latin. Where German and American schools valued science and technology, we in Britain tended to look down on these subjects, and on industry, as somehow grubby, or, if not grubby, then somehow uncultured and even anti-intellectual. I'm afraid, C. P. Snow, that nothing much has changed. If anything, science and engineering are even more looked-down-on today.

What I learned at the RCA, from brilliant men and women who knew instinctively how to wear their own considerable learning lightly, was that intelligent enquiry can be delightful as well as productive. For me, this revelation began with encounters with my first-year tutor, Bernard Myers, a walking one-man encyclopaedia who taught design. He was especially keen on the meeting and mingling of art, science, engineering and design and on the value of technology. In fact, he taught Industrial Design Engineering at both the RCA and to engineers at Imperial College. Bernard was a fundamentally serious person. During my first face-to-face tutorial with him he said, 'When you design something, everything about it has to have a purpose. There has to be a reason.' I looked around at the best designs of the time like Issigonis's Mini, the new architecture of Norman Foster and Richard Rogers, and at the radical designs of the American inventor Buckminster Fuller, and I saw that Bernard was right. I have based all my design on this ever since – honest, purposeful design reflecting its technology and engineering.

At the same time, this was very much the era of anything goes, a time of a brilliant flowering of new art, design, fashion, colour and music. Non-conformity was celebrated. I began to grow my hair, wear flowery shirts and bell-bottom trousers made to measure inside Kensington Market, which smelled of incense and patchouli and was packed with colourful clothes.

Kleptomania in Carnaby Street, the place in those days to buy clothes, proved to be just one edgy peak in the phantasmagorical world shaped by art college graduates letting rip in London in the mid-'60s. As it happened, many of them were from the RCA, showing how young people, encouraged to be creative, could shine whether as artists, designers or entrepreneurs, or, in fact, all three simultaneously. The fusion of original design, innovation and entrepreneurialism is something that has fascinated and driven me on ever since. I was lucky to follow David Hockney, the fashion designer Ossie Clark and film director Ridley Scott at the RCA at the time austerity gave way to a new prosperity and new freedoms in the early '60s. Pink Floyd played at the college, our education was free, student power was expressing itself in protests while Mods and Rockers gathered at seaside resorts. Unlike our parents, we hadn't endured the Second World War. We felt a sense of new opportunity.

Jon Wealleans, in my year at the RCA, and user of a dry powder hair shampoo that would inspire me later on, designed Tommy Roberts's multi-storey Mr Freedom shop in Kensington, where you could buy textiles by Jane Hill (RCA), Jon's PVC and fake fur False Teeth Chair and wild menswear by Jim O'Connor (RCA) coveted by Elton John. Downstairs, the Mr Feed'em restaurant was decorated with pop graphics by George Hardie, another fellow RCA student, who went on to design the Hipgnosis album covers for Led Zeppelin and Pink Floyd.

King's Road was increasingly fashionable. By 1967 the street was, in parts, a colourful parade of one-off boutiques – Aquarius, Bazaar, Chelsea Girl, Garbo, Granny Takes a Trip, Hung on You, I Was Lord Kitchener's Thing, Just Looking, Kiki Byrne, Lord John, Mates, Quorum, The Squire Shop, Take 6, Top Gear, Topper – along with Just Men, where I got my hair cut. What was so interesting was the way in which RCA students revelled in this psychedelia, experimentation and pure fun while also being keen on what Terence Conran was doing with

Habitat in Fulham Road and on industrial design and the unflippant furniture of Le Corbusier, Charles Eames and Jo Colombo. Before he took on the design of Mr Freedom in Kensington, Jon Wealleans had worked for Norman Foster. Roger Dean, another RCA contemporary, designed highly imaginative new furniture, although he is best known today for his sleeve designs for albums by the prog rock group Yes.

An earlier graduate, Allen Jones, a contemporary of David Hockney, who had designed erotic furniture, was refused a degree by the RCA because he would not submit written General Studies papers for exams, arguing, along with David Hockney, that they were artists, not writers and that, after all, academics didn't have to paint paintings. Years later, as Provost of the RCA, I begged the college's academic board to give Jones an honorary doctorate. To no avail. The painting professor dismissed his work as mere 'Pop Art' in spite of there being a major retrospective exhibition at the time for Allen Jones at the Royal Academy. A wrong was not righted.

Lapping up everything the RCA had to offer, I was certainly fascinated by what my contemporaries were up to and quite aware of Vietnam and student protests, but I was never a renegade. I did, though, trade in wine to make ends meet, supplying staff and the Junior Common Room, and I worked nights at a petrol station. My parking space in those pre-yellow-line days was in Hyde Park. It witnessed a succession of second-hand cars – an Austin Healey 100/4 that took Deirdre and me to France and Spain, a split-screen Morris Minor and a Mini van. Between cars, I switched from Furniture to Interior Design, which, then as now, is not the same thing as interior decorating. It was more like architecture.

RCA students were experimenting at the time with new materials like acrylic, PVC and polyester while making cardboard chairs that could be folded into shape and disposable paper dresses. And, in keeping with the open-minded spirit of the times, they moved, as I did, from one department at the RCA to another. Art, design, fashion and even architecture were seen as a continuum.

I think one thing we all learned in those heady days in Kensington is that art and design could be inventive, functional and exciting, seeking new ground without compromising in terms of quality. Perhaps subconsciously, I also learned that what passes for experience in many professions, disciplines and walks of life is often an attitude that becomes blinkered over time. At Dyson, we don't particularly value experience. Experience tells you what you ought to do and what you'd do best to avoid. It tells you how things should be done when we are much more interested in how things *shouldn't* be done. If you want to pioneer and invent new technology you need to step into the unknown and, in that realm, experience can be a hindrance.

Back at the RCA, the tutor who encouraged me to shift from Furniture to Interior Design and Architecture was Sir Hugh Casson, famous for his role as Architectural Director of the 1951 Festival of Britain. I was mesmerised by Sir Hugh's ability to sketch so very lucidly on blackboards and by his engaging manner. Again, he made me think that anything was possible which, of course, was something he had proved on the South Bank in 1951 when he had commissioned buildings and architectural sculptures as fresh and as daring as Ralph Tubbs's Dome of Discovery and Powell and Moya's Skylon, and realised them in truly tricky circumstances, what with shortages of time, money and materials, lousy weather, labour strikes and a mostly hostile press. Sir Hugh, with one foot in the highest ranks of the establishment, the other in contemporary design and architecture, had a talent to amuse and enthuse. He cut a dash, driving a Mini about town and an ancient Rolls-Royce from his country home by the Solent.

His Interior Design department was much concerned with architecture. There was no Architecture department as such when I was at the RCA, but there was an Industrial Design department led by Misha Black whose Design Research Unit had been the design consultants for London Transport's sleekly functional new Victoria Line, the first new Underground line in fifty years. Opened by the Queen in September 1968, its subdued grey and brushed stainless steel aesthetic was the polar opposite of what RCA graduates were exploring in Carnaby Street and on the King's Road.

One of my course teachers in Sir Hugh's department was the brilliant structural engineer Tony Hunt, then in his mid-thirties, who did more than anyone at the time to turn me on to engineering and to make the connection between design, engineering, art and science. Tony was as passionate about the aesthetics of structures as he was about how they worked and were made. Actually, these things all went together. For Tony, a great innovator, already working with Norman Foster and Richard Rogers on early hi-tech buildings like their radical Reliance Controls factory and office in Swindon, concept came before calculation. There were exciting new ways of creating structures and designing buildings that could be imagined, invented and then worked out with maths, log tables, slide rules (the way calculations were done before the calculator was invented) and, increasingly from the 1960s, with computers.

Tony taught us that structure *was* architecture. Most enduringly-good modern buildings of the past fifty years, like the Pompidou Centre in Paris or the Lloyd's Building in London, as well as medieval cathedrals and ancient designs like the Pantheon in Rome, are defined by the structure that holds them up rather by cladding or style. I was also really taken by the work of Buckminster Fuller. Hot news at the time in London – he had come to work with Norman Foster in 1967 – he demonstrated, especially with his geodesic domes, that architecture and structure were indeed synonymous.

One of 'Bucky's' most enthusiastic supporters at the RCA was my fellow year student Anton Furst, an impulsive and hugely creative talent, who went on to design the sets for Neil Jordan's *The Company of Wolves*, Stanley Kubrick's *Full Metal Jacket* – recreating the hell of Vietnam on the site of Beckton Gasworks in east London – and Tim Burton's gloriously dark *Batman*, for which he won an Oscar. Anton's mentally unstable and alcoholic father died when he was at the RCA. Anton himself, dependent on Valium from then on, committed suicide in 1991, jumping from the roof of a Los Angeles multi-storey car park.

Fuller himself was an eternal optimist who would have liked to have lived for ever, his way of thinking as wonderful as it was contrary. He aimed to turn conventional thinking concerning architecture, homes, cars, land use and the way we might live upside down and inside out. He drew heavily for inspiration from the aircraft industry that necessarily sought lightness in its machines. 'Just add lightness' became Fuller's mantra. This led him to create his lightweight geodesic domes. Patented in 1954, these hemispherical lattice-grid structures composed of triangular elements, offered the greatest volume of covered space for the least surface area. Strong, wind- and snowproof, they found favour with the military, polar research bases and exhibition organisers, although they were never mass-produced as homes as Fuller would have liked them to be. I suspect this is principally because most people, including mortgage lenders, prefer conventional homes.

Through Tony Hunt's teaching, Buckminster Fuller certainly opened my eyes as to what was structurally possible and to just how exciting pure structure and design engineering could be. I could see that structural engineering would come to dominate architecture. I also understood that this would be true of products, too, that their technology and engineering would become more important than industrial design casings.

My Interior Design course at the RCA had, through happy accident, tilted me ever more towards engineering. By chance, I was introduced to Joan Littlewood at an event in Clerkenwell, a district of London that was anything but fashionable at the time. Joan – widely known even then as the 'mother of modern theatre' – was the founding director of Theatre Workshop that had taken up residence in the early '50s at the dilapidated Theatre Royal in Stratford – east London that is, not upon Avon. She enjoyed critical success with the London debut of Brecht's *Mother Courage and Her Children*, which she directed and starred in, and won both popular and critical acclaim with two early '60s musicals, *Fings Ain't Wot They Used T'Be* and *Oh, What a Lovely War!*.

When I met her, Joan was keen on building a new children's theatre at Stratford. I came up with a mushroom-shaped structure, its spherical surface formed by aluminium tubes constructed in triangles, and much influenced by Fuller. It was fun to do, and secured planning permission, but it never made it off the drawing board. It did, though, introduce me to Jeremy Fry, the inventor and engineer, who, more than anyone else, encouraged me to think for myself and to 'just do it'.

I had based my design on the Triodetic structural system patented by Vickers, the engineering conglomerate. The company's London headquarters was Vickers Tower, the much-acclaimed steel-and-glass skyscraper on Millbank near the Tate Gallery. Designed by Ronald Ward and Partners, it had opened in 1963. Here I was shown a black and white film of the construction of a Triodetic factory roof. I asked who the man in the film was, lifting the entire aluminium-tube roof of his new factory on his own with a set of pulleys. He was Jeremy Fry of Rotork.

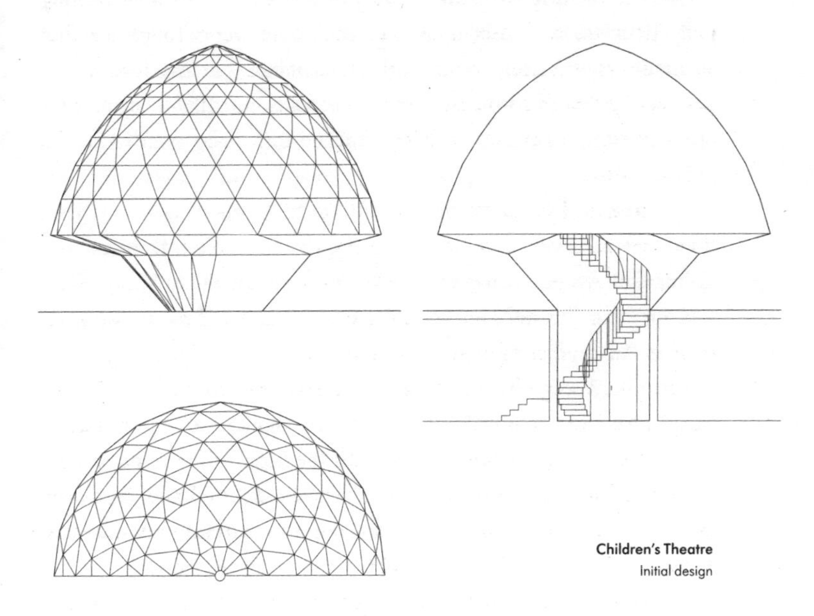

Children's Theatre
Initial design

In 1967, I engineered a meeting with Jeremy through a friend of his. I drove down with my model of the theatre in the Healey 100/4 to Widcombe Manor, Jeremy's glorious early Georgian house in Bath. Over whiskies, my charismatic host said he wouldn't help finance my theatre but suggested I might like to consider doing some projects with him. The first of these, for the theatre and film director Tony Richardson, was to design the seating and auditorium of the London Roundhouse at Chalk Farm for its first ever event, Richardson's production of Nichol Williamson's *Hamlet.* My Joan Littlewood mushroom theatre may not have come off, but here I was designing another circular auditorium for a high-profile production.

Opened in 1966, the Roundhouse was far from an ultra-modern lightweight steel structure. It was a former engine shed commissioned by the London and Birmingham Railway. The 'Great Circular Engine House', however, was a brilliant and radical design of its time, a highly functional circular structure entirely free of ornament or decoration shaped in white Suffolk brick and enclosing twenty-four tracks for locomotives clustered around a central turntable. A ring of twenty-four cast-iron columns forming a circle – a polygon, really – around the turntable supported a conical roof crowned with an open-sided lantern to emit steam and smoke. The logic and engineering of the design were impeccable, yet the building was quickly superseded. For more than a century it served as nothing more than a warehouse, if memorably so for Gilbey's gin. A friend, Torquil Norman, an eternally optimistic entrepreneur, has since bought and brilliantly restored the building as a performing arts venue.

Back to Jeremy. The first project he asked me to do was a pair of 'Jesus floats' – floating skis – for his 8-year-old daughter so she could walk on water. This is harder than you think because skis on water want to do the splits. Because your feet are spaced slightly apart, the downwards force of your weight divides outwards, and if there is no 'grip' on the floor, or water in this case, your feet simply slip sideways. I solved this by connecting the skis with elastic cord.

The next project was a circular pedalo made of a bicycle with paddles brazed to the spokes of the rear wheel. The front wheel was removed and replaced with a rudder. The hull was balsa wood, on the underside of which Deirdre created a painting of a turtle with glorious colours. We tried it out on Pampelonne beach at St Tropez where it did indeed turn turtle. Many times. In 1968, Pampelonne beach was a glamorous place to be, even though there was only one café and that barely had a roof. Next, I made a roof from plywood box sections to be bolted together on site, made in Gloucestershire and driven to Provence by Deirdre and me for a conversion of a long barn.

One evening Jeremy asked me if I'd like to engineer an invention of his, a high-speed, flat-bottomed landing craft, or Sea Truck, that could lift itself over the water riding on a layer of bubbles. We called this an 'air-lubricated hull'. Jeremy had built some plywood Sea Trucks, but these rotted while being eaten by beetles. Jeremy wanted to make a version in glass fibre, a more suitable material for a boat hull. The problem was that plywood, which flexes without breaking, is a much more suitable material for a large, flat platform like the Sea Truck. Boats are normally curved and egg-shaped, deriving strength from this form, and glass fibre is a highly suitable material. It is, in fact, not dissimilar in concept to an eggshell. The problem was how to provide the necessary torsional stiffness to the large, flat shape of the Sea Truck while keeping it very light, an essential quality for the high speeds envisaged. We solved this by having a light cruciform-shaped chassis between the hull and deck.

Jeremy asked me to engineer, prototype and sell the glass-fibre Sea Truck. An art student just beginning his penultimate year at college, I was offered a job as head of a new marine division of Jeremy's successful engineering company Rotork. I jumped at the opportunity. Jeremy, however, insisted that I complete my degree, which is why my third-year project at the RCA was the engineering design of the Sea Truck. This was a curious project, perhaps, for a student supposedly studying architecture and interior design. As open-minded as ever, however,

Sir Hugh Casson allowed me the plain sailing I needed to complete the design of the boat and, in my final year, was quite content for me to work on another project for Jeremy, during which time the prototype of the Sea Truck was being built. This new project was the Wheel Boat, a boat with drum-like wheels that provided the propulsion and flotation, instead of propellers, and which would allow it to cross land and sea in a rather different way from Christopher Cockerell's hovercraft.

The origin of the idea was that boats, including the Sea Truck, lived on water and were unable to travel on land. This meant that they were liable to wear and tear by the constant corrosive effect of water, particularly salt water. If, however, the craft lived on land and only went into water when necessary, corrosion would be limited or even non-existent. While amphibious vehicles immediately come to mind, the performance of these was, and remains, compromised. They are slow on land and slow in water.

Meanwhile, it is well known that the most effective form of propulsion at sea is a paddle wheel. On a calm day, a paddle wheel can outperform a boat's propeller in terms of efficiency. It is a different matter when considering speed, where a paddle wheel is slower, or in rough seas where the paddle wheel regularly comes out of the water and loses thrust. Jeremy wondered whether we could make a vehicle that could both float on its wheels, like a jeep with giant wheels, and be propelled by them. That was the starting point. I took this on as my final-year project, while I was supposed to be designing buildings. Charming Sir Hugh Casson was indeed both lenient and encouraging.

Jeremy and I had this image of a high-vantage jeep-type body connected on suspension arms to four very large wheels providing the flotation with paddles incorporated to provide propulsion. The first task was to consider the size of wheel necessary to provide the flotation and then to develop the propulsion method for those wheels on a scale model. The test tank was to be the garden 'pond' at Widcombe Manor. It was actually more of a small lake than a pond. I did the calculations for the flotation, knowing the weight of the engine, propulsion system, vehicle body and load, and discovered that the wheels would have to be about 3m in diameter. I then built a scale model for testing purposes.

The test rig was a tripod that would have the lake bottom as its foundation. There would be a boom arm that would rotate around the centre of the tripod, with the wheel at one end and the drive motor at the other. The drive motor would be connected by a drive shaft to the wheel, the boom arm would be the support for the drive shaft and the drive motor. The motor would be a Briggs and Stratton lawnmower engine with V-belt reduction to the drive shaft. The wheel would be made of marine plywood for its sides and plastic for the circumferential surface hitting the water. Paddles would be screwed onto the wheel circumference.

I was able to install the steel tripod on the lake bottom and weigh it down with stones. Every day I would take out the boom arm on a small plastic boat, having filled the engine with petrol and fixed the paddle configuration for the day's test. However, I had to paddle the boat back to the bank to moor it, otherwise it would have been in the way of the boom arm as it rotated around the tripod. This left me to swim back to the tripod to pull-start the motor and conduct the tests with a stopwatch. I had to edge delicately around the tripod behind the boom arm as it rotated, otherwise it would have knocked me off. Fortunately, it was April, and not too cold by the time we started the in-water tests.

These were conducted using various loads on the wheel housed in an adjacent wooden box, and with various depths and widths of paddle as well as different numbers of paddles, and determining the speed obtained with a fixed speed and power of motor. At this stage the paddles were radial, although this was to change dramatically when the development continued thirteen years later. The results were sufficiently interesting, first for me to obtain my post-graduate degree, having studied the wrong subject, and secondly for us to revive the project in 1983.

Following our convocation at the Royal College in 1969, my contemporaries went off to practise as designers. Although the talented and successful yacht designer Jon Bannenburg had offered me an exciting design job, I had other ideas. I would not practise as a designer. I had engineered the Sea Truck at college and would now leave the RCA to make and sell it.

CHAPTER THREE

Sea Truck

It still seems amazing that Jeremy Fry had entrusted me with the task of developing and selling the Sea Truck. How well-equipped was I as a designer and engineer? Jeremy Fry was to teach me, without having to say anything, that every day is a form of education. My own education, in fact, owed as much to mentors like Jeremy and the open-mindedness of Hugh Casson at the RCA as it did to serendipity. When I set up on my own in my late twenties, I had to teach myself an awful lot about so many things before I made a success of my dual cyclone vacuum cleaner. You might say that if I had studied science and engineering at school and university, I might have succeeded long before I did, much as I loved making things like model aeroplanes at home. But even though I went to a good school, excellent in sciences, engineering, maths and technology, my family history dictated an arts, indeed Classics, education. It never occurred to anybody that engineering might be my thing.

On my last day of term at the Royal College of Art, I drove down newly opened stretches of the M4 from London to Bath. It was time to start work. I was now general manager for Rotork Marine with a salary of £2,500, an Issigonis Morris 1100 as my company car and a job to make and sell the Sea Truck. While I knew about making things, I felt I knew nothing about selling. Jeremy Fry, though, had done his charming best to convince me that I was the right young man for the job. 'You know the Sea Truck inside out, every nut and bolt. You've made it. You're the best person to sell it.' Which is what I did for the best part of the next five years.

It seemed remarkable to me then that this young man, dressed in floral shirts, flared trousers and a purple raincoat from Just Men, should be making and selling high-speed landing craft to army brigadiers, seasoned construction executives – the word 'executive' was all the rage in business circles and on badges of new cars with fake leather seats seen dashing down the M4 – and hardened oil company managers.

You might well ask, as I did, too, what I was doing making and selling rather than designing. I had, after all, spent the past five years training to be a designer. The theory I formed for myself was that I would be a much better-equipped designer if I experienced the vicissitudes of manufacturing and the feedback of selling viscerally. There was, though, more to it than this. For half my time at the RCA, I had made up my mind that what I really wanted to be was a manufacturer. I wanted to make new things – things that might seem strange, and not things you make because you know they will sell. The ultimate challenge, I suppose, was to design, make and sell inventive and wholly new products. To do this you need to be more than a designer or engineer, no matter how well educated. You need control over the whole process just as my exemplars Soichiro Honda, André Citroën and Akio Morita, creator of the Sony Walkman, had done.

While my decision to work for Rotork Marine with the title 'General Manager' might have appeared at the time to be a waste of my degree and a way of jettisoning a possibly golden future in the world of design, I had observed that designers work on what manufacturers and clients ask them to design and this held no appeal. Rather grandly, I had decided I wanted to be the one developing the technology, engineering and design of a product and making and marketing it, too.

How extremely fortunate I was to be able to discuss with Jeremy, a genius of a mentor, how our heroes like André Citroën had been able to design revolutionary new products and make a success of them. We also discussed how important design was in that success. Jeremy's friend Alec Issigonis and his revolutionary Mini featured, too. Issigonis had produced a cult car that won rallies, a timeless product without obvious styling. These discussions impressed me so much that I was left with a burning ambition to emulate designer engineers like Issigonis and Citroën in my own small way. I was even bold enough to declare to Jeremy that I would work with him for five years before setting off on my own.

Even so, I was well aware that manufacturing and, perhaps even more so, selling were looked down on in Britain as grubby trades. These were the worlds of 'metal bashing' and the 'oily rag' on the one gnarled hand and of the sharp-suited, smooth-talking, snake-oil spiv at the wheel of a Ford Cortina on the other. These nether worlds were clearly unsuited to sniffy middle-class graduates. This perception of the making of things as being somehow dirty and selling as unseemly underlined much of what was wrong with the British economy when I started work at the beginning of the 1970s.

In some ways, precious little has changed. When people ask why companies, including Dyson, manufacture outside Britain, the answer might be complex and yet it boils down to the fact that there are places and countries around the world, from Germany to Singapore, where manufacturing is encouraged and not least because it is seen as both worthwhile and exciting. There were to be several good reasons why, years later, I was to move Dyson manufacturing to Malaysia and, ultimately, its headquarters to Singapore, yet the sheer enthusiasm for making things there, expressed by both governments and entrepreneurs, is a wonderful antidote to the British way of looking down on the world of factories and production.

Inventions, though, no matter how ingenious and exciting, are of little use unless they can be translated through engineering and design into products that stimulate or meet a need and can sell. It might be fun to look at books illustrated with fascinating, if all too often enigmatic, impractical and rather mad inventions and designs that never happened, yet even the most worthwhile and world-changing inventions, from ballpoint pens to the Harrier Jump Jet, need to be a part of the processes of making and selling to succeed.

I happen to find factories and production lines romantic places. They are truly exciting. When we opened our vacuum cleaner factory in Chippenham, I worked on the production line for a fortnight not just to understand the process and to check that it made good sense from a technical point of view, but because I enjoyed it. I like making things, and to be a good manufacturer, you really have to. There was, though, little romance in mainstream British manufacturing in the 1970s, an era of obstreperous communist shop stewards, lacklustre management and poor-quality manufacturing.

Under life-or-death pressure, Britain had proved to be very good at making inventive machinery and products during the Second World War, yet somehow lost interest in peacetime. Perhaps the country was returning to pre-war form when there had been all too little incentive to innovate during the years in which Britain had an easy time selling run-of-the-mill manufactured goods to its imperial territories. Covering a quarter of the globe, these were captive markets.

As Britain's overseas reach declined, innovation and high-quality manufacturing mattered very much indeed, yet there was precious little of it. Why was this? One reason, following a spate of innovation from the Industrial Revolution, is that the wealthy sons of what had been mostly artisans or lower- to lower-middle-class, first-generation industrialists, aspired to be gentry. The scions of industrial magnates were shipped off to public school, immersed in Classical culture and taught to hunt, fish and shoot and to look down on the very world that had allowed their ascendancy from makers to squires.

Anything to do with politics, the military, the law, the established Church, the arts and, above all, making money from money with as little effort as possible, was good. Making anything by hand and, far worse, by machinery in factories, was very bad form indeed. Inventive and highly successful engineers from public-school and officer-class backgrounds like Sir Nigel Gresley, of record-breaking *Flying Scotsman* and *Mallard* fame, were as rare as British cars that didn't fall to bits in the 1970s. Imagine what Gresley's contemporaries and teachers at Marlborough College must have thought when this determined young man opted for an apprenticeship at Crewe Works with the London & North Western Railway rather than a science degree at Cambridge. When I was at school, one of the worst things teachers could say to you, as some did to me, about a lack of academic progress was, 'You'll end up working in a factory'. Boys who struggled academically were exiled to the Technical Drawing shed. That was the culture.

But think of the story of Paul Magès, inventor of self-levelling hydro-pneumatic suspension, around which the revolutionary Citroën DS of 1955 was designed and engineered. Magès's invention, developed for Citroën in secret in Nazi-occupied France during the Second World War, led on to oleo struts for aircraft (patented in the USA in 1960) and to gas-filled shock absorbers, two unsung inventions that have made our lives safer and more comfortable than before yet are taken wholly for granted.

Magès joined Citroën in 1925 at the age of seventeen as a very junior assistant. Twelve years later, he had worked his way up to become a technical draftsman. It was this skill, in tandem with his questioning mind, that led to the new suspension technology that improved people's lives and sales of Citroëns. The DS remained in production for the best part of twenty years, with sales reaching their peak in 1970, fifteen years into its production run. Paul Magès, who also pioneered speed-sensitive steering with the Citroën SM, launched in 1970, was hardly a 'thicko'.

Magès did, though, work in Citroën's venerable Paris factory that by 1970 was markedly behind the times in terms of comfort and cleanliness. And yet, what could be less honourable and grubbier than making nothing – not so much as a single oleo strut – other than money, and especially when this demands as little effort as possible though all but guaranteeing social esteem? Britain has been a nation of plunderers and raiders, pirates, buccaneers and chancers for several hundred years. We have a history of the fast buck celebrated in poetry, song, films and romantic histories, whereas the story of what has long been portrayed as the slow, difficult and dirty business of making things is held up to ridicule and shamed even today.

The opening ceremony of the 2012 London Olympics was loved by the British yet contained a running-down of the ethos of British manufacturing industry, a celebration of how we had moved far from the labour-intensive 'satanic mills' of the Industrial Revolution to the brave and squeaky clean new world of twenty-first-century 'Creative Industries', of which the Olympic ceremony was testament. We are good at these things and should celebrate them, but not at the expense of other things, like manufacturing. This smug sensibility served to reinforce the idea that industry – the engineering and making of things – was uncreative. And yet the Industrial Revolution took millions of people out of serfdom to give them homes and to create wealth for future generations. Instead of feeling shame, as the ceremony implied we should, we ought to feel pride and relief that our industrial might was able to produce the technology and equipment to repel the threats of two twentieth-century world wars.

Rather ironically, BMW's much-visited factory in Munich sits next door to the city's Olympic Park. Fifteen minutes from the city centre, people from all over Germany come here to pick up their new cars from BMW Welt, a wildly inventive building designed by the Viennese architects Studio Himmelb(l)au. When it was opened in 2007, members of the foreign press were puzzled by a colourful opening ceremony during which the Cardinal Archbishop of Munich blessed the building as if it was a cathedral. Where the 2012 London Olympics pictured manufacturing as the lowest circle of Hell, BMW likened it to godliness. And, of course, in Germany you may well be introduced to a crisply attired production engineer – Herr Professor Doktor Ingenieur – at some grand corporate do. His British equivalent – Barry or Dave – would not have been invited.

I remember, for example, when Tony Blair, the energetic 'New Labour' prime minister, began talking about the 'Creative Industries' soon after he came to power at Westminster in 1997. He was referring to businesses and professions like publishing, advertising, architecture and design, and as if manufacturing was somehow not creative. And, by the way, those esteemed professions and businesses are not 'industry'. Manufacturing creates real products every day. That is creative and it is industry.

Perhaps this lack of interest in manufacturing has something to do with the fact that news is fast while making things is extremely slow by comparison. It can take five years to launch a new Dyson product and this is a very long time in the life of journalism or, of course, banking or broking. Equally, there is no respectable profession of manufacturing, no long-established lobby group like the Royal Institute of British Architects or Institution of Civil Engineers. It is not by coincidence that the Institution of Civil Engineers, along with the Institute of Mechanical Engineers, have their Victorian London headquarters just a few minutes' walk from the Palace of Westminster. Manufacturers have no institute or imposing building near Parliament. With no organised representation, they fight their own corners individually.

The people who run Britain have had little interest in manufacturing for a very long time. Ministers and Members of Parliament only very rarely come from industrial or engineering backgrounds. Aside from an antipathy to long-term thinking and the discipline manufacturing demands, memories of industrial unrest and the incompetence of large-scale British manufacturers in the 1970s linger on.

If manufacturing has long been seen, in Britain at least, as a lowly pursuit, then selling has surely been seen as beyond the pale even though bankers and others working in seemingly respectable financial services are mostly salespeople by another name. Selling, though, goes with manufacturing as wheels do with a bicycle. It is far more than flogging second-hand cars or contraband wristwatches. Products do not walk off shelves and into people's homes. And when a product is entirely new, the art of selling is needed to explain it. What it is. How it works. Why you might need and want it.

I happen to think that the Morris Minor was much the better car and yet the way the Volkswagen Beetle was sold into the US market was very clever, with the Doyle Dane Bernbach ad agency playing on the fact that this curious car created by Ferdinand Porsche for Adolf Hitler was small, odd and – to American eyes – ugly. Press ads like 'Think Small' and 'Lemon' worked in an era, from 1955, when most American cars were getting bigger, faster and ever more brash. In 1949, when the VW launched in the US market, just two cars were sold. Sales shot up following DDB's memorable ad campaigns. By 1970, Americans had bought 570,000 VW Beetles and DDB had increased its earnings tenfold.

Unlike the Americans, the British rather despise salesmen and the art of selling. This is a pity. It can be noble and exciting work. Jeremy Fry taught me not to try to pressure people into buying but to ask them lots of questions about what they did, how they worked and what they might expect of a new product. Equally, I learned that most people don't really know exactly what they want, or if they do it's only from what they know, what is available or possible at the time. As Henry Ford said, famously, if he had asked American farmers what they wanted in terms of future transport, they would have answered 'faster horses'. You need to show them new possibilities, new ideas and new products and explain these as lucidly as possible. Dyson advertising focuses on how our products are engineered and how they work, rather than on gimmicks and snappy sales lines.

Manufacturing itself is highly creative. It can turn left-field inventions into reliable, desirable products that sell. Jeremy Fry very much enjoyed thinking up new ways of doing things. This is how he had made Rotork work in the first place, before creating a marine division in 1968. A member of the Fry chocolate-manufacturing family, Jeremy had spent his meagre inheritance on Rotork, a small British engineering outfit he bought with his brother David in the early '50s. Rotork was soon producing motorised valve actuators, engineered and patented by Jeremy, for pipelines in the oil industry. It was an ingenious idea, a form of automation that was compelling to oil companies – Shell, BP and Esso were major customers – and, from 1962, when Rotork took an order for a thousand sealed, weatherproof and explosion-proof actuators for a new French uranium enrichment plant, to the nuclear power industry, too.

I still find myself saying and putting into practice at Dyson some of the same things Jeremy said and did when I worked for him half a century ago. As an inventor, engineer and entrepreneur, he believed in taking on young people with no experience because this way he employed those with curious, unsullied and open minds ... as long as they didn't sport beards or smoke pipes. He thought of beards as old-fashioned, harking back, I think, to the days of George V and Tsar Nicholas II. In any case, the '60s and '70s was the age of Gillette and personal male grooming. A beard was widely considered to be unhygienic, while smoking a pipe was, or so Jeremy believed, a sign of smugness. Pipes have long died out while beards and 'designer' stubble have proliferated. Perhaps Jeremy would have changed with the times.

He believed, most of all, in the power of enthusiasm. He sold new and intelligent things to people not through smooth and well-rehearsed sales patter but because, aside from possessing charisma and charm, he enjoyed talking to prospective customers about engineering, for which his enthusiasm was infectious and second to none. As an instinctive inventor and engineer, Jeremy was always looking for 'a better way of doing things'. He talked face-to-face to employees about new ideas. He never sent memos. Where there were opportunities, he didn't see problems. He was a natural teacher. When I told him I didn't know how to weld, he lit up a gas-welding nozzle and taught me the rudiments in little more than ten minutes. Being quick-minded he expected the same of those he employed.

He talked of the need to listen to your customers, aiming to improve products wherever necessary and, if you are an inventor, simply for improvement's sake. This is not to say we at Dyson ask our customers what they want and then build it. That type of focus-group-led designing may work in the very short term, but not for long. Just before the launch of the Mini car, Austin Morris did indeed consult a focus group, and nobody wanted this tiny car with small wheels. So they cut the production lines down to one. When the public saw it on the streets, they were most enthusiastic for it. Austin Morris never caught up with demand, missing out on serious profits.

What was also undeniably special and important, too, I think, was Jeremy's inimitable sense of style. Under Jeremy, Rotork was based initially at Widcombe Manor, the Grade 1-listed Georgian house he bought in 1954 at the time of his marriage. Treading a lively path between convention and experimentation, he installed a bath in his bedroom, covered the library floor with leather and lined walls in copper. He was a close friend of Princess Margaret and the Earl of Snowdon, who stayed there often. Life at Widcombe was glamorous and, to the relish of the tabloid press, sometimes a little risqué, too. This was not a part of Jeremy's life I knew, although I did admire and get to know some of his other friends, like Andy Garnett, who became a non-executive director of Dyson, and his brilliant wife, the writer Polly Devlin, and the opera director Robert Carsen, a lifelong friend.

Jeremy was a great engineer and an inspiring entrepreneur. He was always happier making things in the machine shop than he was sitting around a boardroom table or mingling with high society. He was interested in young people or anyone enthusiastic who wanted to learn. He loathed smugness and experts, by which he meant those who want you to believe that they know everything about a subject when the inventive mind knows instinctively that there are always further questions to be asked and new discoveries to be made.

The Sea Truck was Jeremy's design, although when we first discussed it, when I was in my second year at the RCA, he said, 'This boat of mine. No one wants it in wood. Would you like to do it in fibreglass?' I would, and did, despite knowing little, if anything, about the process. But what made the fibreglass Sea Truck special? It was light, rugged and fast. It was also flat-bottomed with side skegs. The skegs were sledge-type runners about 50mm deep running from the front to the back of the hull. They served as skids when the boat was put on the ground but, more importantly, they trapped bubbles underneath the boat so that when it was planing at speeds above 12mph, the frictional resistance of the water was dramatically reduced, and it could plane at speeds of 40 knots and above using less power than a conventional hull. The Sea Truck is the only craft where, once it planes, you can throttle back, a testament to the efficiency of the layer of air bubbles it generates. Being flat-bottomed, it could negotiate shallow drafts and ride straight up on to beaches. It had no need for jetties and piers.

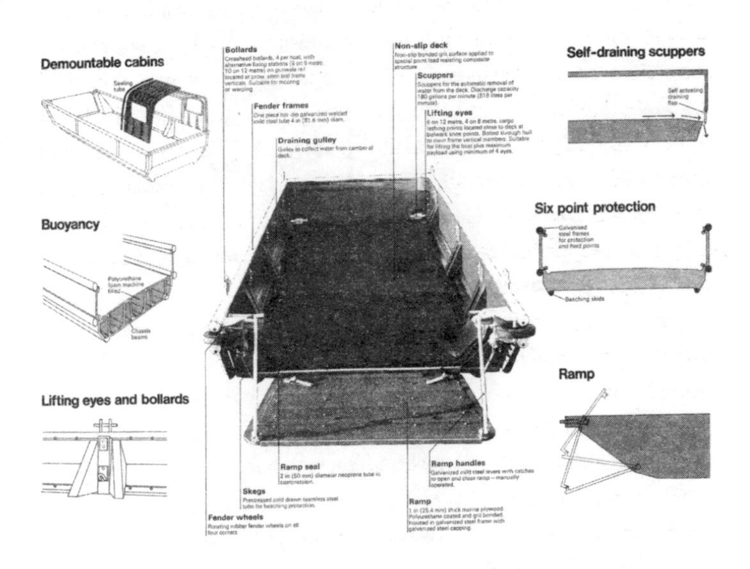

A low-cost modular workboat – cabins and other accessories could be added to the basic craft – the Sea Truck was stable at sea and, without a keel, sailed securely in shallow water. A little shorter and a bit wider than a Routemaster bus, it could carry a pair of Minis, or two Land Rovers, and a heap of equipment, from – as experience taught me – machine tools and cables to guns and ammunition. Initially, we sold one open boat a month. By 1974, I was making 200 a year with cabins, and for clients in forty countries. An invention capable of change and adaptation, there was always, it seemed, a way of making and selling a better Sea Truck. Through the Sea Truck adventure, I learned – on the ground and, in this case, on water, too – how selling and manufacturing are very much two sides of the same coin.

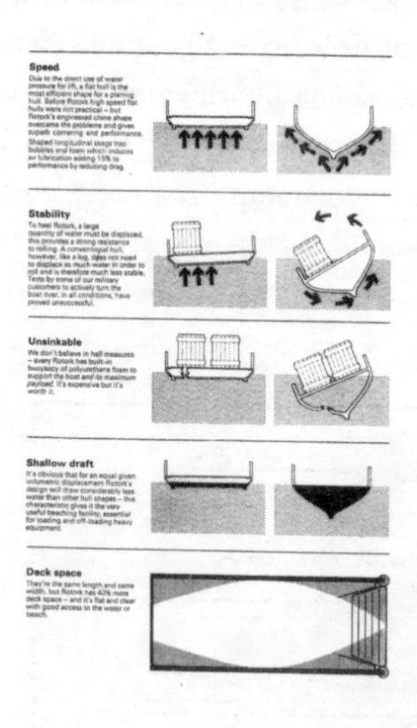

I travelled the world, appointing agents as I went. I sold Sea Trucks in Norway to lay underwater phone lines, Sea Trucks to many foreign armies and navies, Sea Trucks as passenger ferries and Sea Trucks to the Egyptian Special Boat Brigade, who, excited by their low profile – they were hard to hit with gunfire – used them to storm Sinai during the Yom Kippur War. I also sold Sea Trucks to the Israelis. With its bow flap lowered in seconds, commandoes could sprint out onto landing grounds. The Royal Navy said it would buy Sea Trucks if they could be armour-plated, but this would have weighed them down and lost the boats their advantage of stealth and speed. We did, though, meet the Royal Navy's requirements with Sea Trucks commissioned to ferry stores and personnel and to recover torpedoes.

We built a Sea Truck for the London Fire Brigade, a rapid-response tender complete with fire pumps and ancillary equipment. It could tear up the Thames faster than River Police boats. Within a certain weight limit, the Sea Truck, highly manoeuvrable in shallow water, made a good tug. As well as pulling, it was also good at pushing. Demonstrated to the army, the Sea Truck proved itself capable of pushing sections of bridges weighing 45 tons. We also sold one to the University of Southern Florida for work as a marine laboratory, another to South Africa to destroy invasive water hyacinths. The Royal Yacht *Britannia* bought a royal blue Sea Truck with a red carpet to furl out across the ramp. The Queen used it on visits to far-flung parts of the Commonwealth such as the Gilbert and Ellice Islands in the Western Pacific.

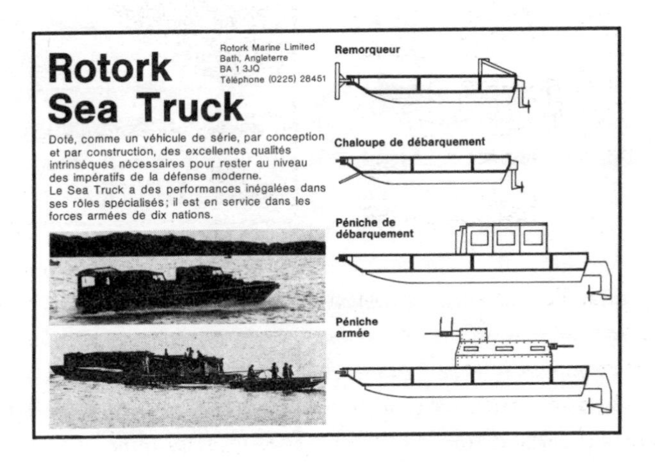

The Sea Truck was at its best perhaps in the months following the terrible floods that smothered much of East Pakistan in the wake of the Bhola cyclone, which smashed across the Bay of Bengal in November 1970. Bhola is said to have been responsible for up to half a million deaths. Widespread criticism of the government of West Pakistan for failing to respond to the disaster led to the Bangladesh civil war (aka the

Liberation War). On its way to becoming the independent Republic of Bangladesh, East Pakistan suffered horribly from atrocities committed by West Pakistani armed forces. Floods, hunger, streams of refugees and the sheer horror of the situation led to fundraising events in the West – most famously 'The Concert for Bangladesh' organised by George Harrison in New York's Madison Square Garden and, on Rotork's part, the dispatch of 100 Sea Trucks to Bangladesh. Jeremy Fry spent several months there managing our service depot and supervising the distribution of supplies.

A true Jack and Jill of all trades, the Sea Truck was part of an unofficial school of engineering and low-key design that includes the Land Rover (a machine that so often went wheels-on-deck with the Sea Truck), the Swiss Army penknife, the Citroën 2CV, the Bell 47 helicopter and Alec Issigonis's Mini. What I liked so much about these machines – and my affection for them remains undimmed – is their ingenuity and the fact that the power of invention invested in them made for designs that reimagined and revolutionised their market sectors and even created wholly new markets. And yet, for their all their functionality, each is a highly individual product with a character and charm of its own.

What is equally interesting is that these radical machines made use of pre-existing ideas and components. With the Mini, for example, the features that made it so revolutionary had been tried at various times beforehand. Before the Mini, which made its debut in 1959, there had been cars with transverse engines, notably the Saab 92 in 1947, a wind-cheating, rally-winning saloon engineered by a small team led by Gunnar Ljungström, an aero-engineer and keen sailor, and stylist Sixten Sason, a former Swedish Air Force pilot and designer of Saab military aircraft. In fact, the transverse car engine dates back to 1899 and the Critchley Light car made by Daimler in Coventry. There had been cars in the mid-'30s with front-wheel drive like the Citroën Traction Avant, engineered and styled by André Lefèbvre and Flaminio Bertoni, to whom we owe the 2CV. There had been boxy cars making as much room as possible for driver and passengers and there had been forms of rubber suspension, too. Issigonis brought these ideas together in a tiny car that, despite its very modest dimensions – just 10ft long and 4½ft wide – could seat four adults and their bags and drive like a go-kart. A whole 80 per cent of the Mini was given over to its occupants. Issigonis was particularly pleased with the big, bin-like door pockets. The bins, he said, held twenty-seven bottles of Gordon's gin and one of vermouth. His tipple.

Like the Sea Truck, the Mini led many lives in its forty-year production run. It was a family saloon, a car for stars of stage, screen and recording studios. It was a tiny estate car (my mother also owned one while I was a schoolboy at Gresham's), a Monte Carlo Rally winner, storming up icy Alpine passes in snow, much to the consternation of the French judges, who tried very hard to find reasons to ban these English terriers, and a racing car capable of taking on and beating fearsome, twin-cam 3.8-litre Mark 2 Jaguars at Goodwood in the early '60s. The Mini was a delivery van, a pick-up truck, a post van, a fashion statement and even a police patrol car.

Although I never met Alec Issigonis – of course I wish I had – he was a close friend of Jeremy Fry and the pair raced minuscule, lightweight hill-climbing cars together with John Cooper, who famously made the Mini into a fast track and rally car. With fellow engineer George Dowson, Issigonis had developed a tiny monocoque, or stressed-skin, aluminium and plywood-framed 750cc hill-climbing racer in the 1930s, culminating in a lithe, minimalist car as light on its wheels on the ground as a fly on its legs on water. It featured rubber rather than conventional steel suspension. It enjoyed a successful career.

For the Mini, Issigonis had wanted to use a form of interconnected suspension improving on that of the soft-sprung Citroën 2CV, a car that leaned alarmingly, if safely, around bends. For reasons of technical and financial expediency, the first Minis were fitted with rubber cone rather than interconnected suspension. Issigonis's friend Alex Moulton, another innovative engineer who lived near Bath and was also a friend of Jeremy Fry, devised the Mini's rubber suspension. The cones were fitted between the car body and the axles adjacent to the wheel. They softened the impact of rough roads, with the rubber cone partially collapsing under load in the same way that a spring compresses. More significantly, Moulton invented the Hydrolastic suspension system that made my Rotork company Morris 1100 such a comfortable and reassuring car, despite its modest size, to drive up and down the M4 to and from Bath. This was a similar interconnected suspension to that used by Citroën in the ground-breaking DS model of 1955. The idea was that interconnecting the suspension between the wheels allowed suspension at other wheels to assist when one wheel was deflected. The system also aided cornering by spreading loads to all wheels.

Alex's family were in the rubber business. His father John Coney Moulton, an army officer who served in India and Singapore during the First World War, had gone on to direct the Raffles Museum in Singapore. An expert in South East Asian cicadas, he was on leave at home in Bradford-on-Avon, when serving as Chief Secretary to Charles Vyner Brooke, the last White Rajah of Sarawak, when he fell ill with appendicitis and died. Alex was six at the time. I got to know him quite well through Jeremy Fry. Curiously, given my own situation, Alec Issigonis's father had died early – he was fifteen at the time – as had Jeremy's while he was at Gordonstoun. The three of them had very different characters, despite their overlapping career paths, but all three shared an independence of mind, a hard work ethic and a drive to succeed.

Mechanically gifted, Alex had built a steam car while at school at Marlborough. After reading Mechanical Sciences at Cambridge, he was apprenticed to the Bristol Aircraft Company. Like Alec Issigonis, Jeremy Fry and Buckminster Fuller, he sought lightness in everything he made along with an efficient use of materials. Just as Jeremy made a better valve actuator and Issigonis a better small family saloon when the alternative was a bubble car, Alex wanted to create a better bicycle.

Impeccably dressed as always, he set to work devising one in the stable yard of his Jacobean house. Launched in 1962, his famous Moulton bike, with its compact lightweight frame, small 16-inch wheels and rubber suspension, proved to be truly revolutionary. It made cycling, especially in town, fashionable all over again. The Mini of the cycling world, it was a darling of fashion editors, featuring in numerous photo shoots in Sunday supplements and magazines like *Vogue* throughout the '60s. It turned ideas about what exactly a bicycle, or any other product, ought to be on its head. As the critic Reyner Banham wrote in the *Architectural Review*, 'bicycle thinking can never be the same again, and there can be no more nonsense about permanent and definitive forms, for even the Moulton is capable of improvement'. The Moulton was also an excellent bike to ride. It remains in production. I own a wonderful extra-lightweight 'Pylon' space-frame model.

Although several companies have cribbed the wheel size and foldability of the Moulton, none have come close to matching its pure and refined engineering. Neither have they bothered with the suspension necessary to smooth out the bumps created by small wheels or the high tyre pressure needed to achieve minimum rolling resistance. The tool kit is a single Allen key slid into a socket under the comfortable saddle.

Something I might have learned from Alex Moulton during my Rotork years, though, was the story of how he had first thought of getting an established bike company to manufacture his radical new design. In 1966 Raleigh brought out a bad copy that hit sales of the Moulton original. Moulton sold out to Raleigh the following year with Alex retained as a consultant. It was not an easy partnership. When sales slumped in the early '70s, Alex reacquired rights to his design and began manufacturing the bike, and versions of it, on his own at Bradford-on-Avon. Alex's story reminds me of how important it is for inventors to keep hold of patents and rights to their designs and, if possible, to run their own show.

The other thing I was learning was that an invention might be a brilliant idea yet either unsuited or irrelevant to the market it needs to sell in. A design might be considered ahead of its time and, sometimes because of this, even ridiculous. The hugely successful Sony Walkman was dismissed when first launched because who could possibly want a tape recorder that couldn't record? And it was received knowledge until Volkswagen and, later, Honda crossed the Atlantic with the Beetle and the Accord that Americans were wedded resolutely to big cars.

Alec Issigonis was a genius of sorts who knew how small cars could be big. He was also an independently minded man who was never going to be able to fit comfortably into the giant, staid and highly bureaucratic British Leyland conglomerate that, in 1968, swallowed BMC, owner of marques like Austin and Morris, who made the first Minis. Aside from anything else, Alec's view that 'market research is bunk' and that one should 'never copy the opposition' was hardly likely to appeal to his new boss, the hard-nosed lorry salesman Donald Stokes, who replaced the much-loved Morris Minor with the hapless, market-researched Morris Marina and smothered the roofs of lithe Jaguars in vinyl. The bestselling British car of all time is the Mini. If market research had ruled Issigonis's roost at BMC, it would never have existed. Inevitably, Alec Issigonis was side-lined. A thorn in British Leyland's tough hide, he was, however, feted by fellow engineers, being made a Fellow of the Royal Society in 1967.

I was given an original Morris Mini Minor by Dyson engineers for my sixtieth birthday. It's kept at the Dyson campus in Malmesbury. It's been sectioned lengthways, so its interior and workings, from the transverse engine, rubber suspension, weight-saving sliding windows and those big door bins are on display. It's an everyday reminder of how ingenious engineering design can be.

The Mini is also a reminder of how a radical product can work its way into the manufacturing heart of an established company and succeed to the tune of 5 million sales over a forty-year production run. It was a left-field design that, if never ordinary, became a part of everyday life. The Sony Walkman is another fascinating success story because, at first, its design appeared to defy common sense. Launched on 1 July 1979, in time for school and college holidays, this personal cassette player that allowed people on the move to listen to music through headphones, was hugely popular from day one.

Priced at US$150, the compact silver and blue Walkman wasn't cheap, while within Sony itself it was controversial and brave because it was unable to record, and no one had made a 'tape recorder' that wouldn't do so before. Nevertheless, Sony's Masura Ibuka – one of the Japanese company's founders – hoped to sell 5,000 Walkmans a month. He sold 50,000 in the first two months. By the time production ended in Japan in 2010, more than 400 million had been sold worldwide.

Sony had come up with the right product at the right time, a discreet music player for people taking up jogging or wanting to study or relax at home or while travelling without bothering others and listening to the music of their choice. Cleverly, Ibuka had asked his deputy Nori Ohga, a physicist and musician who had trained under Herbert von Karajan in Munich, to design a stereo playback-only version of existing cassette recorders like the Sony Pressman. Using off-the-shelf components where possible, in 1978 Ohga and his team produced the TC-45, but priced at US$1,000 and, in Ibuka's mind, too bulky – he tried it on business flights where he liked to listen to opera – it was a case of back to the drawing board.

With lightweight foam headphones and no function other than playback, the Walkman emerged. The press lampooned it. Even the name was ridiculous. The Japanese press was wrong, although the market hadn't known it wanted a tiny personal stereo. When it saw the attractive little device and heard it in action, it fell in love with it. By the mid-1980s, the word had entered the *Oxford English Dictionary*. The Walkman was a cultural phenomenon, while, for Sony, it was also easy to manufacture.

And then there is the story of Citroën and its inventive and highly distinctive cars, from the front-wheel-drive Traction Avant of the 1930s and the minimalist 2CV of the 1940s to the technologically advanced DS of the 1950s. Under the direction of the highly decorated former military aviator and Michelin director Pierre-Jules Boulanger, Citroën drove its Engineering and Design department energetically towards radical design. Boulanger nurtured a remarkable team, chief among them André Lefèbvre, a former racing car designer and driver, in charge of engineering design and Flaminio Bertoni, an Italian sculptor who gave these unmistakably Gallic cars their distinctive styling. As part of Michelin since its first bankruptcy in 1934, Citroën developed radial tyres in tandem with Paul Magès's self-levelling suspension for the DS.

What Citroën proved with these cars – as indeed did Alec Issigonis with the Mini – is that a large public is not afraid to buy highly advanced products. Famously, Citroën took 12,000 orders for the DS on the day it was unveiled at the 1955 Paris Motor Show.

Even so, Citroën was to make a key commercial mistake. Between 1955 and 1970, the company launched no new models. It had been resting on its laurels. And then, realising that the car-buying public was looking for something new, it appeared to panic, investing in and launching a host of new cars. The expense bankrupted the company. Taken over by Peugeot in 1974, it was never the same again, making money yet losing its character and its drive for innovative design, engineering and styling.

There is a lesson here. Rather like the way some sharks have to keep moving to stay alive, innovative engineering-led manufacturers need continuous innovation to stay competitive. Striving for new and better products is often what defines such companies. At Dyson, we never stand still. In a quarter of a century, we have gone from making a revolutionary vacuum cleaner to prototypes of a radical electric car. Invention tends to compound invention and companies need to be set up for this. They can have a DS moment – a prolonged one in the case of the car that in Roland Barthes' words 'had fallen from the sky' – and still fail.

In Britain, though, one of the most revolutionary of all technological inventions, the jet engine, had a rather rocky and painfully slow gestation. It was the invention of another of my heroes, Frank Whittle, whose sheer determination and perseverance have inspired me since even before I first set up on my own as a manufacturer. Those working at Dyson's Malmesbury campus are reminded every day of Whittle's achievement. We have the world's oldest working jet engine – it is also one of the very first put into production – in one of our workspaces.

We fire it up in the car park.

This particular engine, a Rolls-Royce RB.23 Welland, was assembled at Barnoldswick, then in the West Riding of Yorkshire, in December 1943. We discovered its Rolls-Royce ancestry at the time I had wanted to take it to the Goodwood Revival, but was unable to do so because we found a fuel leak. A set of Whittle drawings for the fuel system showed ours was not one of Whittle's original engines, but a version made by Rolls-Royce. Our enthusiastic engineers rebuilt it according to Whittle's original drawings. It now works perfectly.

The point, though, is that this is the forerunner of engines that, in recent years, have powered more than 3 billion passengers a year around the world. A part of what makes the Welland so special is that, as a turbojet, it has very few moving parts in contrast with the Rolls-Royce Merlin piston engine – powering Spitfires and Lancasters – that preceded it. In fact, Whittle's first formula for a jet engine, written and drawn by hand in a school-style exercise book in 1929 when he was twenty-two, is as simple as ABC. He wanted an engine that could power a passenger plane across the Atlantic at 50,000ft and 500mph.

If the Whittle jet had been developed apace, by 1939 the RAF may well have had a range of jet fighters and bombers that would have been a very serious challenge not just to the Luftwaffe but to Hitler's military ambitions as a whole. The tragedy, as is well known, is that the Germans were able to catch up and put jet aircraft into production before the British. The double tragedy is that German engineers were able to study Whittle's design freely. In 1935, the Air Ministry refused to pay the £5 needed to renew his patent. And even before then, Whittle's proposal was free for anyone to study as, believing it to be of little consequence, the Air Ministry had not included it on its 'Top Secret' list.

Against the odds, Whittle succeeded. A working-class boy from Coventry with a passion for flight and for making model aircraft, in 1923 Whittle was apprenticed to the RAF as a metal aircraft rigger. He was one of just five RAF apprentices chosen to fly as officer cadets. Whittle was an excellent, indeed daredevil pilot. Recognising his exceptional intelligence, the RAF saw him through Cambridge and a first-class degree, in two years instead of three, in Mechanical Sciences while at the same time building the world's first jet engine through the Power Jets company. Power Jets was generously and bravely financed by Maurice Bonham-Carter, grandfather of Helena Bonham Carter. It is a matter of disgrace that the government stole the Whittle jet from his company, without compensation and not even allowing Power Jets to compete in its manufacture.

Both the Air Ministry and its experts and manufacturing companies had been sceptical of Whittle's jet. A trickle of funding saw the engine come alive in a rudimentary industrial building in Rugby in March 1937. This was the moment when the Air Ministry gave the project the thumbs-up, although it was really only when the Germans flew their first jet aircraft, the Heinkel 178, days before the invasion of Poland that the vital importance of Whittle's work was fully recognised. With Rover working on the engine and Gloster Aircraft on the airframe, Whittle's prototype jet took to the air in May 1941. 'It just sucks itself along like a Hoover,' said one RAF pilot to another when asked how the E.28/39 flew without a propeller. 'Get me a thousand Whittles,' boomed Winston Churchill. But the production aircraft, with Rolls-Royce-built engines, that Whittle was working on with Gloster was not the E.28/39 – a prototype – but the twin-engine Meteor fighter that entered service in July 1944.

The Jet Age had truly begun and especially so when a Whittle engine was shipped to the US, where it spawned the first American turbojet, General Electric's GE J31. And then, two years after the war ended, Britain's Labour government sold fifty-five Rolls-Royce Nenes, a development of Whittle's first-generation jet, to its friends in the Soviet Union for 'non-military use'. 'What fool will sell us his secrets?' Stalin asked. The Nene was quickly reverse-engineered by Vladimir Yakovlevich Klimov into the VK-5A for what became the highly effective swept-wing Mig-15 fighter US pilots came up against in the Korean War. Today, we need a great degree of secrecy and security to protect our research and inventions. It seems extraordinary that one of the most important of all modern inventions should have been handed on a plate to our Cold War enemies. But that's politicians for you. It also demonstrates how little value they place on home-developed technology and the investment required by companies to achieve it.

If Whittle had been an engineer with Rolls-Royce from the outset rather than a maverick RAF officer working on the fringes of manufacturing industry, might his jet engine have been given a smoother, faster flight through development and production? Perhaps. In any case, when I decided to make my own bagless, dual cyclone vacuum cleaner, I learned – after a number of false starts, especially in trying to get existing manufacturers to take it on – that the best thing for me was to go my own way and build my own factory.

This, as you might expect, was not especially easy. When I started out on my own, factories in Britain were only available on twenty-one-year leases at a time when inflation and interest rates were very high indeed. And even when you are established and need extra manufacturing capacity quickly, the procedures and politics of planning permission and so on can threaten to drag you down, or, of course, prompt you to set up elsewhere in the world where decisions are taken quickly and effectively.

Once you have a factory, how are you going to make your inventive new product? Are you going to make plastic components yourself, or buy them in? And what about motors and gaskets and a hundred other parts that, one or way or other, you have to will into being? Experience taught me that, ideally, a manufacturer – Dyson, certainly – should aim to source as little as possible from outside the company. Those of us who drove British cars made in the 1970s know pretty much exactly why. Poor assembly aside, what often let these cars down were components sourced from poor-quality external suppliers. Electrical failures were legion.

Electric motors spinning reliably at phenomenal speed are at the heart of our products. Nobody made anything other than an ordinary motor. So, we developed our radically new technology that we had made ourselves. This is an expensive thing to do, but it allowed us to revolutionise vacuums. Obviously at Dyson we cannot make absolutely everything on our own, but we can work with suppliers so that they are in tune with us, with our manufacturing standards and our values. Because what we're doing is special and different, we can't go to a company like Foxconn, for example, which, founded in 1974 and employing over 800,000 people in different parts of the world, makes well-known American, Canadian, Chinese, Finnish and Japanese electronic products. Those products are mostly made from off-the-shelf components. We design our own components. We don't buy them off the shelf.

In 1974, I was at a personal crossroads. Life was certainly exciting. Although caught up in my work for Rotork, I was also a father. Our daughter Emily was born in 1971 and our first son, Jake, two years later. Deirdre and I had bought an old stone Cotswold farmhouse that meant the end of my long-distance commuting and the beginning of a lot of hard, physical work that helped spark my idea for an invention of my own designed to make everyday life better. Better, that is, for someone like me restoring a house and having to negotiate paths and muddy tracks with a wheelbarrow that was difficult to steer when loaded and had a tendency to topple on its side. Surely it must be possible to invent a better wheelbarrow?

While Jeremy and Rotork made me think of products that operated, like the Sea Truck, on a big outdoors stage – capital goods like aircraft or excavators or working boats bought by people with plenty of capital – family life helped focus my mind on those that might make some of the simple trials of everyday domesticity that much easier than they were and even, perhaps, enjoyable. It wasn't that I didn't revel in the idea of designing for oil companies and armies, but now I had become interested in designing products for everyday use, designs derived from my own experience. With capital goods, you have to interpret what industries want in terms of the performance of their service, a quite different proposition from what we want at home where we are also the user. This is important because, as a potential customer, unlike with capital goods, you can make the judgement about whether or not you would venture to buy this revolutionary and strange product. What this means is that it is very much riskier to design a revolutionary new and perhaps strange product for someone else or someone whose business you do not know. You are dependent on interpreting what they say and what they might need. And you might be wrong. If, however, you are designing a product that you use yourself, you understand what you need from it. If your new product is radically different and works in a different way, you can use your own feelings and interpretation of the product to decide if it is something that would appeal to you enough to want to buy it. It is easier and safer to take a big risk with a product if it is one that you would use yourself.

I did, though, have another boat in the pipeline. During my years at Rotork I spent creative and very enjoyable months at Le Grand Banc, Jeremy's research centre. This was the remote Provençal hamlet that, abandoned during the First World War, Jeremy had bought in the early '60s and restored – with the help of students from the Architectural Association. The setting, above fields of lavender in the Luberon Valley, with the Mediterranean in the distance, was and remains exquisite. This was to be the natural stage set for Claude Berri's captivating films *Jean de Florette* and *Manon des Sources*, made back-to-back in the mid-'80s. It was a wonderful place to think.

One of these thoughts was for a boat made out of the polyethylene pipes we saw along local roads during the installation of gas mains. The pipes were very hard to break, unlike the fibreglass hulls of Sea Trucks landing on rocky shores at high speed in the hands of overzealous owners. And, even if a pipe were to be damaged on a Tube Boat, it could be replaced. We would have eight or ten pipes lashed together, rather like Thor Heyerdahl's raft *Kon-Tiki* that caught the global imagination when the Norwegian explorer crossed the Pacific in 1947 from South America to the Polynesian islands. Where, though, Heyerdahl used balsa logs, native to South America, I worked with thermoplastic. What the two materials, natural and man-made, have in common is that both are lightweight and durable.

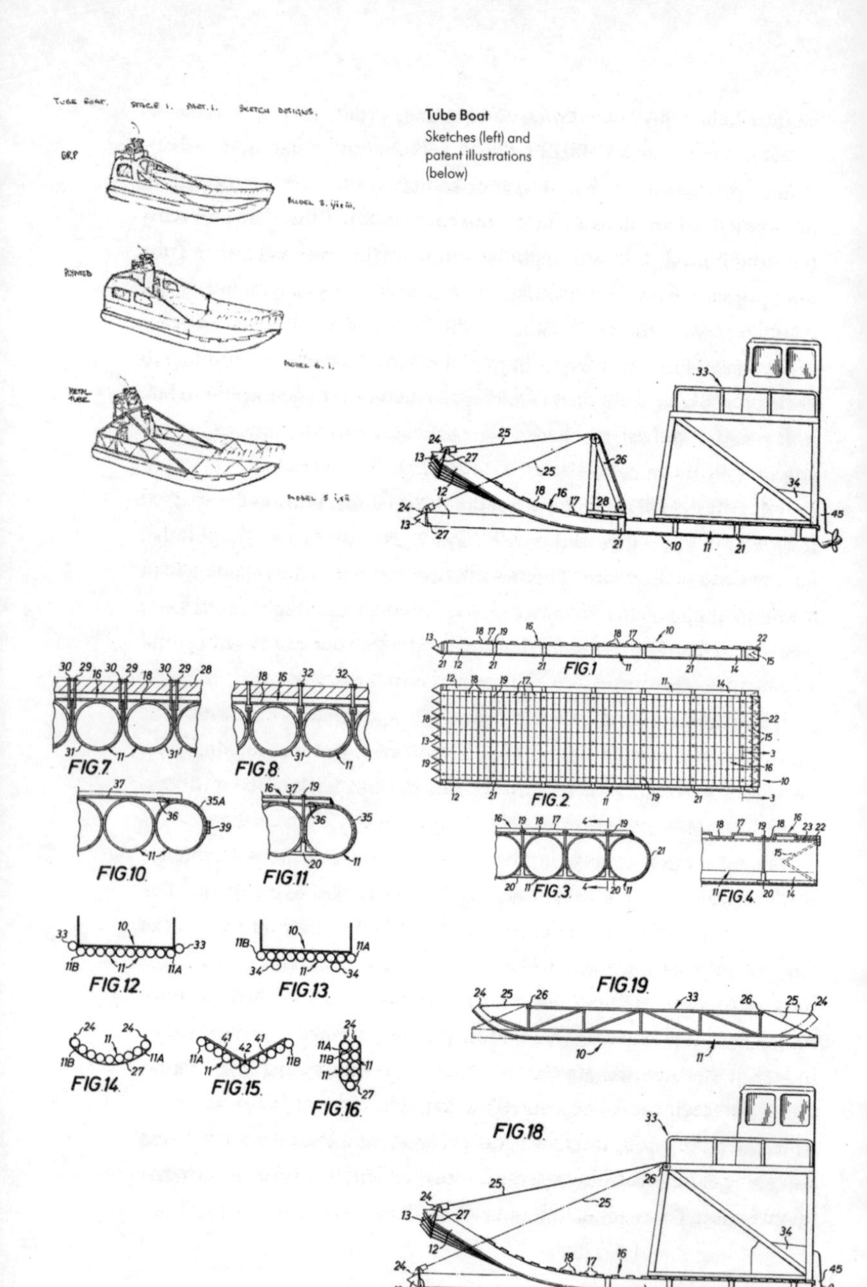

Tube Boat

Sketches (left) and patent illustrations (below)

The Tube Boat would be much cheaper than the Sea Truck. I started making models and testing these on a local reservoir in the Basses-Alpes. Sadly, the Tube Boat never made it into production. Rotork itself had started to lose interest in the Sea Truck venture from 1975 as Jeremy removed himself from the business, but at least one aspect of the Tube Boat's design wasn't wasted. To seal the ends of the polythene tubes, I used footballs. In practice these would have been moulded spheres. What with my frustration with cantankerous wheelbarrows, the ball at the end of the Tube Boat tubes was beginning to give me an idea. This was the Ballbarrow. I hadn't thought of the name yet, but as I sat on a stone wall at Le Grand Banc overlooking the Provençal countryside at its lyrical best, I felt impelled, no matter how much I liked and respected Jeremy and enjoyed working with him, to go my own way. Typically, Jeremy generously offered to back my new venture. This was an offer I dearly wish I had taken up. In 1974, with two young children, a substantial overdraft and a whopping mortgage, I stepped out of an exciting job, a company directorship and a salary and into the unknown.

Deirdre and I had rolled the dice in our minds. As always, her support and guidance were all-important. She said we could bet on my inventive streak. And so we did. For me, though, risk has long been an antidote to inertia. I felt that then. As an artist, Deirdre appreciated what a 'project' or idea was about. It is something you get caught up in. You have to do it, believe in it and trust in a successful outcome. I was very lucky that Deirdre understood and allowed me the rein to plunge into the unknown, a world of debt, risk and potential penury. Back then there were no incentives to start a business of your own, while venture capital was non-existent. This was long before Silicon Valley start-ups. Indeed, it was quite the opposite, with investors earning big money from interest on their capital and unwilling to risk it in manufacturing.

Against the grain, though, I was going to be a manufacturer. I was also going to be something else, and someone I believe is key to successful invention. I was going to be an entrepreneur.

CHAPTER FOUR

The Ballbarrow

Had I been slightly off my head in striking out on my own in 1974? Working with Jeremy Fry had been an exhilarating experience. There had been so much adventure that, just possibly, the reality of the world outside our Rotork bubble had eluded me. Or perhaps was I just too young and driven to entertain the stark realities affecting Britain at a time of political conflict, social turmoil and economic uncertainty.

In hindsight, something intangible had gone downhill from 1970 after the split up of The Beatles and the death of Jimi Hendrix after that summer's Isle of Wight Pop Festival, attended by half a million young people. It was as if a prolonged and creative 'Summer of Love' – my RCA years – had all of a sudden turned to winter. The Cold War and the Vietnam War continued. 'The Troubles' grew in Northern Ireland. The IRA set off bombs at the Houses of Parliament, the Tower of London, on a coach on the M62, in a pub in Birmingham, and at the Biba boutique on High Street Kensington. The Angry Brigade bombed the Post Office Tower – like Biba, a symbol of the progressive '60s.

This violence was accompanied by increasingly demanding strikes by Britain's major trade unions. When the miners went on strike in February 1972, the Conservative prime minister Edward Heath declared a state of national emergency. A ban on overtime working by miners late the following year prompted Heath to impose a 'Three-Day Week' on the country. This meant severe restrictions on the use of electricity from New Year's Day 1974 – Deirdre and I learned how to fry bacon and eggs over an oil lamp – and a further decline in the economy. Televisions at home, with their three channels, had nothing to show after 10.30 p.m. These odd, dark days overlapped with the 'oil crisis', when petrol was severely rationed and 50mph speed limits were imposed on British motorways following the decision of OPEC countries to restrict the supply of oil to countries that had supported Israel during the Yom Kippur War.

Unemployment had already topped a million, the highest it had been since the 1930s, and in January 1974 Britain was officially in recession for the first time since the end of the Second World War. There were other major changes to life in Britain in these uncertain years. The currency was decimalised and, after years of trying – and a 356–224 vote in the House of Commons – the United Kingdom joined the European Economic Community. A general election held in March over the question 'Who Governs Britain?' – the government or the unions – led to a hung parliament for the first time since 1929.

Harold Wilson was returned to power with a minority Labour government. By then, the stuffing had gone out of the British manufacturing industry. This was the era of British Leyland, a huge car company best known for its management fighting the unions and for producing horrible cars. On the one hand you had Derek Robinson, or 'Red Robbo' as the media dubbed him, a communist trade union convenor leading 523 walkouts from British Leyland's Birmingham plant and workers setting up beds in factories so they could sleep during their shifts; and, on the other, a management parachuted in of whom none were car men.

None of them had a clue about what would make a good car. There was no interest in design, no passion for cars in their blood at all. If engineers managed to get a potentially decent car into production, the standoff between unions and management spelled appalling production quality. I blame the management more than the unions because, if they had produced successful cars, the public would have gone along with British Leyland. The company could have made money and dealt with the unions.

Instead, it lost public support, lost money and went bankrupt. The company was nationalised. Relying on state handouts made dealing with the unions even more difficult because these came with obligations to them. It seemed pretty clear, yet again, that the people who ran Britain, whether in Westminster, Whitehall or boardrooms of large companies, had no real liking for industry or for making money by making things.

Inflation, meanwhile, had risen to 16 per cent in 1974, peaking at 24 per cent the following year. Interest rates also peaked at 24 per cent, making it difficult to pay back even the interest on loans. This was hardly the time to think of borrowing money and setting up as an entrepreneur in the manufacturing sector, especially as, in those days, there was no help for small businesses. Nothing at all. No one even thought of it or saw the need for it.

Without entrepreneurship, an inventor may not be able to bring their radical or revolutionary products to the marketplace or at least not under their own control. Without becoming an entrepreneur, they have to license their technology, putting them at the mercy of other companies that may or may not have a long-term commitment to a particular new idea or way of thinking about the future. Sometimes a change of director or vice-president within a licensee company can scupper a new product launch, which may sound a minor matter, but it can ruin the venture. Sometimes the company just changes its mind and drops the idea. And sometimes the licensee company is taken over and the venture dropped. I have experienced all these hazards.

To those of us coming from art school backgrounds, the entrepreneur – if we understood the term at all – was something to do with dodgy wide boys at one end of the spectrum, property developers somewhere in the mix and buccaneering playboys at the other, these stereotypes milking the system for as much 'moolah' as they possibly could. In our naïve and inaccurate way, we thought of an entrepreneur as someone who exploits other people. The true French meaning defines a combination of builder and architect, which I rather like.

This, though, was the era of big corporations, whether privately or state-run, and there were, as far as I was aware, very few individuals setting out to manufacture new and interesting things. Jeremy Fry, however, taught me that if you have a good idea for a new product, you engineer, prototype, manufacture, market and sell it. This makes you an entrepreneur. Jeremy showed me that, far from being pirates, entrepreneurs could be creators and makers of better products, however odd or outré.

In spite of the economic woes of the mid-'70s, Jeremy had given me the self-confidence to abandon his happy ship and to sail off, as entrepreneurs will do, into uncharted waters. And, despite the general doom and gloom of the 1970s, there were plenty of reasons to be optimistic, especially if you happened to be keenly involved in and excited by the world of innovation and engineering.

There was, after all, Boeing's Jumbo jet, Concorde, NASA's space programme, British Rail's HST (High Speed Train, or InterCity 125), the first programmable microprocessors, Kodak's digital camera – not taken up at the time – and the first email. The need for British invention at the time was certainly acute. These were the years Japanese manufacturers began exporting cars to Britain and the public took to them with alacrity. Britain did launch a few truly successful cars while I was at Rotork, like the Range Rover, but the Morris Marina and Austin Allegro seemed painfully awkward compared to the new Honda Civic, Renault 5 or VW Golf. Not that my eyes were focused on anything as bold as the design and manufacture of a car at the time. I had something rather more modest and far earthier in mind: an improved wheelbarrow for gardens and building sites.

Deirdre and I had bought an old Gloucestershire farmhouse near Badminton. The house needed doing up and we needed to create a garden. My weekends were focused on building walls and lugging things about. I used a navvy barrow in anger and its limitations became increasingly clear. Cement slopped out of it. Its tubular legs sunk into the ground. It was hard to steer. Its sharp edges damaged doorframes. The more I used it, the more I realised that nobody had really thought about these problems or bothered to fix them, and for a very long time. In fact, they had barely changed since the Roman design. They still followed the form of the original, with straight wooden shafts from the handles to the wheel axles and planks mounted on the wooden shafts to hold the contents. I wanted to change all this and to rethink the wheelbarrow from scratch.

I started out by making the bin of my design out of glass fibre. This reflected the shape but wasn't the right material. The ball, though, was the problem. I made this in glass fibre, too, at first. I had yet to find out whether or how I could make a plastic pneumatic wheel ball. To complete the assembly, I welded up a tubular steel frame. I was able to do all this in my crude workshop in a barn outside our home. I now had a prototype that I could try out in the garden. It seemed to work quite well. The ball didn't sink into soft ground while the dumper-truck-shaped bin retained sloppy cement well.

The next job was to work out how to make the bin and the ball for production. I went to see the laboratories at ICI, then Britain's premier company for plastics. As a result, we determined that the bin should be made of low-density polyethylene as cement wouldn't stick to it and, being flexible and tough, it would be very hard to break. The ball was trickier. ICI had a material called EVA (ethylene vinyl acetate), an artificial rubber rather like car tyres. This looked promising although ICI doubted whether it would work. I took a big risk and decided to go with EVA.

ICI did have a point. It wasn't as easy as I had hoped. Nobody had tried replacing rubber tyres with a tubeless EVA ball before. Even so, if I could get it to work, my ball would be easier to make than a rubber tyre. It would also be possible to repair punctures with nothing more than a lighted candle or cigarette lighter. To give it pneumatic attributes, I would fit it with a Schrader valve (invented by August Schrader, a German-Austrian, in 1891), the sort you see on car tyres.

The next question was how to put the bin and the ball into production. Plastic comes in various raw-material forms. Sheet plastic can be vacuum-formed into shapes. Granules of plastic can be heated to a melt

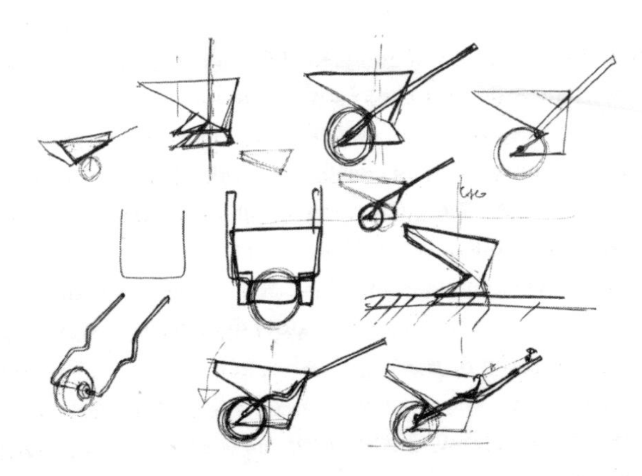

Ballbarrow
Sketches (above) and
patent illustrations (below)

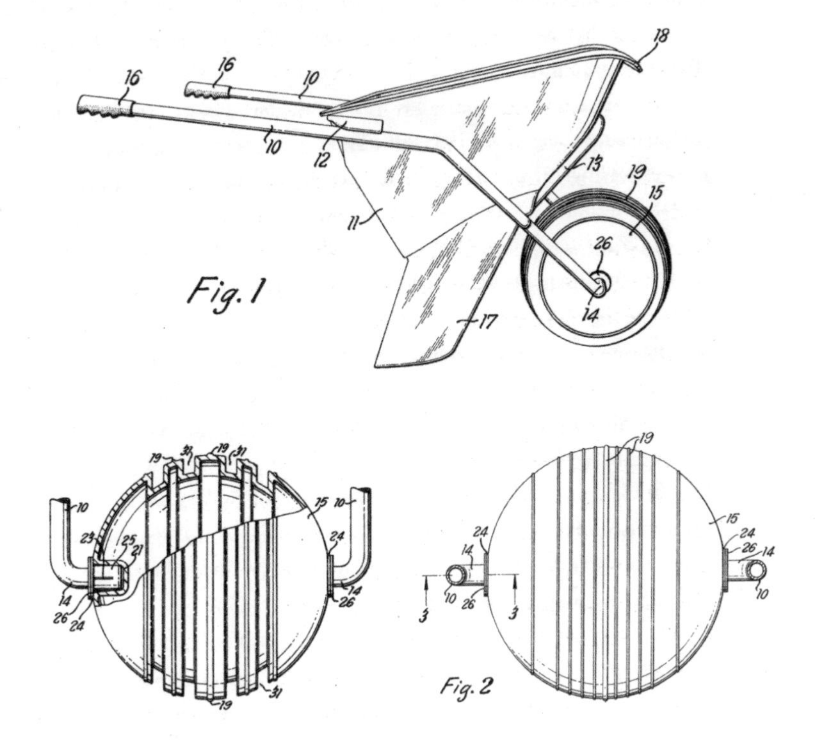

point and then injected into mould cavities. This is called injection moulding. Or powder plastic can be tumbled into hot moulds until it melts and lines the mould evenly. This is rotational, or blow-moulding. I had little choice about the method I decided on. Sheet plastic wouldn't form the shapes I needed. With injection moulding the cost of the moulds was prohibitive, in the order of £100,000, although the moulding was very accurate in high volumes. This left rotational moulding, where the tooling cost was much lower but the cost per unit was higher due to the slower process.

I visited a company in South Wales that specialised in rotational moulding. They could mould both the bin and the ball. It turned out that moulding the ball in a round mould was the most cost-effective way of making it. I commissioned the tooling still not knowing whether my ball would behave like a pneumatic tyre. With tooling completed, we did some mould trials. At last, I had a bin in polyethylene and a ball in EVA. I inserted a Schrader valve, inflated the ball and pushed it around on the barrow. It appeared to work! There were many other details, such as the bearings of the ball, which were injection-moulded nylon caps inserted into the sockets of the ball and on which ran the steel frame stub shafts. No oil or grease was needed. For the steel tubular frame, I went to Birmingham where companies produced steel tube and could fabricate the frame for me.

I now needed to raise some money to set up the company and put my Ballbarrow into production. I went to see my lawyer, Andrew Phillips, now Lord Phillips. He was very enthusiastic and suggested I approach my brother-in-law, Stuart Kirkwood, who he also advised. Stuart was prepared to guarantee an overdraft as long as I did the same. Deirdre generously allowed me to do this with the house mortgaged against the loan, and with the money duly borrowed from Lloyds Bank and debt high above my eyebrows, I set up my first factory. This was in a cart barn and a row of pigsties at home at our Gloucestershire farmhouse.

Despite my experience selling the Sea Truck, there was a lot to learn. How should we sell the Ballbarrow? Where should we sell it? How much should it sell for? None of these considerations were as obvious as they might seem. There were no nationwide chains of DIY shops, super-sheds or garden-centres at the time. Garden centres and hardware stores were individually owned, so, at first, we had to go round to each of them. They would buy just one or two barrows each at a time, and we could only visit one or two of them in a day. Then we had to get the numbers up to fifty for the wholesaler to be willing to buy these and then distribute them to the retailers we had spent considerable time on the road selling to. The wholesaler took a large cut, so with just a single product rather than a catalogue of products to offer them, there was little money left for us. It was a hopeless business model. It was also pretty humiliating when buyers laughed at my design.

The Ballbarrow did, however, win a British Design Council kite mark. At the time, this triangular black and white badge attached to a product was a sign of officially approved good design. Even then, the Design Council sent me a letter saying that I had only *just* got it. Apparently, the committee was concerned with the colour of the plastic ball. Red, it seems, was wrong for gardens, which are green, although don't tell the roses that.

Perhaps the brightly coloured Ballbarrow did look rather too playful for those used to selling traditional metal wheelbarrows. The explosion of colour through the '60s was, though, important in the development and popularising of a new generation of consumer goods. The Ballbarrow was designed to stand out in a line of old-fashioned rivals on display in garden centres or in the pages of sales catalogues. I still like to add touches of colour to highlight the technology or to draw attention to switches and release catches of Dyson products, so that colour has a functional element as well as visual appeal.

Along with selling to individual garden centres, we also took out small newspaper ads for the Ballbarrow, invariably shoehorned between those for improbable personal medical appliances and cures for baldness. It all seemed rather downmarket and a bit dodgy. Nevertheless, we received a gratifying number of cheques in the post. A bolder decision, to take a full-page advertisement in the newspapers and colour supplements, got good results. It also made us seem somehow bigger than we were, more confident and accomplished, too.

With the Ballbarrow, we also learned to shape our own publicity for magazines and newspapers. We had no money for display advertising but thought that we might get editorial coverage. Word of mouth and editorial remain the best way to tell people what you have done. It is far more believable than advertising and a real compliment when intelligent journalists want to go off and talk about your product of their own free will. If you have new technology and a new product, a journalist's opinion and comment is far more important and believable than an advertisement.

We did a photo shoot one day at Dodington Park, near Bristol. This is the late-eighteenth-century house designed by James Wyatt and built for the Codrington family, who first came that way in the sixteenth century and had owned the estate ever since. By chance I shared a bank manager with the Codringtons and through him I had arranged to photograph our range of Ballbarrows at Dodington. Deirdre pushed a Ballbarrow around the grounds for the camera. We could never have guessed then that, a quarter of a century later, Dodington Park would be our family home.

Selling the Sea Truck had taught me the value of editorial coverage. The Sea Truck had been on television, so I rang the BBC with news of the Ballbarrow. It duly made an appearance on *Tomorrow's World* just before the new invention was launched. *Tomorrow's World* was a popular BBC television series looking at new developments in science and technology. Beginning in 1965, it ran for thirty-eight years with regular audiences of ten million viewers. The lead presenter of the programme was the elegant and articulate Raymond Baxter, a former RAF Spitfire pilot. Baxter and the *Tomorrow's World* team wouldn't allow the inventor or manufacturer anywhere near the camera, though, as the show was really more about the presenters than what was shown.

The Ballbarrow might not have been the most demanding new bit of scientific or technological kit, yet, to be fair, the BBC did help put us on the national map. With sales picking up, we added a grass box extension. This was a large one-piece moulding which fitted on to the top of the Ballbarrow bin and extended the sides upwards by 300mm. This increased the capacity for light loads like grass clippings or leaves. It was very popular and increased sales. A useful lesson about accessories.

BALLBARROW Trebles capacity

EXTENSION Extension top Giant 11.5 cu ft capacity

Simply designed clip-on top. Trebles capacity-invaluable for shifting bulky loads. Available for all models. Be careful to specify exact model.

- Indestructable polyethylene – no rust!
- Light and easy to store
- Fits inside barrow – no awkward clips
- Creates a new dimension to barrow work

Flush with success in our own terms, especially with the publicity we were getting, the board decided that we should move out of the pigsties and rent a proper factory. We should also start selling through wholesale and retail shops to reach a wider public. This meant employing a sales manager and a sales force. I found a new factory on an industrial estate at Corsham in Wiltshire. These new overheads meant extra borrowing from the bank at a time when interest rates in Britain had risen to 22 per cent.

At the same time, we experienced problems with a series of plastic moulders and were forced to mould the bins and balls ourselves. For similar reasons we also started making the metal frames in our factory, bending and welding tubular steel. I learned how to spray the Ballbarrow's steel frame with electrically charged epoxy powder, a dry-coating process invented in the US by a man named Daniel Gustin at the end of the Second World War that does away with the need for wet paint. One problem with the process is overspray. You don't want dry-powder paint flying about your factory. It is messy and dreadfully wasteful. The solution was a big electric fan that sucked the spray on to 8ft-square sheets of calico. It was noisy, sounding like Concorde taking off when we switched it on. It was also incredibly inefficient. Stupid, really. Every hour the calico filter clogged. The production line had to be stopped so that the cloth could be taken down and shaken out. In the process the entire factory got covered in black powder.

I asked around in the trade. What did smart people use? Cyclones, they answered. Cyclonic separators collect dust and particles through centrifugal force without the need for a membrane filter like our 8ft calico sheet. I had seen one in action nearby at a timber merchant in Bath called Hill Leigh. All their machines had extract ducts channelled up into this cyclone. It collected dust all day with none of it coming out and, crucially, never clogging. I went at night with a torch and notebook to see how the thing worked and to draw it. I couldn't take the actual measurements as it was too big, but I was able to sketch the vital details of the construction and to estimate the sizes and proportions.

I found the centrifuge dust extraction principle of the cyclonic separator utterly fascinating. The idea of using a cleverly directed airflow to spin out dust particles by centrifugal force seemed like magic even though the technology itself was hardly new. The cyclonic separator was, in fact, first patented by one John M. Finch in 1885 for the Knickerbocker Corporation of Jackson, Michigan, which among other things made wooden cabinets, but it gave me the idea of a revolutionary vacuum cleaner. First, though, I needed to make and sell the Ballbarrow.

To keep production costs in order, we had begun to make and do everything ourselves. I had been making things by hand since childhood, so for me at least this was enjoyable, as was the process of thinking how things are made and how they might be made better. So, naturally, we made our own cyclonic extractor. An off-the-shelf model, if you could call it that, would have cost £75,000, the price at the time of twenty-five new twelve-cylinder E-Type Jags.

It was through repairing and keeping second-hand cars going, and from making things myself, that I had learned so much about engineering. During the '50s and '60s, knowing how a car worked and how to keep it on the road was particularly important to people on stretched incomes.

You did not dare go far in an average car at the time, even a new one. They were forever breaking down. You needed an AA man on standby.

As a student, I couldn't afford to go to a garage. I had to fix cars myself. During my time at the RCA, I had the early 1950s Austin-Healey 100/4. The engineering design was appalling. In racing car style, the wheels were fitted onto the driveshaft on a taper with just a single large wing nut to hold them on. They frequently became loose and I used foil from cigarette packets to provide effective shimming. The car had a four-cylinder engine with only three crankshaft bearings, giving the crankshaft a worrying degree of flexibility. Late one night I was driving near Royston in Hertfordshire on our way to Norfolk when the crankshaft flew out of the engine casing. I rebuilt an engine from a crashed Healey and, while I was about it, rebuilt my car's gearbox. I installed a new radiator and fuel pump, too, and all this without the luxury of a hydraulic ramp.

While breakdowns and improvised repairs could be frustrating, especially when tackled at the side of the road, they taught me a great deal about the engineering of components, their strength and suitability. Repairing a car like the Healey was a good grounding in basic engineering and fitting. I used to carry a full tool kit and likely spares in the boot, which were often needed. Car owners don't need to do these things today unless they're vintage or classic car enthusiasts and they do it for pleasure as much as purely practical reasons. While few people regret the passing of an era when cars broke down all too frequently, keeping them on the road had once been a course, of sorts, in makeshift mechanical and electrical engineering.

With the cyclone, I worked with two very clever metalworkers and together we made this 30ft-high steel cyclone over a couple of weekends, copying what I had seen at Hill Leigh. We had no engineering drawings to copy. I drew one up and we got on and made it. For me, it was another of those empowering transition points from the designer sitting at a drawing board with cups of coffee to manufacturer and entrepreneur. It was about knowing how to make things and making decisions. And, of course, selling the things you had designed and learned to make.

While we were selling the Ballbarrow quite well to gardeners, we thought we might be able to supply it to builders, to replace the traditional navvy barrow. One of the best-loved features of the Ballbarrow was its moulded dumper-truck-shaped polyethylene bin carrying water to different parts of gardens. The water stayed in the bin without slopping or leaking. This was true of wet cement, too, and even if cement was left to go solid in the bin, it didn't stick to the polyethylene but just dropped out. Polyethylene is non-stick. In fact, there is no known adhesive for it.

For the builders' version, we fitted a 350mm ball instead of the 250mm one. This was better for big loads and over rough ground. It also had a bigger bin. We thought that builders would be pleased to carry a bigger load. This was centred over the ball, meaning that, when lifting the handles, they would feel little of the actual load on their arms. We thought builders would appreciate this. We were wrong on all counts! Construction site workers had no interest in pushing around bigger loads, so the bigger-sized bin was not appreciated. Nor did they appreciate having the weight on the ball and not on the handles of the barrow, and for the following reason: when you come to push the loaded barrow up planks into a skip, it is better if most of the load is on the handles rather than the ball, because pushing a heavy load up a steep slope is much harder than carrying the load up a slope. In fact, the same logic is true of pushing a heavy load across broken ground. It was better to halve the load on the ball and carry the rest on the handles. Carrying is easy, pushing is difficult.

Duly chastened, we redesigned the barrow for the builders' market. We positioned the wheel further forwards so that the load was shared equally between the ball and the handles. We fitted a smaller bin, to make their loads lighter. I am grateful for the builders' criticism because this redesigned version was the best Ballbarrow.

Out of the blue, we were approached by British Aerospace, manufacturers of small missiles. They wanted a missile carrier with Ballbarrow ball wheels as these would roll over soft sand without sinking into it.

The problem, though, was that some military shells explode, scattering quite strong needles. These puncture vehicles' tyres. British Aerospace wanted a puncture-proof ball wheel. I developed and prototyped a version of the ball that was not inflated yet behaved as though it were and was impervious to these deadly needles. This EVA ball had deep circumferential grooves around the centre of the ball where the diameter is at its largest. This provided a hoop strength that stopped the ball from deforming yet gave it a degree of springiness. Finally, we had a puncture-proof ball/wheel.

At this time I had another invention up my sleeve, aimed at complementing the Ballbarrow. This was the Waterolla, a garden roller featuring a plastic drum filled with water instead of the time-honoured – although, in my mind, time-expired – metal drum filled with concrete. The market for the Waterolla proved to be smaller than I had thought, perhaps not least because its very lightness, when drained of water, made it all too portable. It was easy to pick up and put in the back of a car. Friends would lend them to other users, so sales were low.

In 1978, the year our third child, Sam, was born, we started selling the Trolleyball, a ball-mounted trailer to tow boats out of water thanks to its pneumatic ball wheels. On land, the pneumatic ball wheels didn't sink into sand. Of course, they floated the trolley in the water. With other trolleys, when you push it into the water, the trolley disappears from vision so you cannot align the boat with trolley. Because my trolley floated, you just slid the trolley under the boat. The boat was held by seat belt webbing so any shape could be accommodated. In the back of my mind were memories of performing this exacting task with my father in north Norfolk. The Trolleyball was also a way of reimagining an invention, however modest it might at first seem, and, funnily enough, this particular invention would come back into play many years later.

These were early days, yet I was learning much that was to be useful when I set up on a much larger scale with Dyson in Malmesbury fifteen years later. I was also putting into practice ideas I'd learned directly from Jeremy Fry and indirectly from Alec Issigonis. Don't copy the opposition. Don't worry about market research. Both Jeremy and Alec Issigonis might just as well have said 'Follow your own star.' And this is indeed what successful entrepreneurs do. The problem, though, was that I wasn't following my own star, and this proved to be the very reason why my first company, Kirk-Dyson, was not the success it could have been.

In 1974, when I had wanted to do the Ballbarrow, my brother-in-law very generously offered to part-fund it and I had rather stupidly assigned the patent of the Ballbarrow not to myself but to the company. We borrowed £200,000 but, now at 24 per cent, the interest rate was phenomenal. As we borrowed more by bringing in new investors, so my percentage share of the company fell. The business grew to an annual turnover of £600,000. It captured more than half the UK garden wheelbarrow market, but even so we didn't make money from it.

Things got worse when a former employee left to join a US company with whom we had discussed a licence. That company produced a rival Ballbarrow, even using our Ballbarrow for the photos in their brochure. Against my wishes, Kirk-Dyson chose to launch expensive legal proceedings. This put further financial pressure on the company and so the need for yet more investment. While there were quite a few normal disagreements about the direction of the company, what I really wanted to do now was to make the vacuum cleaner I had in mind rather than fight the alleged plagiarist in Chicago, as the board was keen on doing.

I couldn't have been more surprised, though, when in February 1979 my fellow shareholders booted me out. There was no apparent reason for this. I later discovered that the son of the other major shareholder had taken over the running of the business. I had lost five years of work by not valuing my creation. I had failed to protect the one thing that was most valuable to me. If I had kept control, I could have done what I wanted and avoided a big interest bill. I learned, very much the hard way, that I should have held on to the Ballbarrow patent and licensed the company. In the event, I lost the licence, the patent and the company. What made it worse was that because Andrew Phillips was the company lawyer, it was he who did the firing and I was now without a lawyer. I was clueless about compensation for loss of office and my shares were worthless.

In this sense, the Ballbarrow – my first consumer product, my first solo effort – was a failure but one from which I learned valuable lessons. There was a lesson about assigning patents, another about not having shareholders. I learned the importance of having absolute control of my company and of not undervaluing it. I knew how to make and sell, but not how to look after myself. We had charged too little for the product while competing with traditional tin barrows knocked up in sheds with no design input or new expense and low manufacturing overheads. In retrospect, the very idea of selling against a utility product was a mistake. Distribution was inefficient – all that selling to individual retail outlets – while the cost of exporting the Ballbarrow was prohibitive. The product was good, but the commercial proposition was a bad idea.

From now on, though, I was determined not to let go of my own inventions, patents and companies. Today, Dyson is a global company. I own it, and this really matters to me. It remains a private company. Without shareholders to hold the company back, we are free to take long-term and radical decisions. I have no interest in going public with Dyson because I know that this would spell the end of the company's freedom to innovate in the way it does. I want to think about the future and to keep going forward with invention, engineering, design, technology and products even if this means sailing against the prevailing tide and into those uncharted waters I find so enticing.

There is another reason. When in those Kirk-Dyson days I had shareholders, I had to consider, not unreasonably, their points of view and what they might want. It was partly their company and not my own. This led to many decisions being theirs and not mine. When you own the whole company, and especially if you are free of debt, from the early days and for better or worse, all decisions are your own. So you take these very seriously and follow your own view of risk balanced, hopefully, with reward. This certainly sharpens the mind. The lessons I learned I felt viscerally. They were to be deeply ingrained. Today, though, with a band of brilliant managers and a board of the best professionals, we make decisions jointly and collectively.

At the end of the Ballbarrow experience, I was penniless again with no job and no income. I had three adorable children, a large mortgage to pay and nothing to show for the past five years of toil. I had also lost my inventions. This was a very low moment and deeply worrying for Deirdre and me. It was deeply upsetting, too. My confidence took a big knock, and it would take some years to regain it. And, yet, as my godmother said in commiseration, the Ballbarrow cloud did have a silver lining. The idea for a revolutionary vacuum cleaner was already maturing. I had learned from my past mistakes and failures. I struck out on my own. This was a real turning point.

CHAPTER FIVE

The Coach House

In February 1979 I was free to go my own way. In one sense, it was all a bit of a disaster. I had no job, no income and a sizable mortgage to pay off. Financially, things looked bleak. Deirdre and I had recently moved to a half-finished early-nineteenth-century house at Bathford, three miles east of Bath. By now we had three children, the youngest, Sam, six months old. I would, in fact, have no real income, just an escalating overdraft, until the sale five years later of the licence of the vacuum cleaner I invented and made to Amway, an American marketing company based in Michigan.

As for the British economy, things were not exactly rosy in 1979. The year had begun in the depths of the bitterly cold 'Winter of Discontent' – the average temperature for January and February across Britain was below freezing – when lorry drivers, ambulance drivers, railway workers, refuse collectors and even gravediggers went on strike. To the backdrop of IRA bombing campaigns, there was talk of yet another recession. James Callaghan's Labour government appeared out of touch and out of control. The Tories were voted into power with a large majority that spring with the formidable Margaret Thatcher as prime minister.

Strange as it might seem, and although money was a constant worry, life didn't seem as tough as perhaps it might have been. Deirdre and I had a handsome house, even if it did need quite a bit of work, our three lovely children, aged nine down to one, and the family retriever. We grew our own vegetables and had a large garden that the children made good use of. Deirdre made clothes, ran art classes and sold her paintings. We couldn't afford glamorous holidays, and yet in many ways it was an idyllic time. We had huge energy and I knew what I wanted to do and was getting on with it.

This was my cyclonic vacuum cleaner. The idea had been in my head since welding up the giant metal cyclone for the Ballbarrow factory. Now it made increasing sense. Here was a field – the vacuum cleaner industry – where there had been no innovation for years, so the market ought to be ripe for something new. And, because houses need cleaning throughout the year, a vacuum cleaner is not, like my Ballbarrow, a seasonal product. It is also recession-proof. Every household needs one. It seemed to tick all the boxes. In any case, I'd used one since childhood and knew from experience that there had to be a better vacuum cleaner.

For the following fifteen years I lived in debt. This might not sound encouraging to young inventors with an entrepreneurial spirit, yet if you believe you can achieve something – whether as a long-distance runner or maker of a wholly new type of vacuum cleaner – then you have to give the project 100 per cent of your creative energy. You have to believe that you'll get there in the end. You need determination, patience and willpower.

Never for a moment, though, had I thought, *I know, one day I'll set up a business making and selling vacuum cleaners and if it takes me fifteen years to make a profit, then that's the way to go.* What happened was that when I was making the Ballbarrow – in fact not long after I'd built the cyclone for the Ballbarrow factory – I bought a new vacuum cleaner. At home we had a reconditioned Hoover Junior – the classic upright styled by Henry Dreyfuss, the famous American industrial designer, and made at the Art Deco Hoover factory on the Western Avenue at Perivale – but I thought we ought to have a modern cleaner and Hoover had a new model that looked like a flying saucer. It was said to be the world's most powerful.

When I set to work with it one Saturday, it screamed away yet appeared to have very little suction. I knew the bag must be full, but because we didn't have a spare, I took the old one out, opened the end, tipped the contents into the dustbin and sealed the end of the empty bag with Scotch tape. Back it went into the flying saucer. Still no suction. I got into the car to go and buy some new bags. With a new bag installed, suction returned. I wanted to know the difference between a new bag and an old empty bag. When I opened the old empty bag again, the penny dropped.

What I hadn't realised, and should have done as an engineer, was that a vacuum-cleaner bag is not just a depository for the dust. It acts as a filter, trapping the dust inside the bag and allowing the air to pass through the pores of the bag. The 'bag full' indicator is not a 'bag full' indicator at all: it is a 'bag clogged' indicator. It registers the pressure as the pores of the bag clog, announcing that the bag is full when it actually means that the pores of the bag are clogged. And this could occur when there was only a very small amount of dust in the bag. This is very significant: because the airflow has to pass through the bag, becoming clogged from the moment the dust first enters, the airflow becomes reduced also, and this means a drastic drop in suction and cleaning ability. As an engineer, I found this interesting. As a consumer I felt cheated, even angry. This anger festered for several months.

I remembered the same clogging problem on the calico cloth with the powder coating in the Ballbarrow factory and the giant cyclone we had made to solve it. What if I could develop a much smaller version and replace the clogging bag in a vacuum cleaner? I had tried to interest my fellow directors and shareholders at the Ballbarrow company in the concept of a cyclonic and bagless vacuum cleaner with its promise of an end to clogging and loss of suction. No such luck. If it was such a great idea, they said, Hoover or Electrolux would have been making it already. And, oh, by the way, you're fired. As the Ballbarrow company had raised money and issued shares to keep going, my stake in the company had fallen. I was powerless.

Now, in 1979, at home on my own, I was eager to see how a cyclone might perform in miniature. I quickly made one from cardboard of similar shape to the metal one we had made at the Ballbarrow factory, but only 300mm instead of 10m high and held together with gaffer tape. I replaced the cloth bag hanging from the handle of my upright vacuum cleaner with this simple cyclone. As I pushed it around the house several times, it seemed to work, collecting dust, fluff and our retriever's hair in the cyclone, although I had no idea of its engineering efficacy.

I went to see Jeremy to show him my idea. He was very keen to support me and we did a business plan and budget for its development. He would put up £25,000, or 49 per cent of the equity, leaving me to find a similar amount for my 51 per cent. If I could, I would have control of my own company. Jeremy said he had no interest in kicking me out. He also said that because we were both engineers and designers we should restrict ourselves to developing the vacuum cleaner and then selling the idea on rather than manufacturing it ourselves. Deirdre and I raised £26,000 for our 51 per cent of the equity by selling our precious vegetable garden as a building plot and borrowing the rest of the money from the manager at Lloyds Bank in Chipping Sodbury. This was the same manager who had arranged for us to take the photo of Deirdre pushing the Ballbarrow at Dodington Park.

Luckily, I had an old eighteenth-century coach house at home to serve as a workshop. I should explain that one half of the building had been designed to house two horse-driven coaches, and the other half to house the horses, with a hay loft above. It was smelly and rotten at first but, beginning with my building a wooden bench fitted with a vice, it quickly began to look and act the part. I bought a set of antique sheet-metal rollers so I could roll prototype cyclones out of brass before soldering or riveting these together. I could make a cyclone a day, not always a completely new one, sometimes a modification.

One of the really important principles I learned to apply was changing only one thing at a time and to see what difference that one change made. People think that a breakthrough is arrived at by a spark of brilliance or even a eureka thought in the bath. I wish it were for me. Eureka moments are very rare. More usually, you start off by testing a particular set-up, and by making one change at a time you start to understand what works and what fails. By that empirical means you begin the journey towards making the breakthrough, which usually happens in an unexpected way. You do need to have bright ideas or be willing to try the unthinkable along the journey, but trying to rush by testing a brainwave rarely works.

Brunel, one of my great heroes, is said to be the father of modern research and development after he painstakingly developed the ship propeller, making just one change at a time. Luckily, the records of his workbooks have survived at Bristol University. Although exacting, this development journey is exciting. There were to be lots of setbacks over the next five years. This was my personal *Pilgrim's Progress*, with plenty of ups and downs, false turns and many, many failures along the way.

My first experiment – mating our Hoover Junior to my homespun cardboard cyclonic extractor – had shown that my simple cyclone did collect some dust. From now on the serious business of developing the technology of the cyclone would begin. Once the heart of the cleaner was completed, I could move on to developing the remainder of the vacuum cleaner.

I discovered as I started to make prototype cyclones that there were two apparently insurmountable problems. First, the state of the art was clear and confirmed by literature: cyclones were only efficient down to dust particles of 20 microns in size, yet household dust is very fine at 0.5 microns or smaller, similar to cigarette smoke. Secondly, the traditional cyclone shape would not trap or separate carpet fluff or human hair. These simply flew out of the air exit with the exhaust air.

At this stage, anyone watching me at work might reasonably have wondered why Electrolux and Hoover weren't making and selling a vacuum cleaner like mine. With all their resources, surely they could have leaped ahead of me – one man and his dog, as it were, in a rural coach house – and cornered the market between them.

There were, though, at least three good reasons why they didn't even think of pursuing a similar path to me. One, which went without saying, was that the 'No Loss of Suction' vacuum cleaner had yet to be invented. The second was that the vacuum cleaner bag replacement business was highly profitable. And the third, rather to my surprise, was that well-established electrical goods companies seemed remarkably uninterested in new technology. With no outside challenges, they could afford to rest on their laurels. For the moment at least.

I buried myself in my own world of prototyping. It's a part of the Dyson story that I made 5,127 prototypes to get to a model I could set about licensing. This is indeed the exact number. Testing and making just one change after another was time-consuming. This, though, was necessary as I needed to follow up and prove or disprove every theory I had. And, however frustrating, I refused to be defeated by failure. All of the 5,126 prototypes I rejected – 5,126 so-called failures – were part of the process of discovery and improvement before getting it right on the 5,127th time. Failure, as I had already begun to learn with my experience with the Ballbarrow business, is very important. I find it important to repeat that we do, or certainly should, learn from our mistakes and we should be free to make them.

Of course, if like me you're lucky to have someone willing to back you up day to day for five years as Deirdre did, you are far less likely to give up. It was also very much a family affair. The children came and worked alongside me in the coach house. Among other things, we built skateboard ramps, vacuum-forming machines and desk lamps. Making things was an enormously pleasurable part of our lives as well as necessary because we couldn't afford them otherwise. Jake made a sledge that he could sit inside with front skis that steered. Sam made an ironing board that folded very simply and easily without the need for annoying catches. He used gas struts designed to hold up car boot lids. Emily made a spring diving board for the swimming pool I had built in the garden.

In this demanding, yet engrossing and enjoyable process of evolving thousands of prototypes, I gained a great deal of knowledge about

My parents' wedding in 1941 at
St Mary's Fowlmere, Cambridgeshire.
My father, Alec, with swagger stick, wears
Army khaki; Mary, my mother, RAF blue

At the wheel of my pre-owned, pre-war
Tri-ang 'Sports' model pedal car

Swallows and Amazons days: My elder brother Tom (left), me (centre) and my sister, Shanie, at Blakeney Point, North Norfolk, a seaside realm of sand, seals, sails, salt marshes and spectacular bird life

Playing a drunken Trinculo (centre) in *The Tempest* at Gresham's School.
Tim Ewart, the future ITV *News at Ten* anchor is Caliban (right)

With Logie Bruce-Lockhart, my old Gresham's headmaster, 96 at the time, at his home in Blakeney, Norfolk, in 2019

Deirdre and I painting together at the Byam Shaw School of Art, Kensington, London, 1965

We got married in London in December 1967

Honeymoon. Deirdre on the beach at Marazion, Cornwall

(Opposite, top) Jeremy Fry in 1968 powering the first Sea Truck up to speed with two Minis aboard at Buckler's Hard, Hampshire.

The Mini with the L-plate was Deirdre's. Buckler's Hard on the Beaulieu River built ships that fought with Nelson's navy and Motor Torpedo Boats during the Second World War

(Opposite, middle) Driving a Sea Truck, demonstrating its firefighting capabilities with the London Fire Brigade in 1971

(Opposite, bottom) A 12-metre Sea Truck at full pelt at Bergen, Norway, 1973

My favourite builder's version of the Ballbarrow. I own this one

Deirdre at Dodington Park, Gloucestershire, modelling a Ballbarrow for an advertising shoot in 1977.

Neither of us knew then that, one day, this would be our family home. Note our 'ball' logo

STABLE
Feet 150 mm further apart than other barrows.

TIPPING
Perfect balance for tipping.

BROAD FEET
Ballbarrow is designed for rough ground with broad feet to prevent sinking.

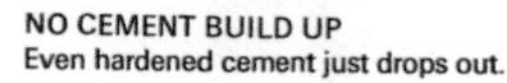

NO CEMENT BUILD UP
Even hardened cement just drops out.

DUMPA TRUCK BIN
Shaped to retain sloppy liquid loads as well as normal loads.

RESILIENT BIN
Takes punishment that would wreck a steel bin.

PNEUMATIC BALL/WHEEL
Because it has three times ground spread, it crosses sites where the old type would flounder.

RUST-FREE
Plastic ball and bin are corrosion free and frame is plastic coated for maximum protection.

As our policy is one of continuous improvement we reserve the right to change the specification of our products without notice.
Ballbarrow and Waterolla and associated products are manufactured in England by Kirk-Dyson Designs Limited. Patents Pending.

PRINTED IN ENGLAND

KIRK-DYSON DESIGNS LIMITED

Leafield Estate, Corsham, Wiltshire SN13 9UD England Phone Hawthorn (0225) 810077 Telex 449740

The back of a brochure extolling the virtues of the Ballbarrow

Making a fishpond, with a Ballbarrow, at home at Sycamore House, with Sam, Emily and Jake in 1982

The Coach House, near Bath, at the time I was developing the dual-cyclone bagless vacuum cleaner

The machine shop on the ground floor of the Coach House

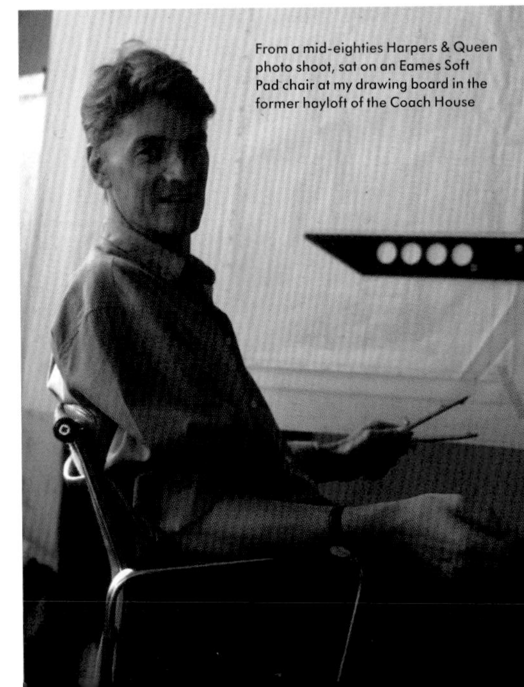

From a mid-eighties Harpers & Queen photo shoot, sat on an Eames Soft Pad chair at my drawing board in the former hayloft of the Coach House

Launched in 1986, G-Force was both a cult consumer object in Japan and a highly effective vacuum cleaner with several technical advancements.

Pages from the manual (below) explained how it worked as well as how to use it

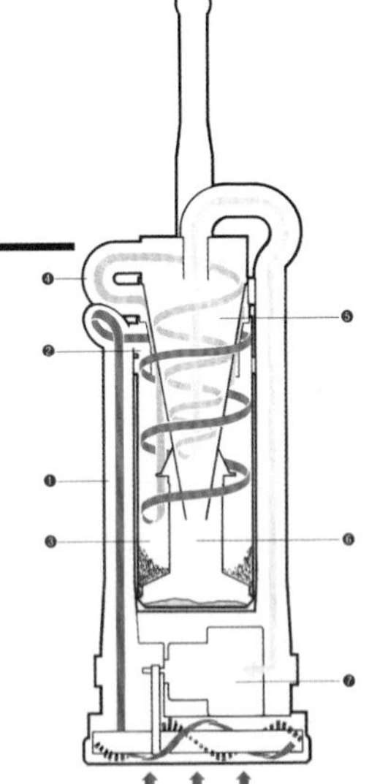

G·フォースは一台二役。
従来の掃除機ではできなかった便利なシステムを、コンパクトな一台にまとめました。

■アップライト方式

広いカーペットもらくらくクリーニング。カーペットにしがみついたホコリやチリを、モーターで回転する回転ブラシでかき出し、吸い込んでゆきます

■ステーション方式

ハンドル部分に収納されている伸縮自在のストレッチホースを引き伸せば、高い所やせまい場所まで、小回りのきくお掃除ができます。

機能だけでなく、デザインも斬新です。

G·フォースのデザイナーは、イギリスのロイヤル·カレッジ·オブ·アート出身のジェームズ·ダイソン。英国デザイン評議会賞、エディンバラ公特別賞など、数々の栄誉を勝ち得てきた、輝けるデザイナーです。「もはや、人が機械に対して恐怖感や違和感で接する時代は終った。これからは人は機械と遊び、楽しむようになるだろう」という彼の言葉には、G·フォースの明るく楽しいデザインコンセプトがうかがえます。

cyclonic technology. However, like Jeremy Fry, I am cautious of experts. While developing the dual cyclone vacuum cleaner, I went to see a scientist, R. G. Dorman, at the notorious Porton Down government research station near Salisbury. Dorman had written a book, *Dust Control and Air Cleaning* (1974), published by Robert Maxwell's Pergamon Press, and was considered the leading expert on the subject of cyclonic extraction. A very nice chap, he told me that the current art was good down to 20 microns, but I knew that for a domestic vacuum cleaner this figure needed to be 0.3 microns.

This, of course, was considered impossible. So, I set out to bring that figure down myself because I needed to and because part of me wanted to prove official experts wrong. Actually, I was helped with the maths for the cyclone by Paul Colombe, my former maths teacher and housemaster from Gresham's, and my godmother's husband, who came to stay on their way to Devon. In Dorman's book there were five different mathematical calculations to determine the efficiency of cyclones of different dimensions at various airflows and particle sizes. Paul Colombe started me off calculating the maths and, not surprisingly, as I worked through the equations, each scientist had a different theorem and different answers! They were no help or short cut. I would have to go through empirical testing to find the answers for myself as well as taking the state of the art to new heights. In the end, after the 5,127 prototypes, we managed to do what the expert view had deemed impossible.

Experts tend to be confident that they have all the answers and, because of this trait, they can kill new ideas. But when you are trying to break new ground, you have no interest in getting stuck in engineering conventions or intellectual mud. In any case, as Jacob Bronowski, the Polish-born British mathematician and historian, said in his 1973 BBC television series *The Ascent of Man*, 'Science is a very human form of knowledge. We are always at the brink of the known, we always feel forward for what is to be hoped. Every judgement in science stands on the edge of error and is personal. Science is a tribute to what we can know although we are fallible.'

This is why I have long had great admiration for engineers like Alec Issigonis, and André Lefèbvre of Citroën who gave us the Citroën Traction Avant, 2CV, DS and the HY corrugated delivery van, among other radical engineering designs that the public responded so positively to. They questioned orthodoxy, experimented, took calculated risks, stood on the edge of error and got things right. And when they got there, they continued to ask questions.

The brief for the Citroën 2CV, set by Citroën's vice-president Pierre Boulanger, himself a distinguished engineer, to Lefèbvre and his team was for a motorised horse and cart that would take a farmer and his wife to market along rutted and muddy lanes crossing ploughed fields with baskets of eggs on board without breaking a single one. They solved this problem by having a very long suspension arm to each light wheel that gave the wheels a long and soft travel. As for the look of the car, this didn't seem to matter, although in fact Citroën's stylist Flaminio Bertoni, a sculptor before turning to industrial design, ensured that it had a very special and endearing appearance very much of its own and one that was smoothed out slowly over the 2CV's long production life. All this required truly original thinking. The 2CV was in production for forty-two years and, like Issigonis's Mini, it has never lost its appeal. These engineers thought freely and independently and had the ability to see radical designs into production and, even if this was not their intention, into the history books, too.

At the same time as visiting R. G. Dorman at Porton Down, I read Jim Slater's book *The Zulu Principle* (1978). Slater was a British accountant and investor who rode the waves of investment finance with a number of great successes, although he had his downs as well as ups. His approach to winning investment, he wrote, was to become a leading authority on a 'clearly defined and narrow area of knowledge'. Following the Zulu Principle, this can be achieved remarkably quickly. Jim Slater's son, Mark, helped research *The Zulu Principle*. I got to know him, admired the investment business he had built up and in 2014 invited him to join the board of Weybourne, the holding company of Dyson operations.

Slater had been intrigued when, after reading a single article on the Zulu people in a copy of *Reader's Digest*, his wife's knowledge of Zulus far exceeded his own. If, he reckoned, she read all the books she could find on the subject and made a quick trip to South Africa to meet Zulu people, she would soon be considered a leading authority. I suppose I did this in a way with the cyclonic technology. Over a period of four years, I had probably built and tested more cyclone dust separators than most cyclone experts. My quest was to increase the ability of the cyclone to trap ever smaller microscopic particles. During the process of testing the 5,000-plus prototypes of the cyclone, I determined the correct angle of the conical section, the best diameter of the cyclone, the best inlet and outlet diameter, the ideal entry-duct shape and the best length of cyclone outlet.

By late 1982, I had a fully working prototype of the critical cyclonic portion of my vacuum cleaner. I now turned my thoughts to the machine as a whole. At the time there were two sorts of vacuum cleaner, an upright and a cylinder one. The upright type had a rotating brush roll in the cleaner head and you pushed the machine, but it had no means of cleaning anything other than carpets, and it failed to clean right to the skirting. Nor did it have much suction. A cylinder cleaner, which you pulled across the floor via the wand and hose, had strong suction with a new bag but usually had just a passive floor tool and was less effective on carpet, though good for cleaning above the floor.

I decided to put a cylinder-cleaner-type motor into my brush-rolled upright, to have good suction, and have a permanently attached stretch hose to match the cylinder version at cleaning the edges of floors and above them. That way my cleaner could compete with either version. The handle for my cleaner could be detached from the body and telescope within the stretch hose reaching over four metres from the machine, with an automatic changeover valve to divert the suction from the cleaner head to the hose and vice versa. No fiddling about, just a quick draw, in western parlance.

Having designed and built my vacuum, I set about patenting the new features and the cyclonic inventions. This is most important, though patents, devised under Henry IV with little change since, are most unsatisfactory and provide scant protection for the inventor. Nevertheless, it was a prerequisite for my next step, which was to try to license the design with all the major British, European and American vacuum cleaner manufacturers. Hoover wanted me to sign a piece of paper saying anything that came out of a discussion with them belonged to them. I didn't sign, and that was the end of my collaboration with Hoover. They did, though, send their European vice-president, Mike Rutter, to appear in 1995 on the BBC's *Money Programme* to say that Hoover regretted not buying my invention because they would have 'put it on the shelf', ensuring that it never saw the light of day. Charming.

I went to see Electrolux, Hotpoint, Miele, Siemens, Bosch, AEG, Philips – the lot – and was rejected by every one of them. Although frustrating, what I did learn is that none of them was interested in doing something new and different. They were, as I had already understood, more interested in defending the vacuum cleaner bag market, worth more than $500 million in Europe alone at the time. Here, though, was an opportunity. Might consumers be persuaded to stop spending so much on replacement bags, which, by the way, are made of spun plastic and are not biodegradable, and opt for a bagless vacuum cleaner that offered constant suction instead? If so, I might stand a chance against these established companies.

Rotork, who had seen what I was doing, took a licence on my design. We had it made by Zanussi in Pordenone, Italy, and sold at home through door-to-door Kleeneze sales reps brandishing product catalogues and at the Ideal Home Exhibitions of 1983 and 1984. This was the pink Kleeneze Rotork Cyclon, its colour chosen to emphasise its difference from all other vacuum cleaners on the market. It cost about £1,000 in 2021 money. Around 550 were made.

I believe that Rotork chose Kleeneze as a partner because they did not wish to be involved in selling the cleaner and also because selling vacuum cleaners door-to-door was a tried and tested method that had long worked for well-established companies like Kirby in the United States and Vorwerk in Germany. Kleeneze was certainly a very well-known company in post-war Britain. It was based on the methods used by the Fuller Brush company in America. In 1982, the suitcases opened by salesmen at front doors were replaced by catalogues, with Kleeneze remaining a successful company for some years to come.

Meanwhile, having licensed the vacuum cleaner to Rotork and while looking for other licensees, Jeremy and I established a company called Prototypes Ltd to work on designs like the Squirrel four-wheel drive, power-steered wheelchair with Lord Snowdon and we revived the Wheelboat project, one that we had explored at much the same time as the Tube Boat with Rotork Marine. By then we had thirteen years' experience selling Sea Trucks and knew that many of them were used as patrol boats rather than as load carriers or assault craft. We thought there would be a market for a fast water jeep on wheels for the army, navy, police and anyone who might use a jeep-type vehicle on water.

By now, instead of working from the coach house at our home, we had bought a coach house for £45,000 at the rear of the famous Royal Crescent in Bath, the licence agreement with Rotork providing the funds. It was an L-shaped Georgian building around a large, cobbled courtyard. This time I built something more resembling a test tank rather than relying upon the pond at Widcombe Manor. I had the same design of tripod and boom arm as that used thirteen years earlier, but the tank was a doughnut-shaped trough made of plywood and lined with a large plastic sheet to make it watertight. The motor was a big electric drill with the power cable going up to a first-floor window. I could measure the power drawn by the electric motor, the speed of the wheel, the load on the wheel and the speed of travel. The wheel was a one-sixth scale model of the full-sized 3m version.

I repeated the same tests from the lake thirteen years earlier and made some progress in efficiency. However, our model water jeep was still slow. I had tried many different paddle shapes, yet I could only travel at speeds far slower than the Sea Truck's. Facing failure while eating my lunch one day, I recalled that the fan blades of a vacuum cleaner, instead of using a paddle-type action and shape to trap and accelerate air, actually had backward and curved-shaped blades, as though these wanted to spill the air in which they moved. However, these were efficient and worked at high speed. I wondered, *What if we had the blades sloping backwards doing the opposite to paddling the water? What would happen?* I set this up and started the test. I was astonished to see the wheel, with its heavily sloped backwards blades, lift out of the water and skip across the top at high speed! Now it could rival the Sea Truck for speed and might be faster. I did further tests to improve the efficiency and speed and filed a patent.

The quick response from the Patent Office was that all documents relating to this invention should be kept in a strongroom safe within a safe. We were not allowed to exploit it commercially as it was considered to be of military significance. The same thing happened to Sir Christopher Cockerell with his hovercraft, and this was a bone of contention between him (and later his widow) and the Ministry of Defence.

I now designed and built a full-size version of the wheel for the water jeep that I had developed in the courtyard. It had a 3.3m wheel with a giant 'inner tube' made of a flexible Kevlar-reinforced polyurethane skin (Kevlar being a less expensive version of carbon fibre) around a light aluminium hub. The tripod and boom arm would be mounted on a Sea Truck anchored in Poole Harbour. The engine driving the drive shaft, hung under the boom, would be a Ford 1,600cc petrol engine with a chain reduction to the drive shaft. The single wheel, representing one of the four wheels of the vehicle, would drive round and round the anchored Sea Truck, testing the best paddles from the tank tests, but at full size. The result was that the 3.3m wheel could drive around at 30mph carrying a 250kg load using only 15hp. It was a most exhilarating sight. This proved that we could build a full-sized jeep, weighing 1,000kg and with a 60hp motor, that would travel at 30mph across the water. Unfortunately, at this very point, Rotork, having sold off their marine division, were no longer interested. The project came to an end.

As the Wheelboat project ended, another began. Lord Snowdon, Jeremy's friend, had long bemoaned the design of motorised wheelchairs and in particular their unsuitability for indoor use. They brought mud and other unmentionables from the outdoors into the home. They were clumsy inside the home and they were ungainly contraptions to look at. Snowdon, who had polio as a child, had long championed the cause of the disabled. A renowned photographer, he was also an inventor. He had previously been the instigator of a motorised platform that he designed with an engineering company. The idea was that it remained indoors, and you could fix your favourite chair onto the top of the platform. The controls and steering were on a simple upright arm. Working with Snowdon, our brief was to design an excellent indoors chair that would perform equally well outside on roads and pavements and without the ungainly contraption-like look of conventional motorised wheelchairs.

The project began with the size of the wheels. The bigger their diameter, the bigger step they could climb. Big wheels, though, are too bulky for indoor use. We knew that if you lay a wheel on its side at, say, 45 degrees, it can mount a higher step than it will do if wholly upright. This is because the diameter described by a wheel at 45 degrees is much larger than its actual diameter. You can easily demonstrate this by taking a plate and rolling it along a floor towards a book placed in its path. The plate will find it hard to climb the book. If you then set the plate rolling at a 45-degree angle, it will mount the book more easily. So, we put the wheels at 60 degrees to the upright. This meant that they could be housed under the platform and hardly noticed.

We also noticed that powered wheelchairs had two big motors to drive them. These were specially made, expensive and heavy. We chose much cheaper and much smaller windscreen-wiper motors and mounted one on each of the four wheels. They could be smaller because we had four of them instead of the normal two. We now also had the benefit of four-wheel drive.

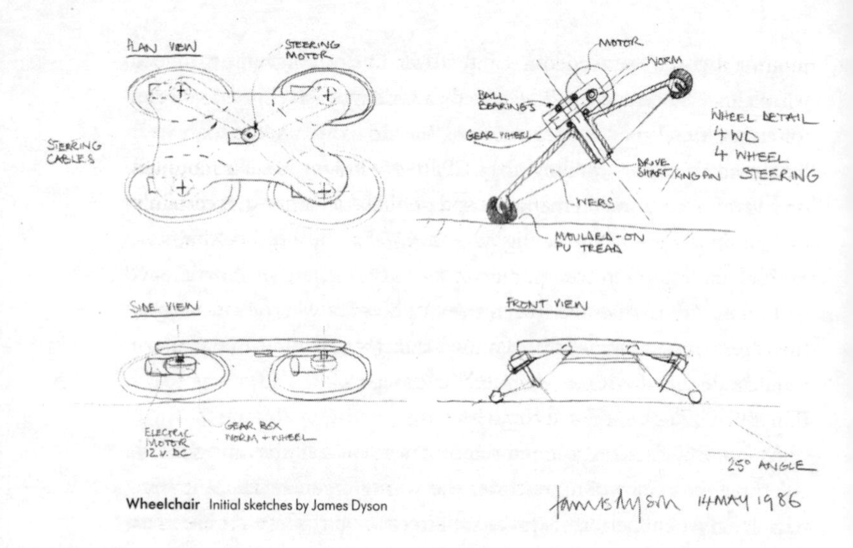

Wheelchair Initial sketches by James Dyson

Conventional wheelchairs have two driven fixed wheels and two castor wheels. This makes their steering somewhat unpredictable, rather like a supermarket or airport trolley. We made all four wheels steer, which gave the wheelchair predictable steering together with a very tight turning circle. It could rotate on its own axis. Since the inside wheels have to turn more slowly than the outside wheels, a fact carefully controlled on a car, we would do the same with the added complication of full four-wheel steering. The electronics for this and the steering geometry with angled wheels was complex and an interesting pioneering step.

The batteries were mounted on top of the platform and the chair on top of the batteries. The chair I designed folded up for transport. The platform with the angled wheels made a second unit and the heavy batteries the third. These were all easily put in the boot of a car. Which is exactly what I did when taking it by British Airways to Nice and from there in a Renault 5 hire car up to Jeremy's Provençal village high in the Basses-Alpes. There it was to be tested by a wheelchair-using friend of Jeremy's. Everything went well until he wanted to drive up a rocky mountain path. He needed a Land Rover to do that, not an indoor wheelchair! However, I still believe that Lord Snowdon's original brief for an indoor chair was the right one. We had made a wheelchair with barely visible wheels, looking like a Charles Eames-type chair mounted on a low platform with remarkable and predictable steering. It certainly had no similarity to, nor the stigma of, normal powered wheelchairs.

I left the project to concentrate on the vacuum cleaner while Jeremy and others focused on the wheelchair. Afterwards, when they changed the wheels back to conventional ones, I thought this was going to make it much the same as other wheelchairs. Although it did go into production as the Squirrel, it failed commercially.

Commercially, the Cyclon upright vacuum cleaner was not a success either. Unfortunately, Rotork chose the wrong man to run the operation. He happened to be its financial director. One of the problems as an inventor licensing other companies is that, while they might be keen when you sign the licence, they may decide, for various reasons, to drop it later on. It is a most frustrating experience for a licensor. I needed to look for other licensees. Black & Decker expressed an initial interest, as did several other American companies including Amway, a marketing company from Ada, Michigan. Founded in 1959 by Jay Van Andel and Richard DeVos, its name is short for 'American Way'. They sent a really nice Australian vice-president to see us in Bath.

He was charming – I still correspond with him in retirement in Australia – spun a good line and in April 1984 we licensed Amway, who were to get a company named Bissell to make the vacuum cleaner in nearby Grand Rapids, Michigan. After I had spent time there with Amway and Bissell, the manufacturer, and handed over all the drawings, prototypes, knowhow and confidential information, they decided to cancel our agreement and to instigate a fraud lawsuit to get back the money they'd paid us.

It would have been very hard and potentially disastrous from a financial point of view to fight Amway in the US courts, during which time we would have been unable to license it to anyone else. Amway had a very tough Bronx lawyer, a barracuda, we were told, with an education. In 1984 Jeremy said, 'Settle it. I don't want to fight a lawsuit.' We gave all the money back to Amway and paid our legal expenses. This put me heavily in debt. At the same time, Jeremy's financial adviser, who thought the vacuum cleaner a complete waste of time and money, encouraged him to sell out. I was able to fund all this and to buy Jeremy out by selling the coach house behind the Royal Crescent to Chris Patten, the Conservative MP for Bath, and borrowing yet more money from the Chipping Sodbury Lloyds bank. Although I now had 100 per cent of the shares, I was sad that Jeremy had decided to cease his involvement. I understood the scare of the lawsuit and particularly that he had more to lose than me. After all, it was my invention and my headache. I was extremely grateful for his support and we remained very good friends.

And, then, serendipity played its part. TWA's in-flight magazine had placed a juicy photograph of the pink and lavender Cyclon on its inside back cover to accompany a feature article opposite. This is what had got the vacuum cleaner noticed in the States. By the same token, a tiny Japanese company called Apex that imported high-end design from Italy and Switzerland along with the Filofax from England saw a solitary photo of the Cyclon in a glossy product-design book. Come and see us, they said. In January 1985 I bought a cheap ticket to Tokyo, flying there by Aeroflot, the Soviet airline, via Moscow.

Japan really was another world then. When I first went to the Apex office, the girls pointed at my nose. 'You have a nose like Eiffel Tower.' This, I learned, was a compliment, although it seemed such a bizarre thing to say, especially when meeting someone for the first time, even if my English nose was much more prominent than theirs. After a few weeks, one of the girls asked me if I'd like to come out in the evening to drink beer with them, which also seemed an unusual thing to do at the time. I was looking forward to it, but just before it was time to go, the chairman asked me to play Scalextric with him at the office instead, a less interesting social experience.

A most refreshing experience, however, was Apex's approach to the vacuum cleaner. They loved it, appreciating each component and the fact that it was different. They took it apart, studied it and understood it. Theirs was a real love of technology, of making of things and artefacts. It was great to be appreciated. I spent a lot of time in Japan redesigning the cleaner head of the vacuum cleaner and improving the separation system by adding a shroud after the first cyclone to trap hairs and fine carpet fluff. Apex wanted to keep the pink and lavender colours as these would make ours stand out from existing vacuum cleaners, and they happened to like them. It was put into production by Silver Seiko Ltd, a company that made well-known Silver Reed typewriters and knitting machines. Apex called it G-Force. It went on the market in 1986 and sold for 250,000 Yen, something like £2,000 today. A status symbol in Japan, it became an instant design classic. Because it was licensed in Japan and accounting there was truly inscrutable, we never knew how many were sold, but it was clearly much liked, continuing in production until 1998.

G-Force Patents for G-Force

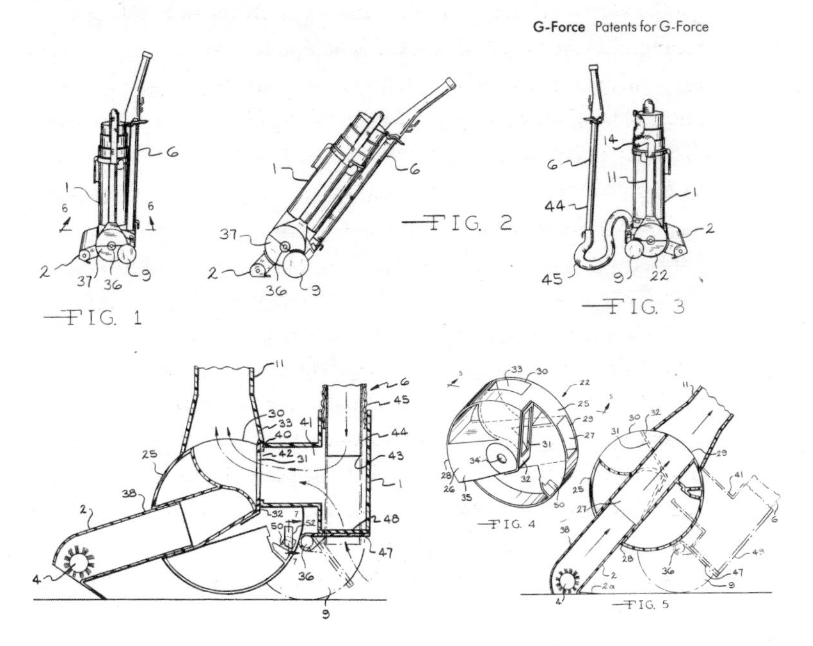

It was while I was away on that first trip to Japan that a squirrel gnawed through a water tank pipe in the roof at home. Water cascaded through the whole house and the ceilings came down. Deirdre and the children had to deal with it in rough camping conditions. I got back home without, as yet, a licence agreement – that came later – to find a wrecked house and, for all my efforts and travels, absolutely no income. It was terrifying at the time – sleepless nights, the lot. But the agreement did finally come through and, with it, upfront money that saved us from ruin.

Not long afterwards, I first met Paul Smith at his Covent Garden shop. He wanted to sell the G-Force. I imported machines from Japan for him so that he could sell them in his British shops. Paul displayed the G-Force in between clothes in a long line across his shop windows. He was my first UK customer and he quickly sold the lot. I only imported 200 machines. With hindsight we should have imported many more. It was my fault. I didn't want to be an importer at the time.

With the Japanese G-Force in production, I went to the USA and spent some time trying to license American and Canadian companies. In one way it was really interesting to learn how the market for vacuum cleaner and suction carpet cleaners could be quite different in other countries, but on the other hand the tricky negotiations and legal aspects of potential deals were often excruciating.

On one flight back from the States I was sitting next to a businessman called Jeff Pike and we discovered that we had shared interests, not least the latest Fay Weldon novel we both happened to be reading. Jeff's main business interest was oil exploration in Canada, but he was also chairman of a company called Iona Appliances, with headquarters in Welland, Ontario. I went to visit Jeff at his office in Toronto and this led to our first licence agreement with Iona (later Fantom) for a dry-powder carpet shampoo machine that we developed. It used Milliken's 'Capture' moist powder product. You loaded dry powder (it is actually slightly damp powder) into the hopper on top of the cleaner head to deposit the powder – not as easy as it sounds: we had to vibrate it to shake up the clumps and to spread it evenly. You then switched to 'brush-in' mode. Finally, you switched to pick-up or vacuum mode. All controlled mechanically from the top of the handle via a Bowden cable.

Iona sold it to Sears. I had an interesting meeting with the Sears buyer, who would only buy it in all grey. I spent a day charming and cajoling her and managed to secure a bold dash of blue near the top! It worked well and we sold some in the UK through a catalogue of new products run by a charming man called Clive Beharrel. Our experience of dry shampoo, apart from Jon Wealleans' hair, is the reason that we sold a Dyson dry powder called Zorb-It-Up and a stain remover called Dysolve alongside our vacuum cleaners.

We went on to negotiate a licence agreement with Iona for vacuum cleaners, and the deal was about to be signed when Amway launched a copy of my vacuum cleaner. Iona had heard about this copy even before it came out as Amway had been to Sears, then the largest retailer in the USA, to try to sell it to them. When Iona went to Sears, they said they had already seen the same product. I was stunned by this shocking news. Amway had not only cancelled an agreement, but it had now copied my technology. This was pretty exasperating, even more so because Iona negotiated my licence deal downwards in the light of Amway's copy. We agreed to share costs to fight Amway through the courts.

When you have developed a new technology or created a radically different product, have beaten the sceptics, established awareness and battled to create a market for it, to discover a similar product from a company that had cancelled a licence agreement is sickening, as though you've been punched in the solar plexus. You feel outraged by the personal theft, and helpless.

By now, 1987, Peter Gammack and Simeon Jupp had joined me in the coach house. They were both graduates in design engineering from the joint RCA–Imperial College postgraduate degree course. I had made a point of keeping in touch with colleges and going to see their degree shows at the end of each academic year. It was a natural way for me to find innovative young design engineers. Pete had done a degree in engineering at Imperial College, followed by two years at the RCA to learn design. At his degree show he exhibited an interesting slot machine to dispense newspapers. It was ingenious and cleverly executed. He hadn't chosen an obviously showy project, instead one that was unusual and well-engineered. I liked that willingness not to be fashionable but, rather, to do a thoughtful project. I was able to afford to pay Pete and Simeon through royalties from Apex in Japan and through various consultancy fees. What was exciting is that, although our main focus was the vacuum cleaner, our thinking was that of a tech company. How else could we evolve cyclonic technology? What other uses could we put it to?

In 1987 we reached a licensing deal so Iona could make and sell dual cyclone upright vacuum cleaners in both Canada and America under the name Fantom. I am afraid that as things turned out it was rather too prescient a name. I was also approached by Johnson Wax in the guise of a delightful Scott Johnson (no relation). They were interested in the industrial cleaning applications of our cyclonic system. I did a demonstration to a group of their top managers where I stamped on a glass of water, creating many shards as well as a puddle. I then vacuumed it all up with my cyclonic vacuum. Unlike a bagged machine it was unaffected by either water or glass.

A senior executive, a most energetic Australian named Ross Cameron, leaped over a table to try the same demonstration himself. He had been an engineer with Austin Morris in Sydney but became a champion within Johnson Wax and we did a deal for them to sell in the industrial arena. Sadly, it didn't work out as the chemical-background vice-president in charge failed to understand the vagaries of machinery and they gave up. What came out of it was better still. I later offered Ross the task of starting and building Dyson in Australia, which he did magnificently. He is now retired. We are the best of friends.

Meanwhile, although I offered Fantom our latest designs and technology, they wanted to go their own way. As this turned out not to be very successful, I offered to buy them out. We were in the final throes of negotiating when they went bankrupt. A quirk in Canadian law is that, although under our legal agreement I had only licensed them with the patents, the patents should have reverted to us. Instead, the receiver started to sell off our patents to the highest bidder. We bought the main ones back, but competitors outbid us for others. Although this was all most frustrating, it did give Dyson the golden opportunity to enter the North American market, then the biggest in the world, with our latest technology.

Because Iona was licensed to make and sell our vacuum cleaners in Canada, but not the US, I had gone to see Conair, the big hairdryer company based in Stamford, Connecticut. Conair had been founded in 1959 in a garage in New York and by now was very successful. The owner was Lee Rizzuto, son of Sicilian immigrants, and he seemed really nice, as did the company. We got close to the point of agreeing terms on a vacuum cleaner when they asked if they could look at the patents. In came the same unpleasant, pipe-sucking executive who had rubbished our patents at Black & Decker. I went to see Lee Rizzuto and said, 'If he is the person examining our patents, we are unlikely to do a deal. He thinks our patents are weak, but he is wrong.' The pipe-smoker killed the licence anyway.

At this point in 1991, Vax came into the picture. A British company based in Droitwich, Worcestershire, it was founded in 1977 by Alan Brazier, a former Rank Xerox salesman. Brazier had created his own shampooing and cleaning vacuum cleaner from a shower pump and milk churn after winning cleaning contracts from motorway service stations and finding existing cleaners unsatisfactory. For a short while Vax was successful. They then wanted to do an upright vacuum cleaner and we designed one for them but found them rather overbearing. They kept rejecting our designs until finally they said they would bring in other designers, so we walked away. Because they were delaying and not going into production, I cancelled the agreement. Now we had a threat of court action at home while at the same time coping with the costs of the US lawsuit with Amway. Fortunately, we had our licensing agreement with Fantom so we could just about keep our heads above water although the end of each month, when salaries needed to be paid, was a worrying time.

The lawsuit against Amway, like all US lawsuits, was long and arduous, five years in all and a constant expense. There is a process whereby, in this situation, you can get an early injunction to stop an alleged infringer of your rights, but judges rarely give them if the damage can be recompensed by money. All very well if you have the millions to fight a five-year lawsuit and subsequent appeals. We embarked on the lengthy lawsuit, taking on two very large US companies with huge resources, Amway and Bissell, in the USA.

Amway had teamed up with Bissell to make its version of the vacuum cleaner. Bissell are based in the same city as Amway and had been a vacuum cleaner and carpet sweeper manufacturer for many years, known in the UK as Bex Bissell. Since I had explained the technology and knowhow to Bissell during the licence agreement with Amway, as well as handing over all the drawings, we had to sue Bissell for their part in the Amway version.

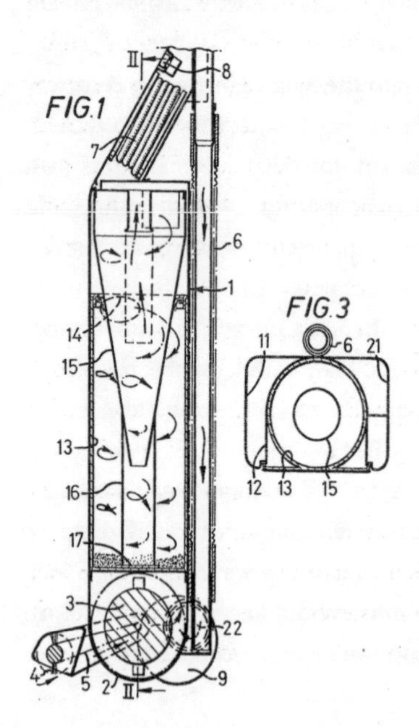

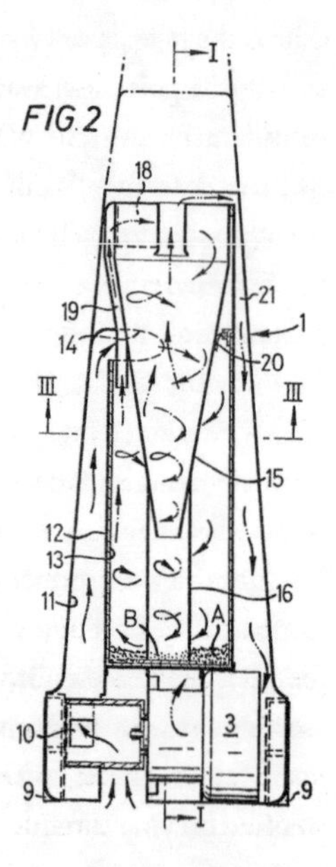

Patent asserted against Amway

Of course, I couldn't afford to fight on my own. The terms of my licence agreement with Iona/Fantom were drastically reduced while they agreed to pay half the costs and take half the winnings, if any. Our litigator, Dick Baxter, agreed to reduced fees with him also taking a chunk of any winnings. This is what is called a partial contingency case. Then the real work started. We filed our claims in detail and they filed their defence. Then both sides called for 'Discovery'. This allowed the other side to ask for every document and prototype imaginable in order to see if there was a flaw in the case and to work out how to attack and cross-examine witnesses, a long process. While we dredged back through all our paperwork and drawings and models, producing dozens and dozens of boxes of Discovery documents, Amway produced hardly anything in response to our requests, which was less than credible.

As the case progressed, and with legal bills mounting, it was clear that we had both a good infringement case and a good misappropriation of confidential information case. However, a classic defence on a patent case is that the patent itself is invalid or the invention is obvious. Thus, throughout the case I, as the patent holder, was defending the patent from attacks upon its validity. This is strange because the time should be spent showing why the infringer has copied the patent. However, just because a patent has been granted, it may not mean that it is valid until tested in court. This is not very helpful, as the cost of filing and annual renewal fees to the Patent Office in each country is very high.

In fact, I twice took cases before the European Court of Human Rights to have Patent Renewal fees declared illegal, on the basis that non-payment of the renewal fee, essentially a massive income to the government, results in the creator of the art losing their rights. This loss does not happen to any other creator of art. Twice the case was lost on the basis that the fees were reasonable, presumably according to the European Patent Offices! Yet when I was a penniless inventor, I was shelling out tens of thousands of pounds every year on renewal fees in just a few countries, which I struggled to pay on borrowed money and so often had to surrender patents.

As a part of US lawsuits, the opposing lawyers get to cross-examine the main witness – in the case of the Amway/Bissell dispute, me – from 9 a.m. to 5.30 p.m. day after day for three weeks. In Grand Rapids, I was faced across a large table by five lawyers from a big Chicago law firm acting for Amway and Bissell, asking question after question, with them passing Post-it notes between them about how to trip me up. In a patent and misappropriation of confidential information case, words matter more than anything. Inventions are described in patents by words rather than by drawings. Although drawings are included, the claims that are the substance of the patent are words. Words can be twisted, or the wrong synonym used, and under cross-examination the use of a wrong word can cost you the case and the validity of your patent. I underwent three weeks of fraught cross-examination by aggressive lawyers, which I would not wish on anyone. Most unpleasant. The Amway CMS-1000 was quite clearly based on my drawings, patents and confidential information.

After five years of this bruising and bankrupting lawsuit, I was ready to pack it all in and settle the legal case. Deirdre, however, stood firm and told me that I mustn't give up. This was well judged. Shortly after my conversation with Deirdre, I was boarding a flight from Heathrow to Detroit for yet more cross-examination when I decided to call Dick Baxter, our charming litigator, in Grand Rapids. Dreading the flight and more cross-examination, I was elated by the news he gave me: 'Amway have indicated that they are willing to settle.' I didn't need to fly. I could go home.

Dick Baxter and Fantom, my licensee, shared the settlement with me. My share amounted to $1 million. More importantly than this, I could regain my life. The black cloud that had hung over our family life for all too many years was lifted. We weren't rich – in fact, the award didn't even cover my share of legal costs – but at least the costs had stopped. Instead of worrying whether or not we were going to be bankrupt we could now think differently. We were more in control of our future. I wouldn't have to spend so much time dealing with the case, reading documents and making trips to the US to see lawyers.

Despite all we had been through, and the settling of our case against the Americans, we were all frustrated at how things were progressing. To date, the North American market had proved to be hard work while the experience with Vax was upsetting, too. Almost all of our licence agreements had failed, and precious time had been wasted. Instead of relying on other companies to make our technology, why didn't we reverse the decision Jeremy and I had made several years previously? We would take on these competitors who were very content selling bags and their loss of suction. We would be free to determine our own future and our own development and designs. Instead of feeling dread about taking on the mountainous task of becoming a manufacturer, we felt liberated.

At the time, the biggest-selling type of vacuum in Britain was the wet-and-dry carpet shampoo machine. We knew that we could make a much better wet-wash vacuum cleaner than Vax. I already knew a great company called Linpac who had injection-moulded the plastic bin for the Ballbarrow and I had kept in touch with their enterprising boss, David Williams. I went to Louth to meet the owner, Evan Cornish, whose son later sold me their excellent Lincolnshire farm. They were really interested in moulding the plastic parts and funding the tooling. We built a prototype and decided to test it alongside a Vax on the carpet upstairs in the coach house. The prototype proved to be pretty hard work. If it was such hard work, why were people buying them, we wondered?

It was time to do a bit of market research. We went to the hardware shop in Larkhall on the edge of Bath, which we knew well, and asked them if they knew a customer who might like to come down to the coach house to take part in a test. Pete, Simeon and I watched as this nice lady set to work with our wet-and-dry carpet shampooer. What do you think, we asked? 'It's really hard work,' she said. 'You know, I wouldn't want to do that twice.'

We abandoned the whole venture on the basis of what this one kind woman said. We knew in our bones that we were barking up the wrong tree. The Vax and those types of carpet shampooers were all hard work.

This was the moment we decided to go against market trends and to concentrate on an upright cyclonic vacuum cleaner that we would make ourselves. It would be a considerable advance on the one we had licensed to Iona, the Canadian firm. There would be no more struggling with other companies, their patents, executives and lawyers. We were going to do it our way, alone and independently.

CHAPTER SIX

DC01

Of course, it wasn't that easy going our own way. While in early 1992 we continued to work on what would eventually become the best-selling DC01 vacuum cleaner in the coach house – a lathe, mill and work-benches downstairs and desks with three computers in the hayloft – I went out to try to borrow the money we needed to set up in manufacturing. Venture capitalists proved to be no help. I went to see Adrian Beecroft, Chief Investment Officer at Apax Partners. As well as Tie Rack and Sock Shop, they were funding Bob Payton, the colourful American restaurateur who had founded the Chicago Pizza Pie Factory in London and was now planning further fast-food joints. These were businesses that promised a quick return on capital, which is what venture capitalists were looking for. In for a few years and then out, maybe.

I was approaching these investors to raise capital, cash, in exchange for equity, or shares, in my company. With this capital I would not be saddled with the enormous bank overdraft that had plagued the Ballbarrow company. When I showed Apax my prototype, and told them my story, they weren't interested, because it was about making things. In any case, from their point of view, I was an engineer, a designer, rather than a proper businessman. Adrian Beecroft said, 'Well, we might consider your application, if you got someone from industry to run it.' I said, 'Can you tell me a domestic appliance business in Britain that is being successfully run?' Hotpoint? Hoover? That was it! Neither of these was at all successful. Existing manufacturers hadn't brought out anything new for decades – I couldn't possibly go down that route.

Apax proved to be a dead end. Even if they had funded me, they seemed fixated on my not running my own business. Where else could I look? There were no government schemes at the time to help. As for other venture capitalists, five turned me down – there were no tax advantages for investing in start-ups. Meanwhile, the clearing banks had been receiving bad press for reclaiming people's homes during the then deep recession, which was why I had avoided them. Why, in the midst of a deep recession, would they lend to me, creating yet another house repossession problem? In desperation, and with no prospect of success, I went to the Corn Street branch of Lloyds in Bristol. My overdraft had escalated, and the bank thought it sensible to upgrade me from Chipping Sodbury to their Bristol head office. To my great surprise it sounded hopeful, but, even then, this caused a battle of wills within the bank.

Mike Page, a 'flying doctor' from the bank, came to see us at the coach house. What do you need the money for, he asked? I told him I wanted to make vacuum cleaners to compete with Hoover and Electrolux. 'That's very interesting,' he said. 'I'll let you know.' He came back about a week later and said, 'We'll lend you £400,000, but you've got to sign over your house.' I went off to Bristol with Deirdre and a lawyer and we signed a nasty grey form making everything we had over to the bank in exchange for the money. It was taking the ultimate risk. With three young children, we could end up evicted from our home. It was heady and scary stuff, even more so when Mike Page upped the loan to £600,000. Deirdre and I discussed the awful risk we were taking. We stood to lose everything we had. It didn't bear thinking about and we had no idea what we would do if we failed. However, it was our last chance to make our invention work, to show faith in what we had done. It was extraordinary that Deirdre went along with this last roll of the dice.

Some time later, when I dared to ask Mike Page why he had lent me the money, he said, 'Well, you had fought a five-year lawsuit in America, so I could see you had determination, and I went home to my wife and told her that you were doing a vacuum cleaner without a bag, and what did she think of that? And she thought it was brilliant not to have a bag.' And that was that, although what we didn't know at the time was that Lloyds had turned his request down. He had to appeal to the ombudsman in the bank's head office in Cardiff, who for some reason agreed with him. Sadly, it is not often that banks or bank managers receive praise for the risks they take.

The great thing is that now we had our own money – well, the bank's money – and no need for investors, who I didn't want anyway. The tricky thing, though, is that we set up Dyson Appliances Ltd just as recession hit. The British economy was pretty rocky for the next two years. There were riots over hardship in Bristol and Cardiff as well as in Birmingham, Tyneside and other parts of the country. The recession had been caused not least by high interest rates used by the government as a means of controlling inflation. This, though, led to high mortgage rates and falling house prices at the same time as the pound was overvalued to maintain British membership of the EEC European Monetary System, making British exports uncompetitive on price. This unhealthy recipe saw unemployment rise to 10 per cent in 1992. When Britain left what was now the ERM (EEC Exchange Rate Mechanism) that September, the pound was devalued by 20 per cent and the economy began to recover.

We moved very quickly to get the vacuum cleaner into production. Pete Gammack concentrated on tooling and we signed up a £750,000 deal with various toolmakers in north-east Italy I knew from the Zanussi days. This was all the money I had, having used up the loan from the bank and the money from the Amway settlement. I had no money for a factory or to buy components, and no money to do any marketing. Meanwhile, I looked for a factory. The Welsh Development Agency was giving grants at the time to manufacturers looking for new factories and new business. The Agency itself was a good organisation but to get approval for a grant you had to apply through one of four consultancy firms, who would then put your proposal to the Secretary of State for Wales. This was David Hunt, now Lord Hunt.

We spent £28,000 with PricewaterhouseCoopers in this process, only to be told by David Hunt that we were undercapitalised. Our case was rejected although we could apply again if we were able to raise more money. We were, in fact, better capitalised than a firm of Japanese zip makers that got a grant at this time. Our problem, I discovered, is that unlike the Japanese zip maker we were not foreign investors and so no 'feather in the cap' for David Hunt.

Funnily enough, soon afterwards I went to see Phillips Plastics, an American firm based in Wrexham in North Wales, to see if they might make plastic components for the vacuum cleaner. Their new factory had been built with the help of a grant from the Welsh Development Agency and approved by David Hunt. It had very little work. I suggested that, as they'd be making the plastic components, perhaps we could assemble the vacuum cleaner inside the half of the building that they were not using. Phillips said, yes, that was fine, so we had the tools shipped there from Italy along with components from various sources – a hose and brush bar from the Accrington Brush Company, a motor from a Japanese manufacturer and brush bristles from Changxing in China.

We employed people for assembly work locally in Wrexham. The first production-quality vacuum cleaner to bear the Dyson name started rolling off the production line in January 1993. We had already shown GUS (General Universal Stores) and Littlewoods, two retail giants, a working model and, published in their January 1993 catalogues, we were able to supply them in the third week of the month.

It might be hard now to recognise what big deals these were for us. GUS, Littlewoods' Manchester rival, was owned and run by Isaac Wolfson, chairman from 1932 to 1987, who created the Wolfson Foundation and built colleges in both Oxford and Cambridge. I drove up to their HQ in Manchester to visit their buyer, Brian Lamont, to show him the Dyson prototype, explaining the no loss of suction with no need to buy bags, and then mentioned that we supplied the machine with our dry-powder carpet shampoo called Zorb-It-Up as a much better alternative to soaking the carpet. Zorb-It-Up was a direct result of Jon Wealleans' hair cleaning habit. I was intrigued when he told me at college that he shampooed his hair with dry powder. I had visions of the flea powder we used on our dog. After we had rejected our wet carpet cleaning machine in favour of an upright vacuum cleaner, I had been thinking of a better way to shampoo carpets. Jon's dry shampoo seemed to be the answer. It could be used with our upright vacuum cleaner, its rotating brush embedding the shampoo into carpets.

Brian Lamont was very sceptical, convinced that our machine would fail to clean up the popular blackcurrant drink Ribena. I rushed out to the nearest petrol station to buy a bottle. Back in his office I spilled some of the deep crimson juice on his fine carpet, applied the dry powder and removed the stain. He then asked me why he should remove a well-known brand such as Electrolux or Hoover from his pages, to put in this unknown Dyson with frightening talk of cyclones and 200mph centrifugal speeds. My response, that his catalogue was 'boring', met with silence for a while, until he finally agreed to buy from us.

By the 1980s, Littlewoods, famous for football pools as well as its retail catalogues, was both the largest family-owned company in Britain and Europe's biggest private company. Its headquarters was the monolithic Art Deco building designed by Gerald de Courcy Fraser, opened in 1938, given over to wartime production the following year. Because GUS was buying from us, they also agreed terms.

All that we had achieved had been accomplished in what I suppose should be described as record time. It was a bit makeshift, though, in that there was a production line in half a factory at Wrexham and HQ was the coach house where we engineers worked along with Bob Bedwell – in charge of accounts, purchasing and logistics – two salespeople and a PA. We serviced our first vacuum cleaners in the coach house.

Three months later, in April 1993, Phillips, who had no other work, decided to double the price of the plastic parts. I said that if they did I'd take the assembly of the vacuum cleaner elsewhere. If you do that, they said, we'll triple the price. Within a fortnight, I'd rented a former Post Office depot in Chippenham and we were up and running very quickly. Deirdre, for one, was pleased. If the family business was finally leaving home, it meant that the coach house was free for her to use as an art gallery to display and sell her works, while the driveway to the house was no longer full of cars.

Even so, moving from Wrexham to Chippenham wasn't quite as easy as it sounds. Aside from the sheer rush, I had to get a court order to remove our tooling from Phillips, and when we did get it back we found there was quite a bit of damage to rectify. In spite of their David Hunt grant, Phillips quit the UK. And, of course, I had to find a plastic moulding company there and then. I found one, Choudhry, in Birmingham.

Our makeshift factory was not exactly ideal. The building had twenty-nine roller doors and was icy cold in winter. When we needed more space, we bought a huge second-hand tent the size of a car park. Local cab drivers called it the 'Madonna tent' because of its twin peaks. On cold days it sweated inside. Hired containers provided further space. This was a big and uncertain moment for me, and yet in July we began production with the DA001 and, soon afterwards, the improved DC01. Within eighteen months, this was the biggest seller in the UK market.

Our first sales were through hefty mail order catalogues. These devoted a few pages to vacuum cleaners. We were among the last pages, at the bottom, with a small, square picture of the DC01. There was no room for descriptive copy, just the tag 'no bag' and the opportunity to pay £1.99 a month instead of £199. Ours was the most expensive cleaner in these catalogues by some margin and they were not the sort of place you would expect expensive items to be sold. Both we and the buyers at the catalogue were, in fact, astonished that DC01 did so well through their pages, with repeat orders coming in. I have never, though, believed that someone's income is a bar to them wanting to buy the best product and a vacuum cleaner is an important purchase. They may have to save or borrow or sign a hire purchase agreement, as people did then, but they will go for the best if they can.

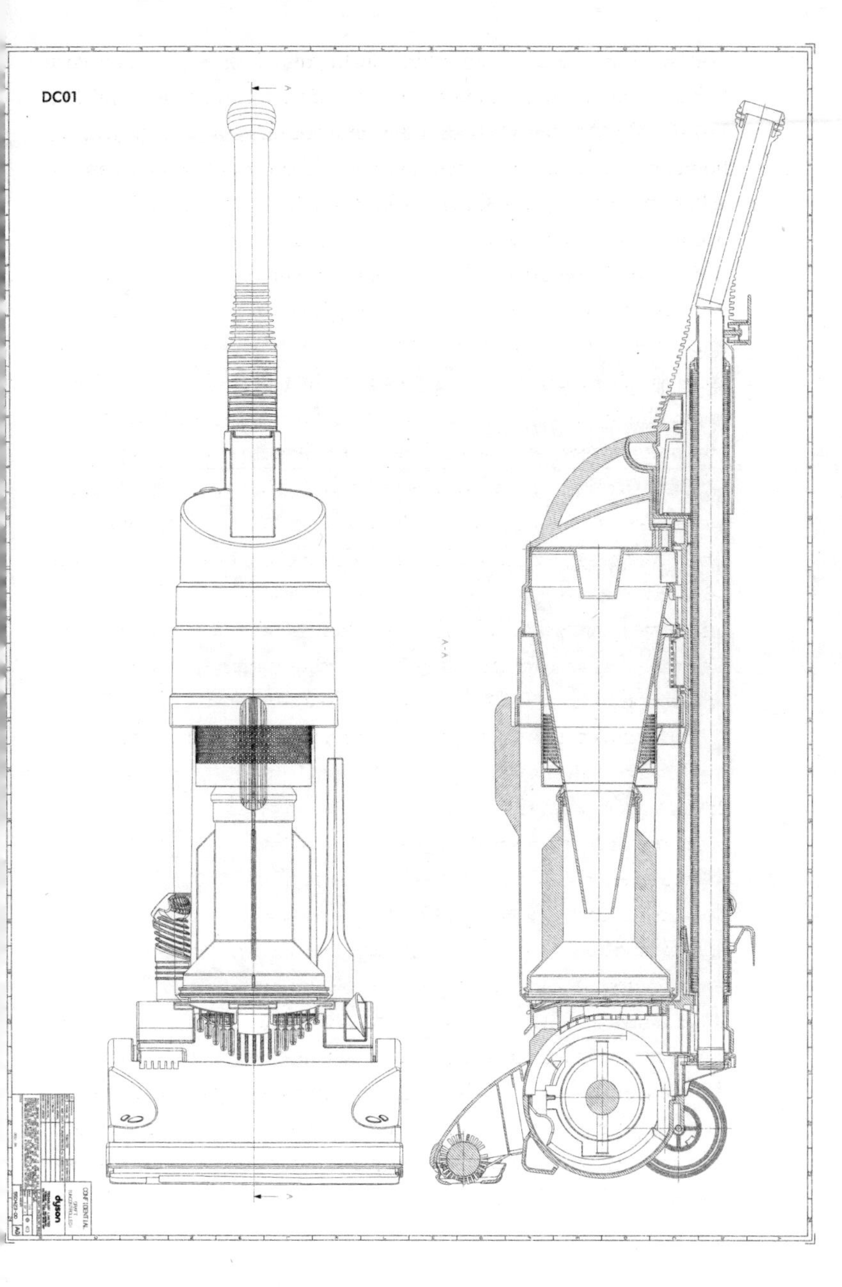
DC01
A-A
CONFIDENTIAL
dyson
A0

This initial success led to the DC01 being taken up by our first retailer. This was Rumbelows, a national chain of shops selling electrical goods. John Lewis, an excellent department store with twenty-five branches, followed, although they insisted on a trial. This was successful, although some of their branches refused to stock it. Now, though, we were selling quite well across a good spectrum of sales outlets and I knew we would survive. Deirdre and I could relax a bit.

DYSON dual cyclone

Operating Instructions

Please read carefully before assembling or using your Dyson dual cyclone.

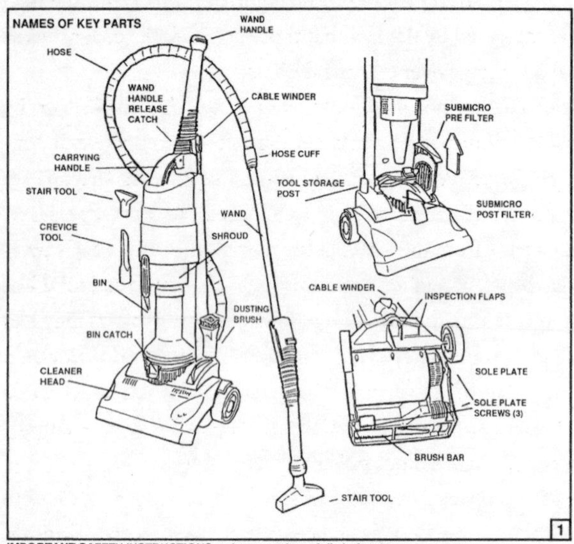

IMPORTANT SAFETY INSTRUCTIONS - ***please read carefully before use.***

When using the Dyson dual cyclone vacuum cleaner please adhere to these simple precautions.

WARNING: TO REDUCE THE RISK OF FIRE, ELECTRIC SHOCK OR INJURY:

1. Do not leave the Dyson dual cyclone with the plug connected to a mains outlet. Unplug when not in use. **Always remove the plug from a mains outlet before carrying out maintenance**
2 Be careful to make sure that hands, feet and fingers, particularly of small children, are kept away from the cleaner and especially from the brush bar.
3 Do not use without Bin correctly fitted and all Submicro filters in place
4 Do not fit any other type or make of filters other than Dyson spare parts, otherwise guarantee may be invalidated.
5 Electric shocks could occur if used outdoors or on wet surfaces do not use the machine or handle the plug with wet hands
6 Do not use with damaged cord or plug, or if the vacuum cleaner has been damaged, dropped or has come in contact with water or liquids. In these cases contact the Dyson Service number or contact your nearest Dyson dealer.
7 Do not put any part of the body, clothes or any object near or into any of the openings or moving parts of the machine. Do not use the machine with any of the openings blocked with any object that may restrict the air flow.
8 Do not damage the cord by running the cleaner head over the cord, running the cord around sharp edges, closing a door on the cord or putting the cord near hot surfaces.
9 Be careful when using the machine on stairs or when the machine may topple over.
10 Do not use the machine to pick up flammable or combustible liquids and do not use near such liquids
11 Turn off the machine by the switch button before unplugging from an outlet socket

Operating Instructions Page 1

I knew I had various degrees of perseverance, determination, grit and what you might call sheer bloody-mindedness, yet these qualities – I like to think they are qualities – were underpinned by a kind of naïve intelligence, by which I mean following your own star along a path where you stop to question both yourself and expert opinion along the way. A willingness to keep questioning the validity of an idea, or indeed a product, might sound naïve in a world of slick global business, but I can say it worked for me and Dyson as I think it can and will for inventors, engineers, designers and makers of the future. I had been warned, for example, that at £200, or at least three times as expensive as most other vacuum cleaners, the DC01 would prove to be too expensive. It sold really well. The sheer cost of producing it, and as a result the high price tag, was mitigated by the fact that those choosing it recognised its technological advantage over existing designs.

I was also told that no one would want to see dust sucked up by a cleaner inside a transparent container. Simple market research confirmed this. However, Pete, Simeon and I enjoyed seeing the dirt we had extracted in all its gory detail, so we ignored the market research. Curiously, and aside from the fact that the new cleaner was clearly powerful and with constant suction, this is exactly what customers *did* like to see. They were fascinated by the sight of just how much dirt they had successfully cleaned up. They clearly liked our venture into television advertising, too. 'Dyson. No bag. No loss of suction.' It was all a bit odd at the time because we were promoting what we didn't have – a bag – when, traditionally, advertising spoke only in positives.

This negative approach was the result of spending a week with the brilliant Tony Muranka, a freelance creative who worked with us. Instead of talking about the clever cyclones, about no need to buy bags, about the quick-draw hose, the powerful suction, we decided to highlight the Achilles heel of other vacuums – the bag and its shortcomings. It worked, except for one small problem: it became known as the 'bagless' vacuum. This term was coined by our competitors, not by us. I think they came up with it to point out to retailers that, if they tried to sell a Dyson they would, in turn, lose sales of replacement bags, to put them off selling Dysons. However, we would rather a Dyson be known as the 'no loss of suction' vacuum, which is a much more important performance feature than the absence of a bag.

Fifteen years of invention, frustration and determination were, however, beginning to pay off. I was now a proper manufacturer, thrilling to the sight and sound of a production line in full swing. I found it impressive, staggering even. I still do. I worked on the line for two weeks to understand how to make the vacuum cleaner more efficiently and have watched all of our lines ever since. I was able to alter how we assembled vacuum cleaners along the line by moving some sub-assemblies off the line, making it move faster. I learned which components were difficult to assemble and encouraged our engineers to visit lines frequently. Most importantly, this experience helped me to look at all our subsequent products to understand where production inefficiencies lie.

One of the great challenges at the time in Britain, however, had been to win over the wholesale and retail business beyond the GUS and Littlewoods catalogues and John Lewis and to get these to thoroughly endorse DC01. How best could I share my enthusiasm for what I knew was a great and innovative product? Some of the obstacles lying in our path were quite frustrating. I dropped into the Chippenham branch of Rumbelows. They had a line-up of gleaming new vacuum cleaners, but the DC01 on display was full of black dust. It looked disgusting. I asked a sales assistant why it was so dirty. 'Oh,' she said, 'it's because we clean the shop with it. It's the best of the vacuum cleaners.' So I asked, 'Which vacuum cleaner do you recommend to customers?' 'Oh, Panasonic.' Why? 'Because I've got one at home.'

Among key retailers for vacuum cleaners at the time were local Electricity Board showrooms. These disappeared eventually, following on from the privatisation of the British power industry, but they used to be on many high streets, and this is where people came to pay their electricity bills, so they had lots of customers. We went to see the Eastern Electricity Board, where their buyer had been testing one of the vacuum cleaners for a fortnight. He told us, 'It's a really good vacuum cleaner. I like it. It works really well. But I'm not going to take it because you can't afford to advertise on television.'

I did a quick mental calculation and said, 'If you buy two thousand, I'll spend £40,000 on telly. So, for every thousand machines you buy I'll spend £20,000 with Anglia Television.' He said, 'Deal'. And we sold a lot of vacuum cleaners in East Anglia. While it was great to discover how well it was received, and especially by those who chose to use it in preference to existing vacuum cleaners, it was almost impossible to know why some retailers would take it while others wouldn't. Sometimes our dealings with retailers bordered on the bizarre.

We started to sell really well in Scotland through Scottish Power and Scottish Hydro showrooms. One day, we went into a Scottish Power store in Edinburgh to find out how our cleaners were sold face-to-face. A young woman dressed in a Scottish Power uniform did her best to sell us a Hoover. I asked her, why a Hoover? 'Well, it's the best vacuum cleaner.' I asked her, in a nice way of course, if she was really from Scottish Power. 'No,' she replied, 'I'm from Hoover.'

It was all a bit disconcerting, and certainly misleading to consumers. We got on really well with John Lewis, a company that truly liked innovative design, but sales of DC01s in one branch could be very much higher or lower than in other branches around the country. Friends who had gone into the Bristol branch told me that they were telling people not to buy a Dyson, but to buy a Sebo vacuum cleaner from Germany instead. This is known in the retail trade as 'switch selling'.

What I heard is that Sebo was sold only through John Lewis. Their importers went round John Lewis branches training staff on the virtues of Sebo products. Was it true? I drove to Bristol and pretended to be a customer. 'Oh, no,' I was told, 'you don't want a Dyson. It breaks. What you want is this well-made Sebo vacuum cleaner from Germany, which is made of better plastic.' I said, 'Well, I'm Mr Dyson, actually, and, by the way, the Dyson's made of ABS, a tough and expensive thermoplastic polymer and costly polycarbonate, while the Sebo is made of cheap polypropylene used for washing-up bowls.'

I was told that I must make a proper appointment to come and talk to the staff, all of whom, of course, were partners of John Lewis. Two weeks later, I was back in Bristol again. I explained that we used this very expensive material, polycarbonate, that is four times the price and four times as strong as polypropylene. You can hit a Dyson with a hammer, and it won't break. One of the partners said, 'I don't believe you.' I suggested she go and get a hammer and try for herself. And she did just that. 'Hit it as hard as you like,' I said. She walloped the cleaner's bin and the hammer bounced off, as I knew it would. 'Now hit the Sebo.' She did and the thing shattered. She certainly made an impression. Bristol became one of our best-selling John Lewis branches.

This, though, wasn't the end of the Sebo story. Within eighteen months of setting up production in Chippenham, we had 20 per cent of the UK vacuum cleaner market. Deirdre and I bought a house in Chelsea. One Saturday morning, we were in Peter Jones, John Lewis's flagship store in Sloane Square. Deirdre was buying kitchen appliances, while I couldn't help wandering over to the section where they sold vacuum cleaners. A young chap came up to me and I asked him what he thought of the Dyson. Should I get one?

'No, no, no,' he said. 'We get lots of returns of the Dyson. What you want is a Sebo.' I got in touch with the buyer and asked to see the returns figures. What the floorwalker had told me was completely untrue. Funnily enough, Terence Conran had been in Peter Jones around the same time. He said he was interested in buying a Dyson and got the same treatment. He told the sales staff, 'Well, if you're not going to sell it, I will.' I didn't know Terence personally then, but he got in touch immediately afterwards and said, 'We'd like to sell the Dyson in the Conran Shop.' That was exciting because I had tremendous respect for Terence and the Conran Shop sold some of the very best modern design you could buy. Unbeknown to him, I had spent the summer of 1967 working in the Conran Design Group as a student placement, designing chairs for Heathrow Airport. Following this we became good friends, with me chairing his Design Museum and having a joint furniture company.

We did, though, build a very good relationship with John Lewis. One of the things I learned was that we needed to talk to its partners, to explain why and how we were offering something different from existing vacuum cleaners. I gave talks to partners at their headquarters and at their country-house hotel and conference centre at Odney in Berkshire. The Sebo story aside, John Lewis understood good design and knew what their customers would buy.

But it took nearly two years to get our vacuum cleaners into mass-market Comet, Argos and Currys warehouses. For a long while their buyers wouldn't speak to us, wouldn't answer the phone. We needed a break, and we got it when our wonderful local Wiltshire MP, Richard Needham, turned up out of the blue at the Chippenham factory. He happened to be the very energetic Minister of State for Trade in John Major's government.

I started to tell the minister all the things that were wrong about politics, when he suddenly said. 'Shut the fuck up, Dyson. What's your turnover?' Refreshing for an Old Etonian and government minister. 'About £3.5 million,' I said. 'I want it up to £50 million within twelve months. What help do you need?' I explained the problems we were having with the mass-market retail trade. Needham arranged for Geoffrey Howe, the former Chancellor of the Exchequer and Deputy Prime Minister, to come and visit. It all seemed rather odd, not least because Howe spoke so quietly that you had to lean in very close to hear what he was saying. He seemed very interested to know that I couldn't speak to buyers at Currys, Argos and Comet. 'Oh,' he whispered, 'Elspeth's on the board of Comet.'

Elspeth Howe was, I think, chairperson of the Broadcasting Standards Commission at the time. She is also the daughter of Philip Morton Shand, the architectural writer who did much to introduce the work of Walter Gropius and Le Corbusier, both of whom he knew well, to England. He also had a company that was the first to import bentwood furniture by the revered Finnish architect Alvar Aalto.

The next morning, I got a call from Comet. They would like to sell our vacuum cleaners. Currys and Argos followed. From the perspective of the 2020s, it might seem odd that – John Lewis, Peter Jones and the Conran Shop aside – we built up our sales through these mass-market retailers. Aside from the fact that we needed them – they were the major British outlets for electrical goods – their customers wanted a vacuum cleaner that worked really well, and without the need for replacement bags, of course, and they weren't put off by the price.

Yes, you could buy a vacuum cleaner for £40, and the Dyson was a stretch at £199, but we were selling to people who were truly house-proud. They wanted the best. There is an assumption that Dyson products are purely for middle-class buyers. It simply isn't true. Cod psychology might say that interest in vacuum cleaners is in inverse proportion to one's wealth. Our products might be expensive, but that's due to the research and development that goes into them, how they are made and how they perform.

It was the last of these qualities, in fact, that took us a very long time indeed to get across in shops and retail warehouses. For many years, products were distinguished, looks aside, by price alone. Over several years I tried to persuade John Lewis, Currys and Comet to let us say something about why someone should spend £199 on our vacuum cleaner rather than £40 on another in the same display. The breakthrough was with Comet, who allowed us to explain what our vacuum cleaners did on stickers on the clear bin. I wanted to explain, even if in just a very few words and figures, the distinct advantages of our new technology and how our vacuum cleaners, proven by standards tests, performed much better.

Before Christmas in our first year in Currys, Mark Souhami, one of the bosses alongside the founder Stanley Kalms, invited me for lunch with them both. They explained that because of Dyson they were now making a profit in their vacuum cleaner section and he wanted more Dyson products. Why didn't we do different colours and different features to make a range? Henry Ford had previously been my guide, and as early

Fords came only in black, early Dysons were strictly silver and yellow. It was sage advice for which I was grateful and immediately followed.

By 1995, just two years after our launch, Dyson was turning a good profit and expanding rapidly. We had paid off the enormous bank loan and were able to tear up the grey bank guarantee forms. Deirdre and I were enormously relieved. We could keep our home and pay off the mortgage. And our long-standing overdraft was finally a thing of the past. At the same time, we knew that we were caught up in the most exciting adventure of our lives.

We desperately needed a new factory, and wouldn't it be great if we could build one that reflected our thinking and values? I got in touch with Tony Hunt, my Royal College of Art tutor for structural engineering, himself an inventive and successful structural engineer who, by now, had worked with Norman Foster, Richard Rogers, Nicholas Grimshaw and Michael and Patty Hopkins, Britain's pioneering hi-tech architects whose work fused engineering and architecture in seamless fashion. Tony recommended three architects to design the new Dyson factory.

Chris Wilkinson, who, before setting up his own practice in 1983 and had, like Tony, worked for Foster, Rogers and Hopkins, impressed me. Instead of telling me what he thought I needed, he asked lots of questions that helped me think what we really needed and wanted, and drew sketches for me in response to my answers. This, of course, is exactly what Jeremy Fry had taught me to do when selling the Sea Truck. Ask your client what they think they want and then suggest solutions.

Because, though, the planning process would take too long – we needed a new building there and then – we looked instead for an existing factory in the area. I found a site on Tetbury Hill, on the edge of Malmesbury in Wiltshire, ten miles north of Chippenham. I was able to buy it outright through income from DC01. The previous occupant was Linolite, a company founded by Alfred Beuttel, who, in 1901, when he was twenty-one, had patented a double-ended incandescent lamp. Before the triumph of the fluorescent light, strips of double-ended incandescent lamps lit factories the length and breadth of Britain and around the world. They were also used for concealed lighting in Art Deco clubs and hotels and in the interiors of graceful pre-war ocean liners.

Malmesbury, on the face of it a deeply rural market town serving farmers first and foremost, had also been a hub of inventive manufacturing with a global reach. This was especially true during the Second World War, when Linolite equipped the RAF and the British Army with patented clips used in the fuel lines and cooling systems of aircraft and tanks and in de-icing equipment first fitted to the Fairey Battle bomber in 1937.

Rather intriguingly, the first recorded British aeronaut was Eilmar, a monk of Malmesbury Abbey. In around 1010, Eilmar leaped from the abbey's tower and flew across the valley for a few seconds before, presumably, his wings collapsed, and he fell to the ground, breaking both legs. He wanted to try again – as inventors must – but was grounded by his abbot. Happily, Eilmar lived to a ripe old age.

Later on, Malmesbury became well known for its cloth trade and silk mills, while, again during the Second World War, it was home to ECKO, a radio manufacturer that worked on radar. Even though our first factory here was old-fashioned, Malmesbury seemed like the right place to be and after Chippenham, our new accommodation – all 80,000 square foot of it – seemed palatial. Within what seemed little more than five minutes, though, we had filled the old Linolite factory to its gunwales. The DC01 had become a commercial success. Luckily, I was able to buy land next door from Lord Morton with outline planning permission and, in 1996, invited Chris Wilkinson and Tony Hunt to develop our first purpose-built Dyson building.

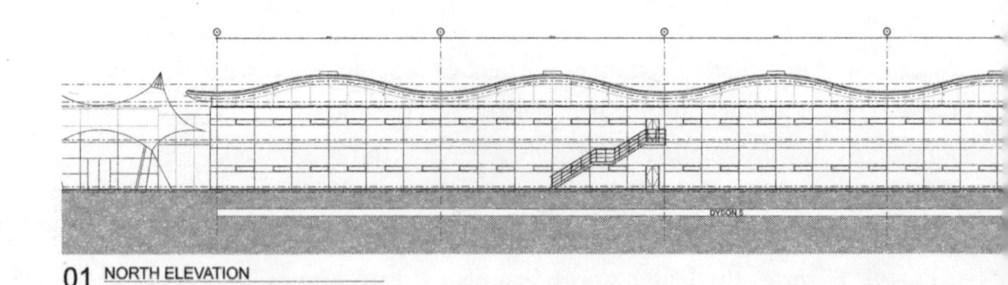

Chris did a wonderful job. Viewed close up, the gentle waves of its long roof are a romantic nod to traditional sawtooth or zigzag skylines of traditional factories. The roof profile also helps in the collection and removal of rainwater. Because this flows relatively slowly down the roof's arches, it needs only small-diameter pipes and, as a bonus, these suck the water away. Chris believes architecture should be a fusion of art, science and nature, something I feel he achieved for us on Tetbury Hill, both in the way the building looks and in the ways it works. He also stuck resolutely to our budget. We've worked together closely ever since.

I wanted the building – a cluster of buildings, in fact – to be an inspiring place to work in and, indeed, a youthful marriage of architecture, art, design and technology. The reception area, a glass cube joining two buildings, is entered over a glass bridge across a pond filled with an installation of fibre optic 'reeds' designed by Diana Wilkinson, Chris's wife and a talented sculptress.

For the courtyard between the two buildings fronted by the glass cube, I sourced copper figures by Peter Burke, a former Rolls-Royce aero-engine apprentice engineer before he turned to sculpture. These comprise forty life-size men and women made from old copper hot-water tanks that had succumbed to corrosion. Peter flattened these by driving over them with a JCB and then vacuum-formed them into figures at Dowty Aerospace. My hope is that they will eventually go green as the copper oxidises.

I like the idea of Peter's figures animating a space like students in the quadrangles of Cambridge colleges. Frank Whittle, my hero, had read engineering at Cambridge while serving with the RAF. He went on, of

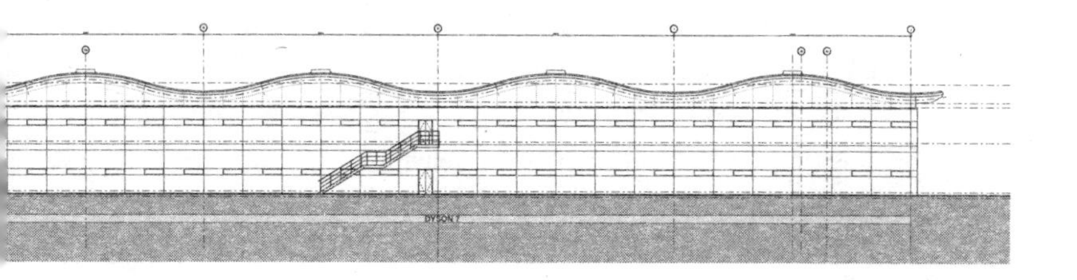

Malmesbury main building Architect's drawing, WilkinsonEyre

course, to collaborate with Rolls-Royce Aero Engines, who manufactured the Welland jet engine on display inside the Malmesbury factory. At this time we were recruiting a lot of engineering graduates especially for research and development. Increasingly, I felt the factory should be like a university campus – although one where products were designed, prototyped, tested and made – and this is indeed what it came to resemble when we further extended it, and especially when, more recently, we added a village for Dyson undergraduates along with a sports centre and cafés. The extra land came again from Lord Morton, himself a self-made entrepreneur, and for whom we have planted a fine Turkey oak by which to remember him.

I wanted ours to be a healthy place to work. When, at first, the M&E (Mechanical and Electrical) contractors proposed a conventional air-conditioning system whereby you recirculate 80 per cent of the air and only let in 20 per cent fresh air, I said I didn't want any recirculated air as this causes illnesses and 'sick building' syndrome. Just the 20 per cent fresh air, please? The mathematics baffled me. What was the point of recirculating 80 per cent of the air? Apart from the massive ductwork and fans necessary, how did the 80 per cent circulation benefit the air in the building? I could only imagine recirculated bad air.

The expert view was that I couldn't have what I thought was right. This led to a two-week standoff with the Mechanical and Electrical consultants. Eventually they complied and the building is much simpler and healthier as a result. Fresh, clean Wiltshire air is fed up from the floor and exits though air outlets in the roof. There are chiller beams under the ceiling as cold air falls, with underfloor radiators concealed around the perimeter. With no unsightly ducts gathering dust, the system is very simple but highly effective and uses less power.

I have always loathed companies that use 'greenwash' as part of their marketing. I would rather reduce our environmental impact quietly and by action. We were, and remain, a company primarily of engineers and because of this we have sought from the outset to use as little energy or materials as possible to solve or complete any one particular task. Lean engineering is good engineering. This is something, in fact, engineers I admire have done since the advent of powered machinery. When, in 2010, I took part in Channel 4's TV series *Genius of Britain*, I was able to show how, by rethinking Newcomen's slow and inefficient steam engine, dating from 1712 and used mostly for pumping water from mines, James Watt was able to increase its efficiency fivefold. The same was true for us at Dyson. From the DC01 onwards, our first-generation vacuum cleaners used a proprietary Japanese 1400W electric motor, while our competitors boasted of 2000W and even 2400W motors. However, we had no loss of suction, therefore we wasted none of our power, whereas the competitors lost suction and indeed wasted so much of their power through their clogged bags.

For me, as for all Dyson engineers, lightness – lean engineering and material efficiency – is a guiding principle. Using less material means using less energy in the process of making things. It also means lighter products that need less energy to power them and are easier to handle and so more pleasurable to use. What I learned right away when making the Sea Truck was that the weight of the boat was critical to its ability to get up on the plane and speed over water at up to 45mph. Lean engineering was, perhaps, built into me right from the beginning – long-distance runners need to be lean, too, of course – and I think it is built into most engineers. It's in our genes.

At Malmesbury we began reducing the weight of the vacuum cleaner from the word go. We thought it ridiculous, for example, that plastic mouldings needed to be as thick as they were. The thicker the plastic the more of it you need. The more of it you need, the more electricity you need to melt and mould it. We had a stand-off with our plastic material suppliers, who insisted we needed mouldings 2.5–3.5mm thick in order to push the plastic through the cavity of the mould. This is what their mould-filling computer programmes predicted. We designed and made one of our clear bins with a 1mm thick wall. Lo and behold, unlike the plastic supplier's computer programme, it worked.

We were tripling production of the vacuum cleaners every year from 1993, so this seemingly simple move saved a considerable amount of material and energy. At the same time, of course, we had to ensure that our machines remained durable over a long period. We set up a large-scale 'torture course' to test our vacuum cleaners to well beyond what could possibly be expected of them even in the roughest, toughest household. Although these consist of many mechanical test rigs, we also have hundreds of people brutally using our products on test courses, doing their best to break them or wear them out. Every 200 actions, photos are taken and notes recorded. This controlled rigour allows us to use fewer materials and make lighter products.

We had refined the workings of the vacuum cleaner from the word go at Malmesbury, too. Each model was very different from its predecessor, always advancing the technology. As Buckminster Fuller said, 'You never change things by fighting the existing reality. To change something, build a new model that makes the existing model obsolete.' Bucky wasn't thinking of change as in the 'built-in obsolescence' of American cars of the '50s and '60s, but of revolutions in design. We were, in fact, to make our own early vacuum cleaners redundant – although old models kept and will keep working – as we researched and developed new technologies and materials.

This is only slightly jumping the gun, as the radical technological changes we have been able to make in the past two decades were possible because Malmesbury, Malaysia, the Philippines and Singapore developed apace as R&D-based manufacturing campuses. This is also why we were able to recruit engineering graduates, who may well have wondered why they might consider joining Dyson when our campus was in rural Wiltshire, the best part of a hundred miles from London, and we made seemingly prosaic things. Why work on the design and development of a vacuum cleaner when you could be working on an aero engine or the latest computer?

I love the fact that we tackled prosaic products, making the vacuum cleaner into a high-performing machine. As we did, we developed Dyson into a 'tech' company driven by the very graduate engineers we took on. I say 'we' more often in this chapter than 'I', because we quickly built up into a team and what we began to think of as campuses of brilliant young engineers who understood that their job was to think freely, ask questions and to challenge what we did to find better design, technology and products.

The atmosphere of what morphed suddenly from a factory at Malmesbury to a research and development centre was and remains attractive to young people. So, too, are our research campuses in Malaysia and the Philippines and our new St James Power Station campus in Singapore. At Malmesbury we might, in London, New York or Singapore terms, be miles from anywhere, yet the campus has its special attractions. With Chris Wilkinson, again, as architect, we were to add The Hangar, a multipurpose sports building, and the Lightning Café, where everyone can meet. Here, you sit on Fritzen Hansen chairs around tables with an English Electric F1.A Lightning Mach 2 RAF interceptor hanging from the ceiling by just three wire cables. The Lightning, a sensationally fast and supremely purposeful aircraft, was restored over eighteen months. It's one of several machines that continue to inspire me dotted about the campus, including the aforementioned cross-sectioned 1961 Mini, a Harrier Jump Jet, a Sea Truck (forty-five years old and bought back from a harbour authority in Wales) and a Bell 47 helicopter.

The stories and the people behind these machines show what's possible when engineers think big, unconstrained by the current state of the art. They can really change the world. Our machines are not museum artefacts; they're here to be touched and understood in terms of the ideas behind them, the engineering that made them possible, and the frustration and failure that went into them along the way.

The damselfly-like Bell 47 was designed by the US inventor and philosopher Arthur M. Young, who spent twelve solitary years as a young man in a workshop converted from stables on his father's Pennsylvania farm, between 1928 and 1940, developing working scale models of what was to become the world's first commercially available helicopter.

Our D9 research building, designed by Chris Wilkinson and opened in 2016, gave us the space and facilities to dig deeper into the research of future technologies and products. For a scientist or engineer it's an attractive place to be. Its two vast floors centred on an atrium and its centrally located labs are hidden behind the building's wrap-around reflective glass walls. The glass walls are 5×3m prefabricated boxed glass panels which, for the first time in a building, form the periphery building structure. They are reflective partly because, increasingly, we needed top-secret laboratories, but also to reflect in the glass the trees that surround the building.

I can never say in public what we're up to here at any given moment because the investment in our research has grown apace as Dyson has moved on and we need to keep a very tight control on our work for future products. The D9 research centre, though, is a great place to work. We've tested new ideas using the very fabric of the building, like the Cu-Beam Duo lights developed by Jake. Suspended from ceilings like satellites, these provide individually adjustable task and ambient light, their LED lamps cooled using heat-pipe technology found in satellites and micro-processor technology and they had a life of 180,000 hours.

We've aimed at making the campus a place of both invention and creative play, with what we research and make reflected in the design of the buildings. When you look down from the top of the stair in the Lightning Café, the spiral form coming up to meet you is a visual reference to the turbines in our electric motors. This is clearly our campus, with a character of its own.

From the beginning we decided that we would create our own publicity materials and advertising. We would not use outside agencies. This is because we want to talk fearlessly about technology, which, of course, is what had driven Dyson into being. Since we have developed the technology, we should know how to explain it to others!

In the late '90s, a Belgian court banned us from talking about vacuum cleaner bags. I didn't realise they could do this, and I would have thought it was illegal. But Belgium had tight comparative advertising laws and our European competitors ganged up together to sue us, arguing that we shouldn't say that we didn't have a bag as this gave Dyson a comparative advantage. While this seems absurd, the court found us guilty. We produced an advert, shot by the photographer Don McCullin, with the word 'bagless' blanked out repeatedly and a strap line that read 'Sorry, but the Belgian court won't let you know what everyone has a right to know'. This got the media interested. We were able to tell them the story of how European manufacturers, as a group, were trying to silence competition.

One of the ways we made Dyson distinctive is by not allowing ourselves to rest on our laurels. Not for a moment. With the upright DC01 selling well, in 1995 we launched DC02, a compact cylinder dual cyclone vacuum cleaner that solved a problem known to every household not just in Britain but around the world: it sat securely on stairs. This was our first cylinder model. Peter Gammack and the other engineers were fully occupied with DC01, so I designed it with Andrew Thomson, who had joined Dyson straight from the Imperial College via the Royal College of Art Engineering course. The clear bin and cyclones were at 45 degrees, with the long handle on top and the motor and wheels underneath them. It departed from the blob or cylinder shape of our competitors. I was interested that, while nobody has ever commented on the design on the DC01, they have on the DC02.

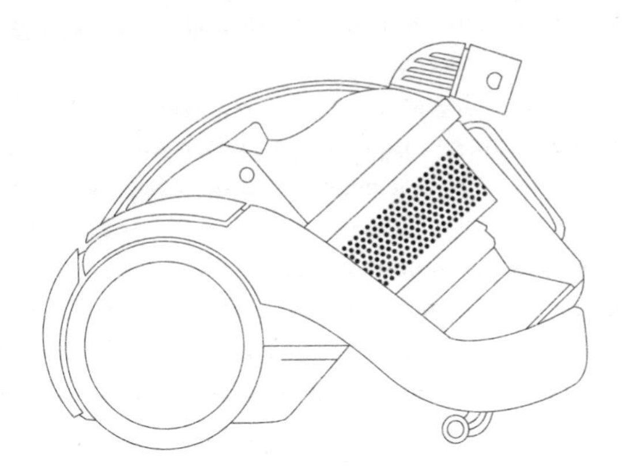

DC02

We made it available in a number of colours and special editions, including De Stijl, our homage to the early twentieth-century Dutch design movement that, to my mind, was the true foundation of modern design some ten years before the Bauhaus. In fact, we made quite a number of special editions of different cleaners, whether to raise the profile of particular causes like the 'green' DC02 Recyclone made of green recycled plastic or to raise money and awareness for charities.

In 1994, we issued the silver and light-blue DC01 Antarctic Solo signed by Ranulph Fiennes, the explorer, the first person to walk without support across Antarctica. We funded the trip to raise money for Breakthrough Breast Cancer. Both my parents having died from cancer, I was anxious to help find a cure. Surprisingly, at that time very little fundraising was going to breast cancer research. I decided to use Sir Ranulph's trip as an awareness-raising exercise. Fiennes had approached me because his wife had sent him off one weekend to buy a vacuum cleaner. He chose a Dyson and had an inkling we might see eye to eye. We did. We also designed a special compass for his expedition.

This was a fecund as well as highly productive period. By 1995 we began to think of altogether new products. One of these was the Dyson Contrarotator, a washing machine. It proved to be a really good washing machine as it happened – this is not just me talking – but it never made money. The problem was that it was expensive to make and the price we sold it at when it was finally launched in 2000, although high at £1,099.99, was too low for it to make a profit.

We started trying to reduce the cost of production, rather unsuccessfully. The marketing team, who I listened to, said to me, 'If you make it £200 cheaper you will sell a lot more,' and I believed them. We made it £200 cheaper and sold exactly the same number at £899.99 as we had at £1,099 and ended up losing even more money. I had made a classic mistake. This might sound counter-intuitive, but I should have increased the price. The Contrarotator was not meant to be a low-cost washing machine. It was a different and well-engineered product made for those who would appreciate it and be willing to pay that bit extra

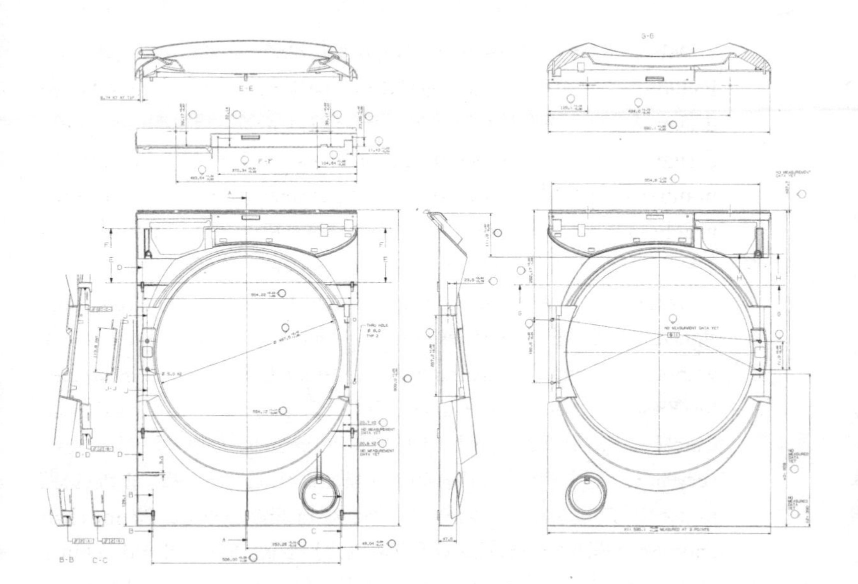

Contrarotator

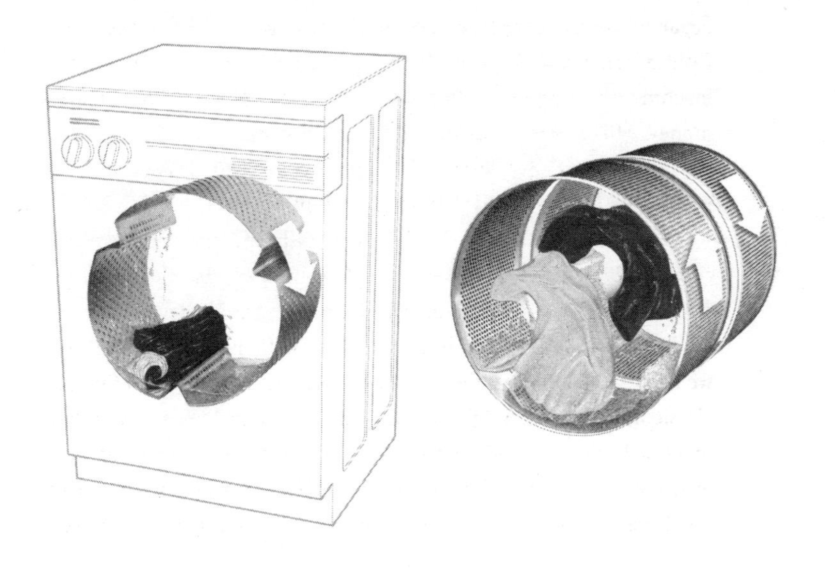

for it. We stopped making it in 2002 and I was sorry to do so, not least because it was a way of rethinking the way washing machines work. We serviced Contrarotators for the next fifteen years. Mine are still running.

Traditional washing machines have a single drum that rotates in one direction at a time, taking a sopping wet and heavy wodge of clothes with it. With so little actual washing action, the clothes need an extended soak in detergent to get clean. Our research showed, contrary to popular perception, that hand-washing was more effective at cleaning clothes than using a washing machine. In fact, we found that a washing machine running for two hours was no better at cleaning clothes than washing them by hand for fifteen minutes.

The Dyson Contrarotator was the first washing machine with a pair of drums that rotated in opposite directions to replicate the movement of hand-washing. The result was cleaner clothes faster, with a bigger load and using less water. The Contrarotator operated at 30–40°C rather than the usual 50–60°C to prevent shrinkage. It was, though, a bit more complicated than standard washing machines: it needed two motors, one for each drum. Because both drums had to spin on the drying cycle at high speed in the same direction, the Contrarotator also needed a clutch and gearbox. And so, round we go in a cycle to the question of manufacturing cost and selling price.

One other product we let go of at Malmesbury was a pet project of mine developed over many years. This was the Diesel Trap, or cyclonic exhaust filter, a hangover from the coach house days. I'd read a great deal about the dangers of diesel exhaust – particulates exhausted are carcinogenic – and I couldn't square what the EU and the British government were saying about the cleanliness of diesel exhaust with serious medical research. It did seem pretty obvious to anyone driving in the slipstream of another vehicle, drawing in their exhaust fumes. You can smell them, and you are breathing them in. And, if you took the train west from London to get to us, you experienced the dirty qualities of diesel train exhaust. Diesel engines left running for lengthy spells under Brunel's great glazed roof at Paddington Station pumped out all too visible and dangerous exhaust fumes. The government, though, cut duty on diesel to promote it following the lobbying of car-makers, EU diktats and consultation with of one of its key advisors, Sir David King, the chief scientific advisor at the time. He later admitted that the government's view on diesel had been wrong.

At the coach house we developed a cyclone for catching diesel exhaust pollution. I got to demonstrate this in a 1993 edition of BBC TV's *Blue Peter*. For this, I drove around Shepherd's Bush in west London in a diesel-powered Ford Transit and showed *Blue Peter*'s Anthea Turner the muck I'd collected. It was impossible, though, to get vehicle manufacturers to adopt it because diesel fuel was significantly cheaper than petrol and everyone was being told that diesel was the 'green' way to go, despite it producing nitrous oxides, sulphur dioxides and hydrocarbons. This, sadly, was a case of banging our heads against an unflinching concrete wall, although we only finally gave up in 1998.

Although this was a disappointment, the Diesel Trap was proof that my cyclonic technology was transferable to other inventions and products. This technology is something that was becoming increasingly important at Dyson. While always on the lookout for revolutionary developments in technology, we could develop and refine what we had come up with to date for use in other devices, as we'll see in the next chapter.

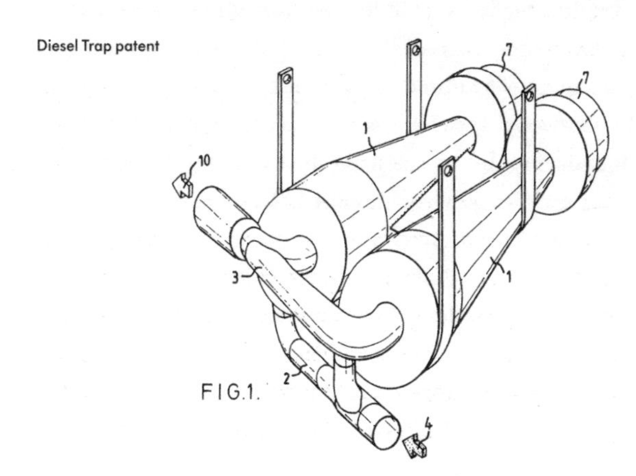

Diesel Trap patent

CHAPTER SEVEN

Core Technologies

I never intended Dyson to be just a vacuum cleaner company. The cyclone was the first of our core technologies to be applied to a product and this happened to be a vacuum cleaner. I always saw this as our first step. Since then, we have focused on pioneering core technologies around which we have engineered better products in a growing number of fields, nailing our colours to ever more masts as we've sailed on and grown.

Ours, though, has been neither a quick nor cheap approach. Investment in new technologies requires many leaps of faith and huge financial commitment over long periods. En route, there are multiple failures, sleepless nights, a great deal of frustration and just a few real breakthroughs. Ours has been more of a pilgrim's progress than a straight path to success. The upright DC01 vacuum cleaner might have been a rapid commercial success, yet we – as relentlessly dissatisfied engineers – immediately and impatiently had ideas to make a much better one. At the same time, we were keen to apply our cyclone technology to different formats.

Upright cleaners had been invented in the USA to deal with fitted carpets, for which they needed a 'beater' brush bar. You pushed them in front of you by the handle at the top, a bit like a lawn mower. They were brought to Britain in the 1930s and gained popularity along with fitted carpets, but not to other countries, most of which use cylinder cleaners,

so called because these were originally metal or cardboard cylinders. You pulled these behind you with a hose and wand extending ahead. They did not have a brush bar because, unlike American houses, other countries didn't have fitted carpets so had no need for it.

We wanted to make a cyclonic cylinder cleaner next. This was the compact DC02 that sat securely on stairs. DC03 was designed to be a very light upright vacuum cleaner indeed, an important antidote to heavy vacuum cleaners. Just 100mm thick, it could be laid completely flat to clean underneath furniture. This flatness and thinness meant it could be hung on a peg against a wall for storage. Obsessed with HEPA filters, we fitted a pair of large ones to the DC03, one before and one after the motor. HEPA filters are High Efficiency Particulate Arrestors of the sort you might have previously found in hospital vacuum cleaners and in hazardous places such as nuclear facilities. They use very fine glass strands in densely packed pleated media. The pleated media creates a much larger surface area and since the air must only pass through the media at a specified low velocity, the surface area of the media is critical. So, too, is the sealing around the filter, which is why every single Dyson fitted with an HEPA filter is tested on the production line for HEPA efficacy.

DC03 was also the first vacuum cleaner with a clutch to disengage the brush bar drive or, if the brush bar were jammed by a piece of Lego, for example, for the clutch to slip and save damage to the drive belt, a major headache for users. We were able to guarantee toothed drive belts for a lifetime. We also produced a Clear version of DC03 but ceased production after Terlux, the polymer we were using for the injection moulding, proved to be a less rugged material than the early tests had shown.

DC01, meanwhile, was selling very well and customers seemed to be pleased with it, but we kept on thinking of improvements and brought out DC04, another upright vacuum cleaner, with more than double the air watts and suction power of DC01. There are quite lengthy ducts inside an upright vacuum to convey dirt to the cyclones and then to the suction motor at the end of the system. There is also a changeover valve that automatically actuates to direct the suction from the cleaner head

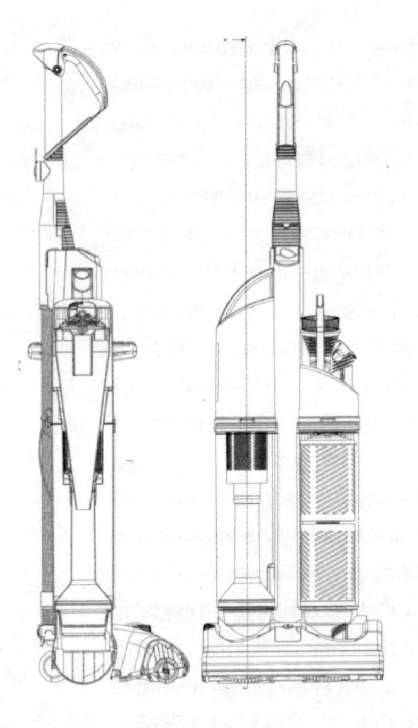

DC03

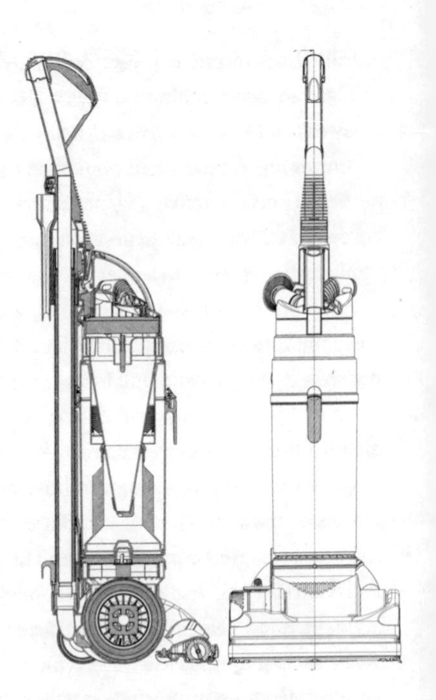

DC04

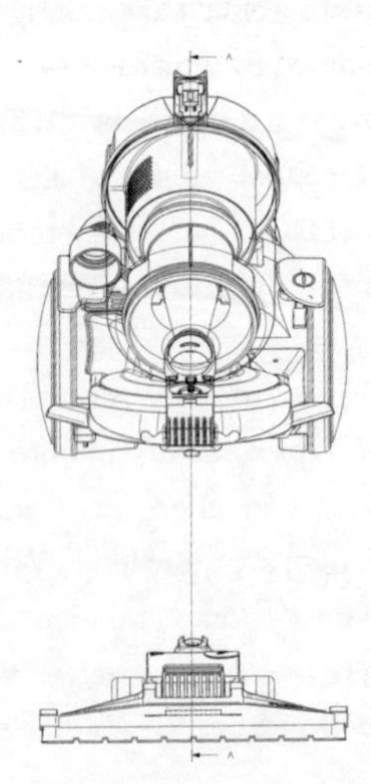

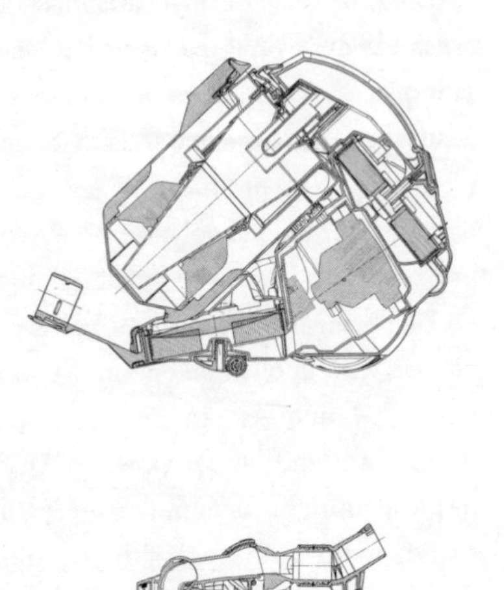

DC05

or from the detachable telescopic hose. We smoothed out all the airways and the valve and improved the performance of the cyclone to achieve a power gain.

We thought that DC02, our first cylinder vacuum, was heavy so we developed a light cylinder vacuum that was also more powerful. I believe that it is crucial to keep on improving and never to relax with a product that appears to be selling well. Permanently dissatisfied is how an engineer should feel. DC05 was our first vacuum cleaner to use a power nozzle, a motorised brush to deep-clean carpets. All conventional cylinder cleaners, including DC02, sit on at least four wheels or castors. This means that you drag around dead weight and have to force them to turn. Instead, we fitted two large rear wheels and put all the weight over them, so when you started pulling the hose, the DC05 did a 'wheelie' back onto its two wheels, making it easier to push and follow you around. It felt so much lighter than other vacuum cleaners.

DC06 was our first attempt at a robotic cleaner, of which more later in this chapter. The DC07 upright cleaner – we're up to 2001 now – was the first vacuum to have multi cyclones, seven in all. These were placed upside down to improve suction power. It also had a clutch that not only saved the drive belt of the brush bar but also automatically declutched the brush bar if you were using the hose. This saved the brush bar from damaging anything when it was not working across carpets. This was the machine we launched in the US market in 2002.

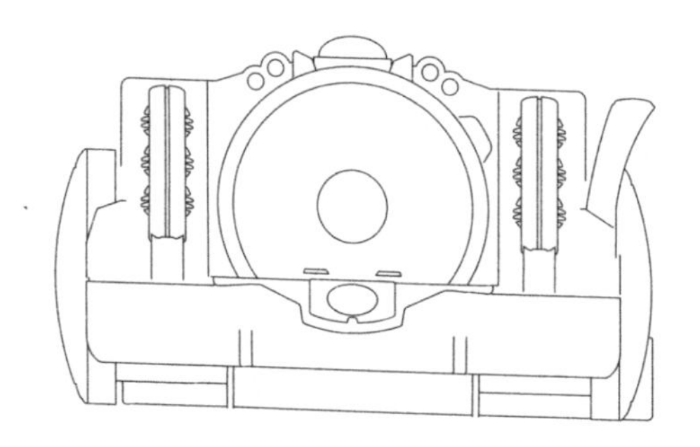

DC06

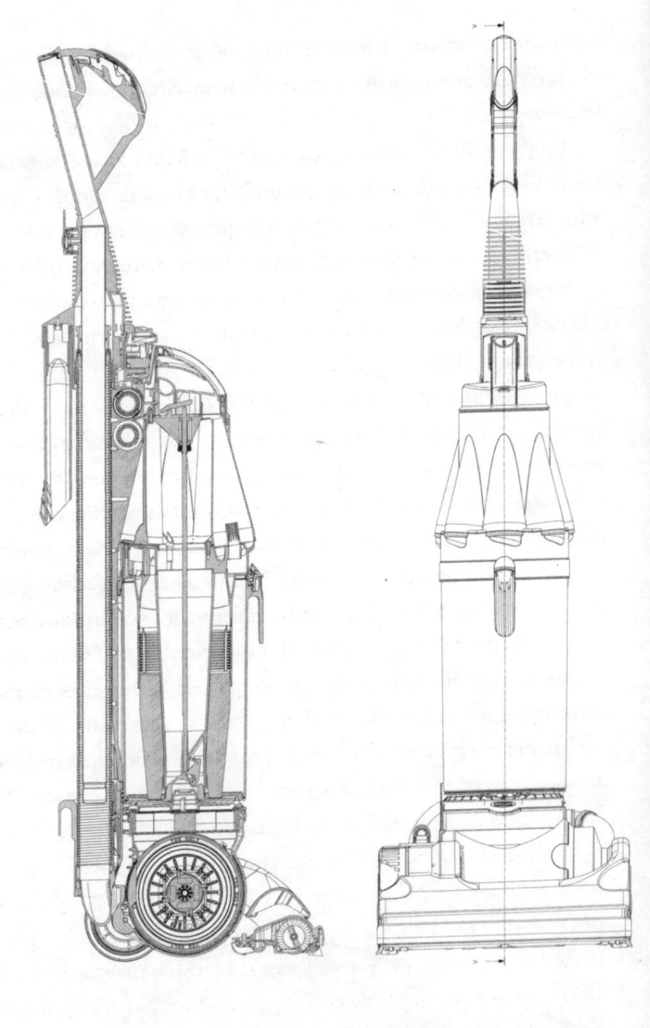

(Right) DC07
(Below) Cyclone technology evolution

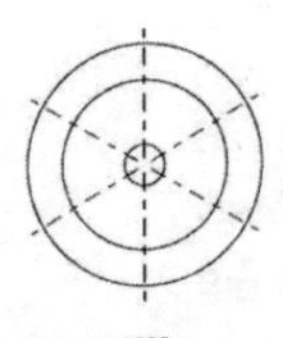

1993
Dual Cyclone™ technology

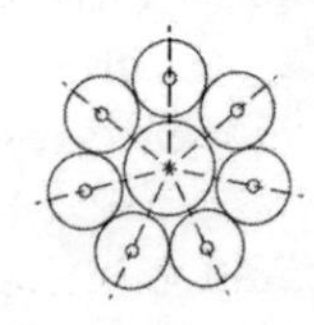

2001
Root Cyclone™ technology

2008
Root Cyclone™ core separator technology

2012
2 Tier Radial™ cyclones

2013
Dyson Cinetic™ cyclones

While we were improving our vacuum cleaners, there were a number of diversions along the way. In 1999, the editor of the *Daily Telegraph*, Charles Moore – now Lord Moore – invited me for lunch at the Savoy, my only one there ever. He suggested I write a history of great inventions in conjunction with the *Daily Telegraph*. This was duly published as a magazine series with the newspaper and later as a book. Moore followed this up by suggesting I design a garden as the *Daily Telegraph*'s entry for the Chelsea Flower Show. He kindly repeated this offer over the next few years until an idea came to me, at which point I couldn't resist the diversion.

I teamed up with a talented garden designer, Jim Honey, whom I had known for many years, to design what we called the Wrong Garden. The idea was to stress the virtue of not being afraid to do the wrong thing. Jim's idea was to have none of the green, pink, yellow or red colours normally found in gardens, but, rather, dark crimson and tan colours. At the front of the garden there would be a water feature echoing the clever Escher drawing where water apparently went uphill. On either side of the water feature would be benches made only of glass, balancing in a seemingly impossible way on a V-shaped structure. Everything was to look impossible and wrong.

This was Jim's first Chelsea garden, too. He had to be on a very fast learning curve as the garden had to be laid out in just two days and moved on immediately after the show ended. My water feature was made in glass and, as in the Escher drawing, there were four ramps – mine were 3m long – forming the frame of a square. The water was to flow up each glass wedge ramp and drop, like a waterfall, into a square pool. From this pool the water would go up the next ramp and so on. Each glass ramp was filled with pressurised water. Pete Gammack and I developed this with Derek Phillips inside the factory at Malmesbury.

We discovered some phenomena which helped us. Firstly, the water, contrary to the visual impression of running up the ramp, actually ran down the ramp. It was forced out of the top of the glass wedge through a slot at the top of the ramp and ran back down the glass ramp. But because of the surface tension at the sharp edges of the ramp, the water did not spill over the edges of the glass but spread evenly across the glass surface. It was hard to see that the water was not running neatly upwards.

The Wrong Garden water feature

Secondly, we made certain that it looked as if it was running upwards by introducing streams of bubbles under the glass ramp inside the wedge-shaped tank, going from the bottom under the glass up the underside of the glass slope to the top. The effect was perfect! The bubbles on the underside of the glass ramp made the downward stream of water on the top of the glass ramp look as though it was instead flowing up it. We also had water running out of the same slot at the top and pouring like a waterfall into the square trough below. The final effect was that you thought you saw water running up the glass ramp and creating a waterfall into the square trough. It fooled everybody.

The mechanics were surprisingly simple. At each corner there was a submerged water pump to push water into the glass wedge-shaped tank, forcing it out of the slot at the top of the ramp to create the waterfall. Then each tank had an air pump to create the trick-effect bubbles starting on the underside of the base of the glass ramp and travelling up to the top. The total effect was of water running up each glass slope, cascading down into the next square trough and so on, continuously around the square frame. It required Derek's gifted engineering expertise to keep it all running smoothly. I did give the secret away to the Queen when she asked how water could possibly go uphill.

The glass benches were made of a long horizontal plank of laminated glass. An inverted 'V' pyramid of glass was glued to the top of the glass plank in the centre with an identical 'V' pyramid under the centre of the plank. The glass benches looked as though they were balancing impossibly and dangerously on the point of the 'V' of glass. The trick was that there was a single stainless-steel cable that ran from a hidden anchor point in the ground under one end of the glass plank up to that end of the plank, up over the central glass pyramid and down over the other end of the glass plank and from there into the ground, where it, too, was anchored. At each point of contact with the glass it was secretly fixed. This stopped the bench from toppling and gave the glass plank the required support when people sat on it. I was delighted when, at the opening of the show, Jerry Hall both bravely and decorously reclined on the bench just as I had intended it to be used as a photographer from *The Sun* happened to drop by.

Most other gardens at Chelsea displayed wildflowers in charming ways; not surprisingly the wrong colours of our Wrong Garden did not appeal to all the judges. Jim was keen that our garden should be suitable for any garden, which would not be the case with short-lived wildflowers. Nevertheless, we were awarded a Silver Gilt prize. The Wrong Garden travelled on to another show, in Birmingham. We had better luck there, winning the 'Garden of the Show' award. Derek was a proud host at the shows as well as engineering, making and operating the garden brilliantly. I was immensely sad when the talented Derek, who had created so much with us over the years, died of cancer a few years later.

We needed a larger boardroom table and, inspired by the theme of glass supported by wire cables, I decided to make a legless glass table. The glass table top, made from two laminated sheets, 5m long by 2m wide, weighed 750kg. Four stainless steel cables came down at an angle (this was important) from their fixing at the ceiling. We were concerned about the weight, so bolted the attachment eyes right through the concrete floor above. The four cables went through stainless steel ferrules in the table top then down to fixings in the floor, crossing over at angle on the way. The crossing-over was crucial to prevent slewing of the table.

The cables were tensioned very tightly. The glass top was remarkably stable, more so than a table with legs. Everyone can sit at any point around the table without a table leg being in the way.

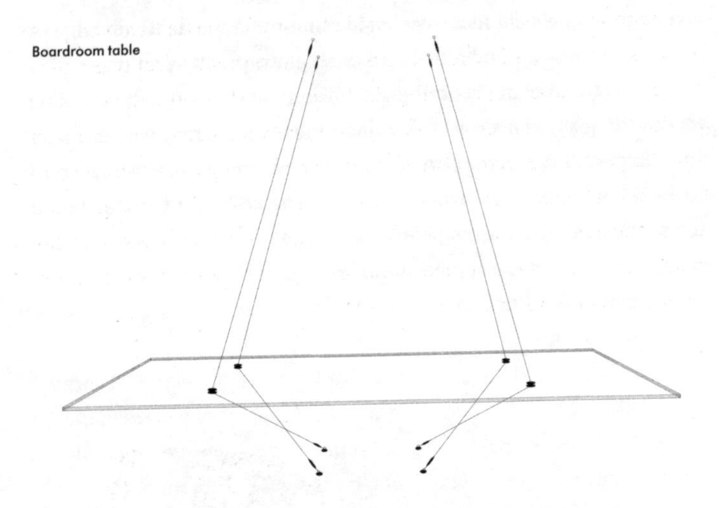
Boardroom table

Another very enjoyable diversion involved bringing vacuum cleaners to the Paris catwalk. On my trips to Tokyo, I would always be invited to visit Issey Miyake. He has only recently revealed that he had been seven when one of the two atom bombs dropped on Japan in August 1945 fell close to his home. Issey is a creative force and was a wonderful supporter to us as we found our way in Japan. I like his clothes, too. In 2002, Issey Mayake invited me to design a show for the latest collection at the Paris October fashion event. Intriguingly, they had designed the collection based on Dyson vacuum cleaners. Some of the shapes of the garments mirrored those of various parts of our cleaners. Our technical drawings appeared on the clothes, all of which were silver and a deep magenta. I was enormously flattered by the idea and longed to see how they would execute the designs. However, I had to wait until the show before this was revealed.

The theme of the show was Wind. Of course, our cleaners used powerful airflow to separate dust from the air so the theme was perfect. They wanted us to create powerful streams of air across the catwalk models as they walked down the runway, as if in a strong wind.

The show was held in a large, rigid tent on the rue de Rivoli edge of Les Tuileries. I placed extremely large fan blowers at several stages outside the tent and channelled the air down giant, flexible yellow plastic ducting of 1m diameter, just like our yellow vacuum hoses but many times larger. These were made to sway to and fro by cables attached to the hoses pulled by very strong assistants. The hoses extended right out across the audience to blast across the catwalk models. It was very hot inside the tent and the audience was enormously grateful for the unintended air conditioning.

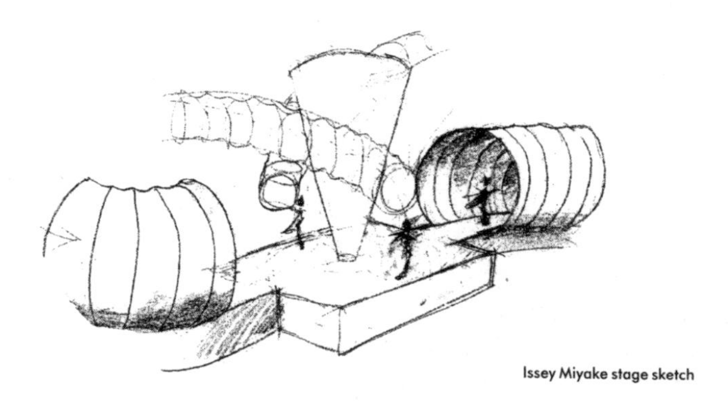

Issey Miyake stage sketch

To complete the flexible hose theme, I had the models appearing out of a 3m-diameter snaking yellow hose, as from a tunnel, at one end of the catwalk. Afterwards, an excited Australian journalist came up to me and gushed that she 'loved the models emerging out of the giant yellow vagina'.

This was the first and only fashion show I have ever been to. I was astonished by the huge size of the audience for what is a short display as well as by how wonderful and imaginative the collection was. It proved to be a commercial success, too. What was I doing designing the backdrop for a fashion show? Well, I couldn't resist it for clothes based on our vacuum cleaners. Back in Tokyo, after the show, some Dyson engineers and Issey Miyake designers created wonderful mannequins from Dyson vacuum cleaner parts to model clothes from the show. I was able to see these exhibited in the Tokyo Design Museum. There were even charming babies and dogs built from the vacuum cleaner parts.

In 1999 a physics graduate from Bristol, Charles Collis, joined us with his idea of a small screen being placed in front on your eye that would appear to be the size of an A4 sheet of paper. You could read a full page of text at the same time as seeing ahead. A camera could be mounted at the same spot, filming in the direction that your head was pointing. The device, which also had a microphone, would be connected back to your ears, with earphones for audio and down to a pocket phone/computer. It was a brilliant idea. You could read emails, or have them read to you, and dictate a response, as well as record and watch films, play music and make phone calls. We built a prototype and were delighted how well it worked. However, we decided not to proceed with the idea and Charles developed robots with us instead. A similar device was later brought out by Google as Google Glass.

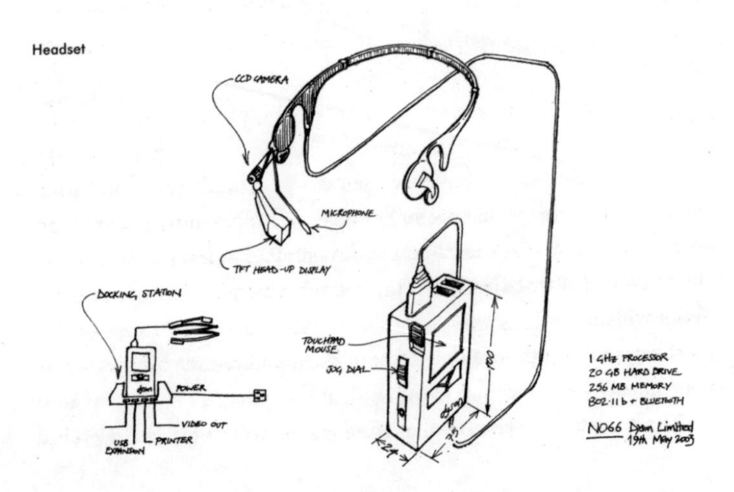

Headset

A more recent digression was the Dyson Symphony performance at the Cadogan Hall, London, on 18 February 2018. Every year at Dodington Deirdre holds an evening of opera arias often accompanied by the Orion Orchestra. One year, over a drink in the kitchen, the orchestra's conductor, Toby Purser, suggested he commission a piece of music that used Dyson machines and newly invented instruments to represent the creative process at Dyson. There would be a competition to find a young composer. Remembering Malcolm Arnold's piece performed with vacuum cleaners at a Gresham's School speech day, he wondered if these could be included. It hadn't occurred to me that Dyson machines could be played as instruments, let alone that anyone would choose to play them, but I jumped at the idea.

As well as using vacuum cleaners and motors, we would challenge our engineers to invent orchestral instruments using Dyson components. Toby Purser and his idea captured the imagination of Dyson engineers and I was astonished at the inventiveness and range of the instruments made by self-selected teams of engineers. They clearly had a passion for music, and each played their own instrument. The largest was an organ with pipes made from forty-eight coloured aluminium vacuum cleaner wands, air generated by eight Dyson fan motors and valves controlled automatically by a computer. This was all housed in a Buckminster Fuller geodesic dome also made from coloured wands. It sounded surprisingly similar to a church organ.

Another instrument was a type of harp with the plucking generated by a digitally controlled Dyson motor. And there was a guitar whose notes were varied by a Raspberry Pi microcomputer connected to a gyroscope that measured how far the player tilted the instrument up or down. This meant that you could play different notes intuitively without all the complex fingering. Even the Dyson undergraduates, up to their necks with exams, contributed a clever pipe instrument. Five of the most interesting and inventive Dyson instruments were then added to the composition.

The composition competition itself was won by a Cambridge PhD student, David Roche. I heard his piece played for the first time in the Dyson Sports Hangar and subsequently in the grander surroundings of London's Cadogan Hall. He had created his work after spending a day with Dyson engineers and acousticians, learning about the development process and its up and downs, with – hopefully – a rousing ending. David and Toby created a novel and most creative piece of music. What could have been no more than a gimmick proved to be thoughtful and enjoyable. I believe the result is still available on YouTube for those who are interested in hearing it.

Getting back to the day job, DC08 was a multi-cyclone cylinder cleaner replacing DC05 and offering a big jump in suction power. It came with a bigger powered brush bar designed to deal effectively with pet hair. Although bigger than DC05, it too had the 'wheelie' feature, which made it also feel light when manoeuvring.

There was no DC09, nor was there a DC10. I felt these designations were owed to McDonnell Douglas airliners and not to Dyson. DC11 was a departure. We were increasingly concerned about the anaconda-like nature of the long and unwieldy plastic hose and the difficulty of storing both the hose and the wand. Although customers have had these awkward hoses and wands since the invention of the vacuum cleaner in 1901 (each country disputes both the date and the country of origin), they told us how inconvenient they are. We needed to try to solve the problem. Listening to what our users say is gold dust and I really enjoy reading or hearing about complaints. We devised a system of reporting remarks heard by customers in stores or by store salespeople from all over the world, so that everyone in the company can see this priceless intelligence.

While on holiday with my family in January 2002, I came up with the idea of wrapping the hose around the machine itself, in a racetrack provided, and telescoping the wand from 1.5m down to 0.5m and clipping it against the side of the machine. As such, DC11 was one neat bundle, easy to carry and to store. This was more difficult to develop than it sounds. We could no longer have the bulky cyclone bin, so

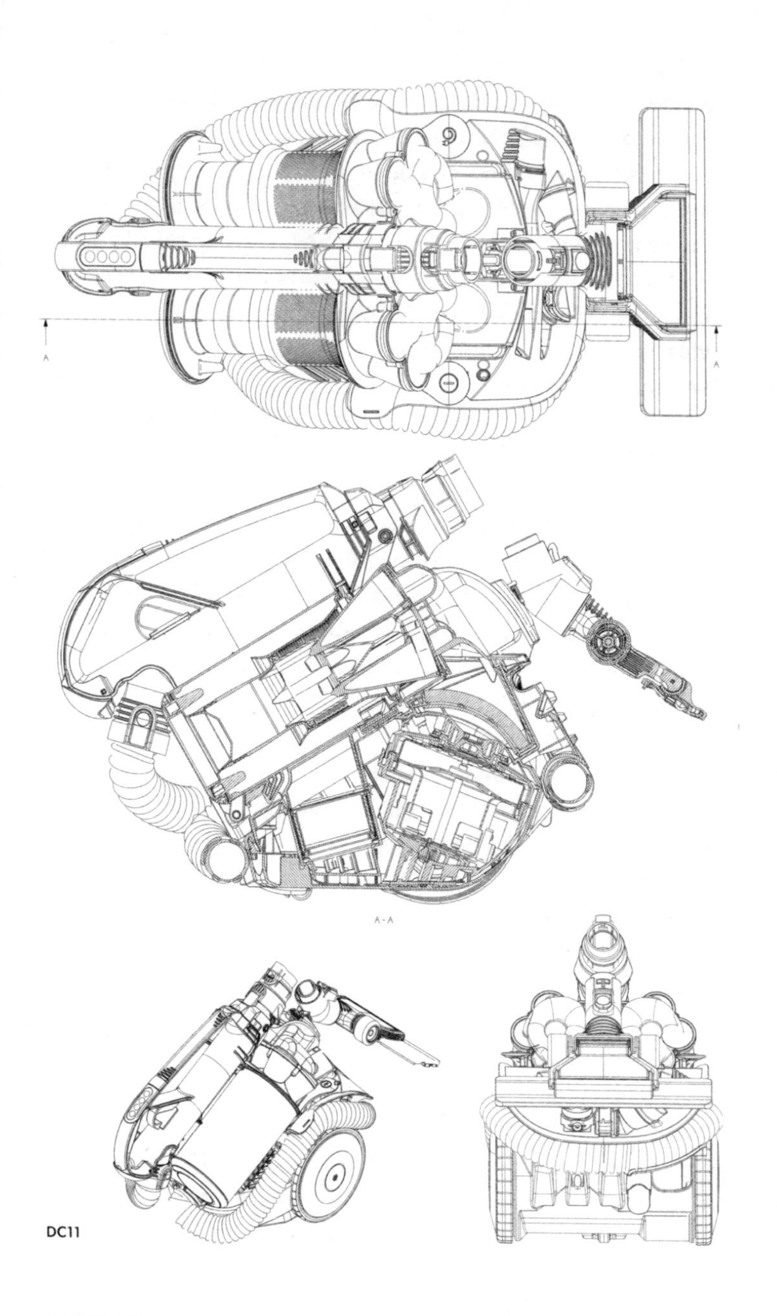

DC11

instead we developed twin bins side by side, which took up less space. Each of the bins had its own multi-cyclones on top.

The telescopic tubes to make DC11 as compact as possible were designed to be easy to use. Normal telescoping tubes, as you find on a camera tripod, require you to release each stage individually as you compress it, and again fix these one by one as you extend it. We developed the wand so that once the first catch at the top was released, all the others followed suit. When extending, all catches automatically clicked into locked mode.

I'll return to DC12 in a moment, for an important reason. There was no DC13. Retailers and consumers in some markets can be superstitious. DC14 was a new upright launched in 2004 with a lower centre of gravity and other design improvements, while DC15 Ball, released the following year, featured the very kind of ball I'd come up with all those years ago with the Ballbarrow. Other vacuums had four fixed wheels, which made the tight manoeuvring necessary for a vacuum very awkward. The ball allowed you to spin literally on a dime, as they say in the States.

At this point we decided that all our vacuums, uprights and cylinders, should sit on a ball for easy manoeuvrability. The ball cylinder cleaner had its centre of gravity on the middle of the ball and, indeed, for turning spun on the centre of the ball. The motor and filters were inside the ball. We wanted to make the first vacuum cleaner that never required any maintenance, with no filters to change or wash and, of course, no bag to replace. The only interaction would be to empty the bin. We developed a machine with thirty-two cyclones. The secret to making them so efficient at collecting fine dust was to add what we call 'quivering tips' at the bottom of each of the cyclones. The tip oscillated, thereby increasing their separation efficiency. We call this model the Cinetic, spelling it with a C instead of a K for trademark purposes.

While we improved every aspect of the vacuum cleaner, there had always been one vital core component for which we had to rely on others: the motor. The conventional motors we used were made in Japan.

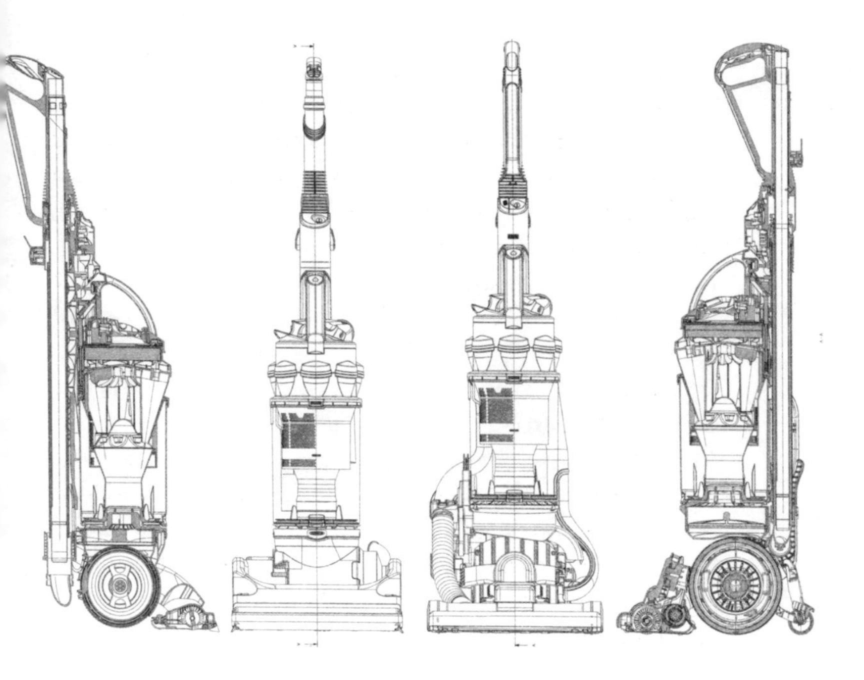

(Top left) DC14
(Top right + bottom) DC15

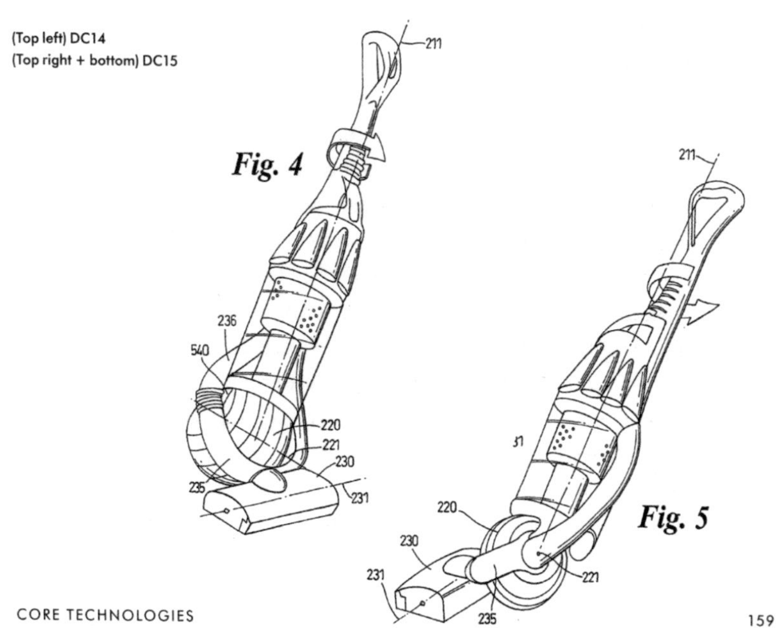

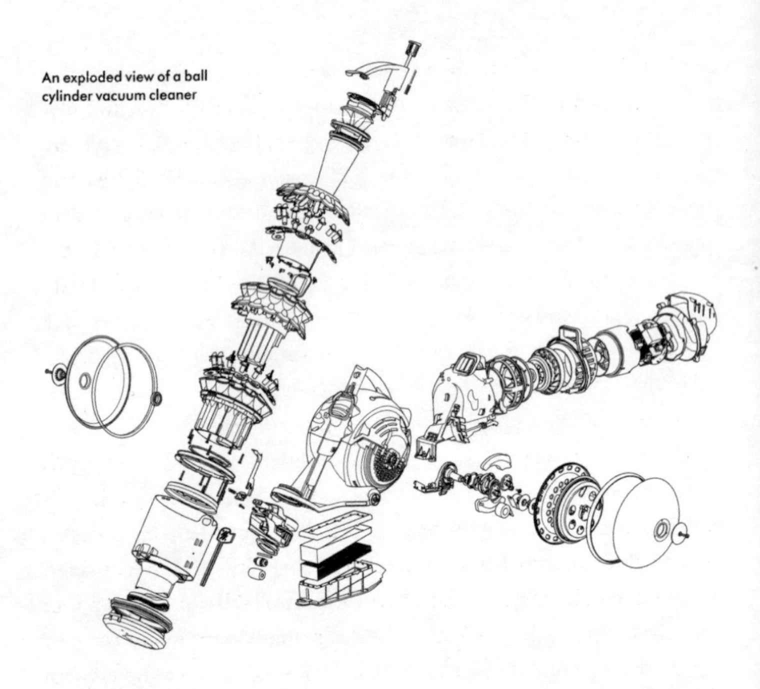
An exploded view of a ball cylinder vacuum cleaner

They were the best we could buy, yet still had inherent faults. They had no form of control or intelligence. They were heavy. Their commutators, connected to the motor's copper windings, were in continuous contact with the carbon brushes that provided the power. The commutator copper segments to provide the switching were fragile and liable to failure and the carbon brushes not only emitted foul black carbon dust but also wore out quickly. Because we were buying from a supplier, any of our rivals could buy them, too, and we had no control over quality. A vacuum cleaner motor drives a fan or turbine mounted on its shaft. This was equally unsatisfactory. It was made of pressed aluminium and stapled together, not the most robust of constructions. It was not designed for high-speed operation, nor was it capable of development for increased output and efficiency.

I'd been thinking about ultra-high-speed motors for quite some time. It was, in fact, a topic I'd discussed with Jeremy Fry during the development of those very first cyclonic cleaners in the '80s. The theory is that the faster the electric motor spins, the more electrically efficient it becomes. Similarly, the faster the turbine spins, the higher the pressure output. With this combination we could dream of smaller, lighter and more efficient electric motors. At the time we had approached an Italian manufacturer to do something interesting in partnership. They liked the idea of a high-speed motor but didn't want to be the first to invest in it. They may also have been reluctant because vacuum cleaner motors were already fast by comparison with others. A jet engine spins at 15,000rpm, a Formula 1 engine at 19,000rpm and a conventional vacuum cleaner motor at 30,000rpm. Why go very much faster?

Although at the time we were neither designers nor manufacturers of electric motors, we wanted to come up with a breakthrough in their design, creating a quantum leap in performance: many times faster, much lighter and smaller, brushless for a longer life and no emissions, more electrically efficient and above all controllable for speed, power and consumption. If we could make this revolutionary motor then, by extension, we would be able to make lighter, smaller and more efficient vacuum cleaners. And maybe other products, too, something that Pete Gammack and I discussed early on.

I began by recruiting brilliant motor and motor drive experts from universities and starting a development programme at two universities. The turbine speed we initially aimed for was 120,000rpm, or four times faster than existing high-speed vacuum motors. We would reduce the diameter of the turbine to 40mm instead of the conventional 140mm because of the centrifugal forces generated at such a high rate of revs. The smaller the turbine's diameter, the lower the centrifugal forces or loads endured. We planned to go smaller still.

There is no doubt that embarking on making our own new-technology, high-speed motor at this stage in our existence was a leap of faith. In fact, at least one board member was against our proposal. His view was that we shouldn't think of competing with established motor manufacturers. It was too risky and too expensive. To us, though, the risk was surely worth it. Perhaps he had been offering good and sensible advice. The proposed investment was certainly eye-watering. But we carried on investing heavily.

Today, Dyson pioneers the world's smallest high-speed motors. These have enabled us to reinvent the vacuum cleaner again with a pioneering new Dyson format. They have also allowed us to improve products in wholly new areas. By 2020, we were manufacturing 24 million motors a year on our own advanced production lines. These are assembled entirely by robots 24/7 in our fully automated factories rated as 'clean-room standard' (i.e. no undesirable dusts or gases inside) in Singapore and the Philippines. Having taken this approach, we were able to control every aspect of our motors' production and development.

From the outset, the motors were designed to perform at the highest levels in all our machines. This was important and exciting because the motors allowed us to make a rapid succession of jumps and even quantum leaps in the products we offered. People often ask if we would supply other companies with our motors. Although it might be profitable to do so, we supply no one other than ourselves. This is because I want Dyson engineers to be 100 per cent focused on our next exciting motor development and not on retrofitting our motors to someone else's product.

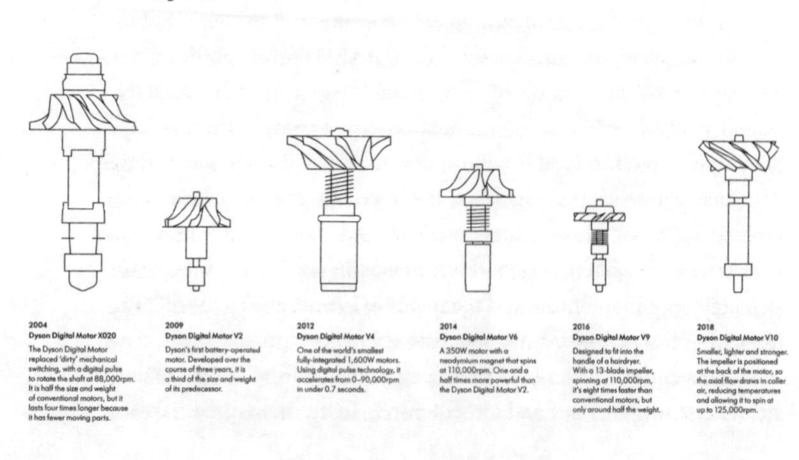

2004
Dyson Digital Motor X020
The Dyson Digital Motor replaced 'dirty' mechanical switching, with a digital pulse to rotate the shaft at 88,000rpm. It is half the size and weight of conventional motors, but it lasts four times longer because it has fewer moving parts.

2009
Dyson Digital Motor V2
Dyson's first battery-operated motor. Developed over the course of three years, it is a third of the size and weight of its predecessor.

2012
Dyson Digital Motor V4
One of the world's smallest fully-integrated 1,600W motors. Using digital pulse technology, it accelerates from 0–90,000rpm in under 0.7 seconds.

2014
Dyson Digital Motor V6
A 350W motor with a neodymium magnet that spins at 110,000rpm. One and a half times more powerful than the Dyson Digital Motor V2.

2016
Dyson Digital Motor V9
Designed to fit into the handle of a hairdryer. With a 13-blade impeller, spinning at 110,000rpm, it's eight times faster than conventional motors, but only around half the weight.

2018
Dyson Digital Motor V10
Smaller, lighter and stronger. The impeller is positioned at the back of the motor, so the axial flow draws in coller air, reducing temperatures and allowing it to spin at up to 125,000rpm.

Because we wanted a highly advanced lightweight brushless motor that would generate power at high speed while keeping cool and spinning as quietly as possible, we had set ourselves a significant challenge. We needed more brains to join the team, specialists in turbines, in plastics, software, aerodynamics and even astrophysics at Malmesbury and Singapore. As the project gathered momentum, the motors team, which started with just four engineers, grew quickly, as did the cost. Very soon, this was in the hundreds of millions of pounds.

There were technical barriers to overcome as our motors became faster. Balance was a major issue as an out-of-balance rotor spinning at very high speed can exert astonishing lateral forces of 30 to 40 tons. Bearings needed to be specially designed and made. Specialist magnets and their attachment had to be developed so that they, too, could withstand enormous centrifugal forces. Shafts had to be of small diameter and highly accurate, sometimes made of ceramic.

The high tensions and forces exerted inside the motor meant that we needed the world's best and most durable materials for production. For example, we chose PEEK (polyester ether ketone), an organic thermoplastic polymer that can reliably withstand operating temperatures of up to 250°C, melting at 343°C. Used in satellites and medical implants, PEEK is also very strong and can be moulded. It is also very expensive.

A conventional motor does pole-switching through the brushes onto different segments on the commutator by brushing mechanical contact. This provides the switching of positive and negative current alternately as well as a magnetic field that excites the motor to rotate. This mechanical system would be far too slow and cumbersome for our new speeds. A new way to drive the motor had to be invented. A main drive chip on a circuit board allows switching 6,500 times per second. The set-up cannot wear out since it relies on digital switching within the chip, rather than anything mechanical, and is capable of extraordinary speeds. The use of the chip as the fundamental drive is why we named this new type of motor a Dyson Digital Motor. We have made different designs of both AC (alternating current) and DC (direct current) electric motors, each working in a different way, improving the design and output each time, becoming smaller and more powerful.

In time, and as Dyson became more global, we also discovered the effect of altitude on the motor. As anyone who has driven a small- or medium-powered car up and across mountain roads knows, performance tails off as the engine draws in ever-thinner air. Might it perform differently in Colorado, the Swiss Alps or Mexico City? In case you are wondering, our motors now use an altitude meter, so that they can adjust their performance depending on their height above sea level. The motor will work just as well in Mexico City as it will in the Zuiderzee Works. It is so accurate, it can determine whether it is sitting on a table or lying on the floor.

The development and testing process was, in its Lilliputian way, rather like Frank Whittle's with his first jet engine. We were soon working with aero-engineers at the Whittle Laboratory in Cambridge alongside Rolls-Royce and its mighty turbo-fan jets. The scale of the challenge, it must be said, was not directly proportional to the scale of the motor.

With each new version of the motor we aimed to double its power output and halve its weight. The motor remains a great focus of ours and we now have a global team working with universities, including Cambridge, and dedicated labs at Newcastle. Creative design is as much a part of solving problems in prototyping, manufacturing, testing, rectifying and improving products as it is in their conception. We do lots of important work with universities, not relying on them for innovation but, rather, using them as an extension of our own research department, and this is as it should be.

DC12 was the beginning of a new chapter for us, since it was our first digital-motor cleaner, designed for Japan, where apartments are often smaller than those commonly found in Europe or North America. We waited some time before applying our Dyson Digital Motor to all our vacuum cleaners, as it was very expensive. In fact, it cost five times more than a conventional motor. This was entirely due to the electronics, not the motor itself. This is often the case when a new technology is first launched. It may take years to bring the costs down to anywhere near the old technology and so its introduction in the first place really is an article of faith. But such faith and perseverance have enabled us to make those leaps forward. In the process of its long gestation, our digital electric motor was about to transform Dyson.

DC12

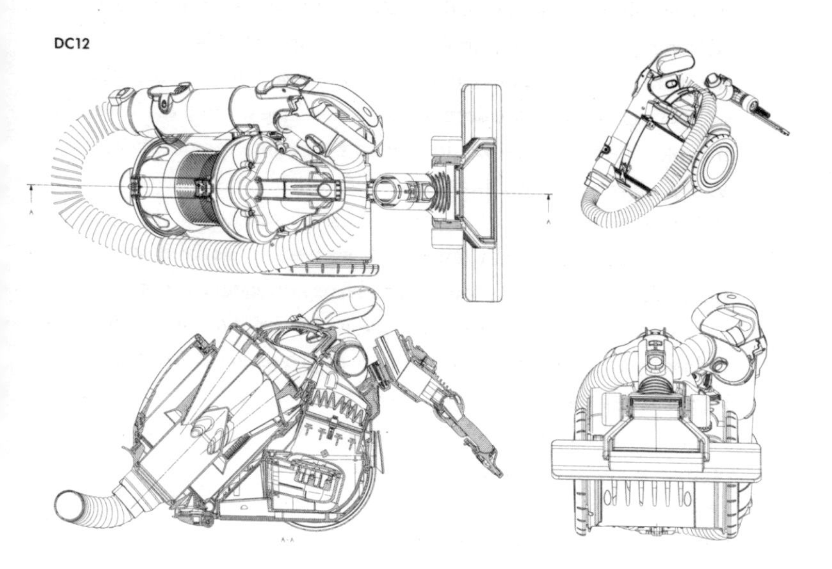

Handheld battery vacuum cleaners had been around for many years for above-floor cleaning, for cars and for boats. They sold well, but suffered from a clogged bag issue. The bags were tiny and presumably needed very thorough washing to try to rid them of the clogging dust, so were probably permanently choked. They also had very little power and even that faded very quickly as the batteries ran down

We knew that we could overcome the clogging problem by using smaller versions of our cyclonic separation system. We also knew that we could overcome the feeble battery power by using lithium-ion batteries even though, in 2012 when we started to do so, these were very expensive and were not used in appliances. As our mantra is 'no loss of suction', we thought we would challenge the idea that battery-powered devices should suffer a fade in power as the battery empties. This fade occurs because voltage in the cells reduces as the charge leaves them.

We devised software and electronics that would maintain the voltage as the charge left. This meant that the cells would 'flatten' more quickly, but at least you have full performance until the end, the end coming suddenly. This was a risky policy as our competitors would argue that our charge would not last as long as theirs. But we believed that someone using a vacuum would want to do the task as quickly and effectively as possible and they would not like hearing the motor gradually slowing down and its suction declining.

This, in fact, proved to be the case when we launched DC16, our first handheld vacuum cleaner. We were criticised for its short run-time, though this criticism has largely gone. We also gambled that battery performance would improve over time, as indeed it has. We think the Dyson DC16 is the first example of a battery-powered product that doesn't fade in use.

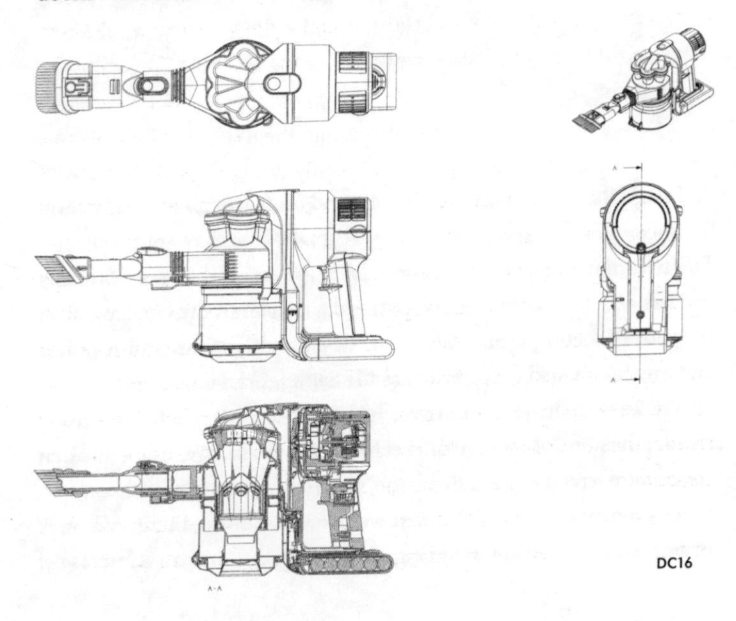

DC16

We were still in the midst of developing motors for powerful, mains-powered vacuum cleaners and it hadn't occurred to us that the future of all our products would be in small, battery-powered motors.

There is a serendipitous quality to invention, as there has been in our quest to engineer better products. We had observed, for example, that handheld vacuum cleaners had the handle at one end and that the entire product cantilevered forward from it, making the cleaners heavy to hold. Bizarre, of course, for 'handheld' cleaners. If you hold something heavy in your hand, it is better that the weight is centred on your wrist. So we put the two heavy components above and below the wrist: motor above, battery below, and with the handle as a pistol grip between the two. That left just the cyclone and the bin, which are light, in front of the wrist. This design avoided the heavy pull forwards of existing handhelds.

After we had been successfully making the DC16 for two years, in 2012 an engineer tried attaching one of our upright cleaner's wands onto the inlet of a DC16 with a cleaner head on the other end. What Pete and I were looking at was something we had never seen before, but which was also 'incorrect' thinking. Upright and cylinder vacuum cleaners are designed to avoid putting the weight of the machine, including the motor, onto the user. In both cases, heavy components sit on the floor. Here we were putting the entire weight into the user's hand. While this seemed like wrong thinking, what suddenly occurred to us was that if the weight could be light, then the entire product would be light, and the ability to manoeuvre and clean would be so much easier.

There were other matters to consider. Pete and I had been worried about the long hoses and ducts on vacuum cleaners, partly because they cause a pressure loss, but also because they are unwieldy and prone to damage. We reasoned that if we could have a much shorter and simpler airway, we could avoid the waste of power involved. Similarly, if we could remove the unwieldy hose, and avoid the large deadweight of the vacuum cleaner, we would remove by far the majority of the materials used.

It goes without saying that removing the cord also saved that tedious business of unwinding, plugging in, unplugging, and all the faff just to move from one room to another. This procedure is frustrating, and we realised it was unnecessary. What we now had was a light, highly manoeuvrable machine that was a fraction of the weight, without the restriction of the cord. Freedom at last!

Together with our high-speed Dyson Digital Motor, this led to our new stick vacuum cleaner, the Dyson DC35 Digital Slim. Pete and I thought this was the future, but we were the only ones who did. I do understand why others hesitated. The motor is battery operated and batteries had not had a good history. And, because it looked so thin, it seemed unlikely to be powerful. Its bin was relatively small, too, and it's not until you take it home and use it that you understand that this changes the way you clean. Little and often, rather than a blockbusting affair, since the Digital Slim is so easy to get out and put away.

From V2 to V6 to V8 to V10, we've made leaps in performance with our Dyson Digital Motor while reducing its weight. The team under Andy Clothier and Tuncay Çelik have brought the greatest advances in electric motors since Michael Faraday. They have a long string of patents to their names. By 2016, we had established a following of people prepared to abandon mains-powered machines, yet the vast majority either didn't know this was possible or didn't believe what our new vacuum cleaners were capable of. This is understandable. I don't think many people follow the development of vacuum cleaners as assiduously as they do that of cars or mobile phones. Our competitors had long boasted of their big 2,400W or 17amp motors, as big in marketing terms spelled powerful. By this point, our own sales were half battery- and half mains-powered cleaners. Even so, we decided to focus our future development on our new invention.

In the process, we ended up killing off our original product – the DC01 and its successors – with a better one because it made sense to do so. Our latest machines pick up more than the bulky full-size variety can while being much easier to use. I actually think this was a far more important development than the cyclone itself, the core technology on which Dyson was built. Our latest invention came about through the

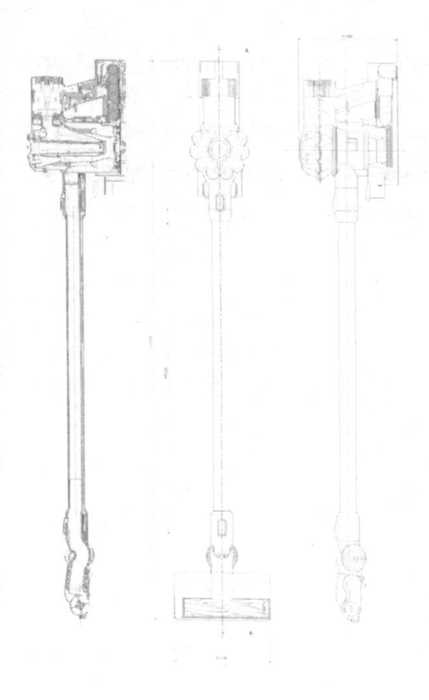

(Left) Cordfree
(Below) DC35

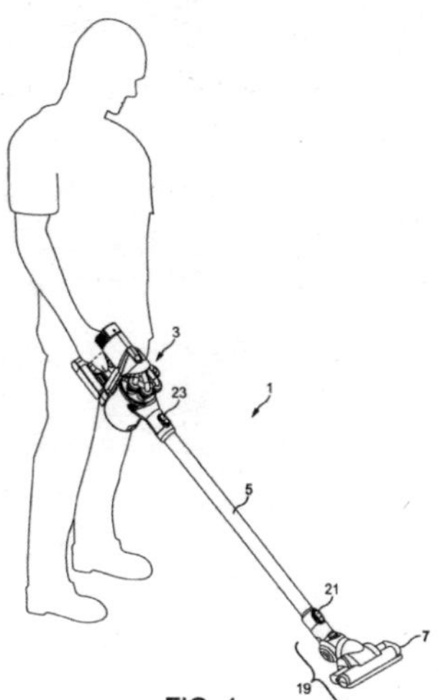

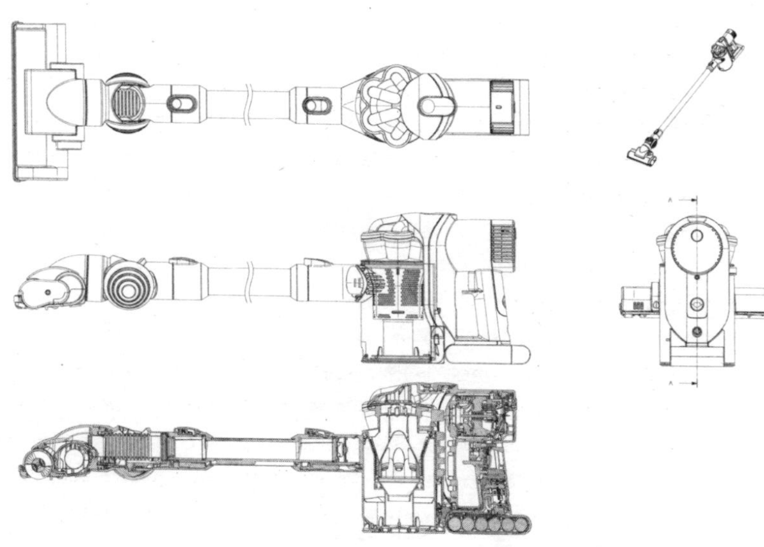

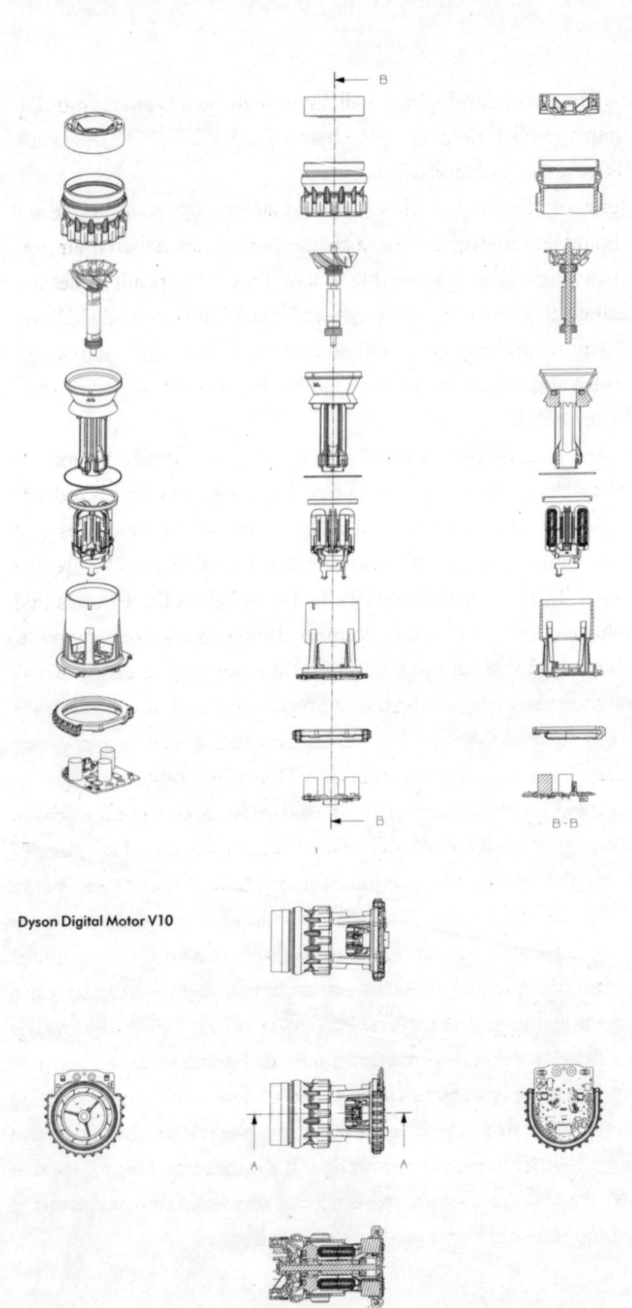

Dyson Digital Motor V10

relentless process of developing small, light, high-speed electric motors, rather than a eureka moment in the bath telling me that cordless stick vacuums were the way forward.

In spite of the Digital Slim cleaners being lighter, using fewer materials and less energy, they have greater performance than their predecessors and are more pleasurable to use. This is the result of design, engineering and science coming together. This, I fundamentally believe, is why scientists and engineers will do more than politicians and activists to solve today's environmental problems. They have more than words. They have solutions.

Our new Digital Slim vacuum cleaners now outsell our mains-powered machines 15:1, and, of course, the usual vultures – hundreds of them – have jumped on the bandwagon, all making machines that look remarkably similar to the one Pete and I saw develop before our eyes. Typically, they waited until we had established the benefits and performance of this new type of vacuum cleaner in consumers' minds, before happily riding on our coat-tails and enjoying the spoils. Many of these rival products are flagrantly reverse-engineered copies. Their makers go further, copying everything, including our imagery, our marketing claims and even our typeface. This is immoral.

At school you can be expelled if you copy someone else's work. In the commercial world it is allowed, even encouraged, under the guise of 'competition'. The senior British patent judge, Lord Justice Robin Jacob, argued (Apple vs Samsung, 2013) that this kind of copying should be encouraged. He was wrong. Copying reduces choice for consumers, rather than encouraging different products working in different ways and achieving different objectives. Plagiarism is lazy, while avoiding the costs of developing and introducing new technology. Patents exist to allow the inventor to commercialise an invention without being copied for twenty years from the date of filing the patent, which in practice means ten to fifteen years of production. If the inventor didn't have that opportunity to make a return on his efforts, why would anyone invest in researching new and better ways of doing things?

Lord Justice Jacob's comment applied to design of products, but if you followed that thinking, artists, musicians, writers could simply copy each other and reproduce each other's most popular works. Surely we want difference and originality, not similar products? We don't all want to listen to the same song over and over again, or look at the same painting, all done by different artists. Quite rightly, the law rules against plagiarism to protect the rights of artists. Why not the same in engineering? After all, patents are described as art – 'Prior Art' is a term for previous patents, and 'state of the art' summarises the sum total of what has gone before. A patent will not be granted if it is 'obvious to one skilled in the art'.

In my view, since the patent system was devised under Henry IV in the fifteenth century and has changed little since, it is time it was substantially revised. For example, the twenty-year life was devised under Henry, yet today it may take twenty years to develop, produce and release new technology onto the market. Patents need a longer life to reflect today's long research and development cycles. The weakening of patents due to obscure and irrelevant 'Prior Art' (something disclosed in previous patents) needs to be overcome as patents are too narrow and easy to engineer around. They need to be made cheaper to apply for and renewal fees should be reduced, particularly for individual inventors and small companies. Finally, one change that was made recently should be reversed. This altered the principle by which an invention should be owned by the 'first to invent' rather than the 'first to apply for a patent', which is the current ludicrous position. An inventor should hold the patent and not a plagiarist who sees it and files first, as sometimes happens today.

If, meanwhile, our future was to be dominated by motors, we needed the right kind of new-technology batteries to power them. This was the start of our urgent quest to make a quantum leap in battery technology. In any case, we were consuming something like 6 per cent of the global supply of lithium-ion batteries, so the need to improve the state of the art was visceral. I look more closely at our battery development in the next chapter.

We had been experimenting for some time with blades of air and working with sophisticated computational fluid-dynamics models for a project that remains secret. We were fascinated by the power of these high-speed 400mph blades of air when one day we noticed that if you directed the blade of air across your hand it would make the skin ripple. We wondered if it would scrape water off your hands, which it did. I was all too familiar with noisy hand dryers that took so long to dry your hands that you ended up walking away and wiping them on your jeans. This blade scraped water off your hands rather nicely, rather like the action of a windscreen wiper, and dried them almost instantaneously. We had accidentally developed a new form of hand dryer. What's more, it didn't need a heater.

Existing hand dryers used big, 3,000W heaters to evaporate water from your hands, turning the water into steam. It's a time-consuming and energy-hungry process. It's also unkind to the skin, removing natural oils from your hands and causing them to chap. Our discovery was that all we needed was the blade of air, and without the 3,000W heater. The only power necessary was 750W for the motor, and because the dry time is reduced to ten seconds, the motor is on for a third of the time of those of conventional heated hand dryers.

In 2006, the Dyson Airblade hand dryer, powered by our very first mains-powered digital motor, went into production without any market research on our part. Hand dryers were under-loved products and we followed our instinct. The Dyson Airblade sold quickly to hotels, airports, schools and hospitals, based on the speed at which it dried hands – with air moving at 430mph, or late-model Spitfire speed – and on its energy-saving properties. It has a carbon footprint six times smaller than that of paper towels. We were slightly worried about the noise and have since made it much quieter. However, an architectural practice that fitted them in their washrooms surprised me by saying they like the noise. Why? Because they can tell whether or not those exiting the washrooms have washed their hands.

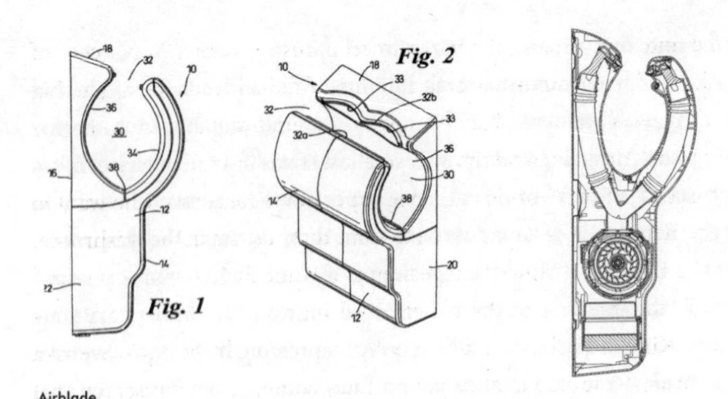

Airblade

Operated by sensors, the Dyson Airblade hand dryer was touch-free, its motor activated by infrared sensors, and fitted with HEPA filters capturing particles, including bacteria and viruses, from washroom air. We went on to make the Dyson Airblade Tap. This is a large-diameter stainless steel tube projecting over the sink, delivering warm water and two blades of air to dry your hands from a pair of projections that look like aircraft wings. This means that you don't need separate locations for washing and drying in washrooms. It is quick, efficient and means that excess water is directed into the sink rather than dripping across the floor. I have these in my bathrooms at home.

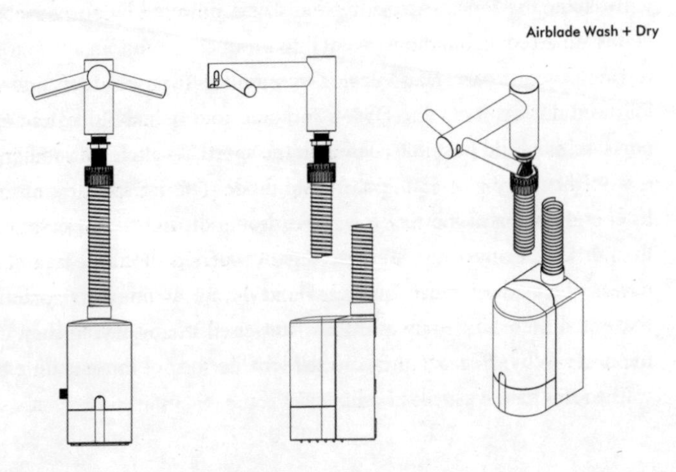
Airblade Wash + Dry

Despite our inroads, the paper towel industry retains 90 per cent of the hand-drying market, worth billions of dollars each year. The big paper players want to defend a highly lucrative status quo. Although the Dyson Airblade hand dryer is small in terms of market share, it has brought us to the attention of the big paper towel consortia, who want to retain their business. This has led to some dirty tactics in the washroom.

The European Tissue Symposium is a Brussels-based group representing the interests of the paper towel industry. It counts corporate giants Kimberly-Clark and Essity AB (previously SCA – Svenska Cellulosa AB) in its membership and has commissioned research that tries to show our products in a poor light. As we have discovered, this research is often based on flawed assumptions that then produce misleading results. These results are then used in an attempt to keep the unnecessary paper towel in the world's washrooms. Although a significant body of readily available scientific research shows the Dyson Airblade hand dryer to be fast, efficient and hygienic, the paper towel industry wants to portray it differently. It has commissioned studies, from two of its long-term consultants, claiming that Dyson hand dryers increase bacteria counts on hands and spread viruses. They promote this research widely, yet to arrive at these bold and inaccurate claims they rely on grossly misleading methodology. Gloved hands are smeared with bacteria and viruses from chickens, and, unwashed, dried immediately with a Dyson hand dryer.

This flawed methodology in no way simulates the real-life experience of using a Dyson hand dryer. Such an unreasonable quantity of bacteria and viruses would never be on someone's hands, let alone on someone's hands when they go into a Dyson hand dryer. In fact, even these absurd test conditions showed that our hand dryers distribute comparatively few particles. Conversely, research by the University of Bradford, among others, shows that Dyson Airblade hand dryers reduce the transfer of bacteria onto hands by up to 40 per cent, while the HEPA filter in our hand dryers purifies air in the washroom. In fact, if you examine the HEPA filter inside a Dyson hand dryer at the end of its life, you can see exactly how many nasties the filter has purified from washroom air. Of course, the bigger problem is the vast quantity of trees, water and fuel consumed in the manufacture and shipping of paper towels. Their use causes unseemly waste, which generally cannot be recycled.

As often happens, our observations during the development of the Dyson Airblade hand dryer led us to the principles used in other products, like our Air Multiplier fans and, in turn, to heaters, humidifiers and air purifiers. One thing we noticed in our research into airflow was the way in which the ultra-high-velocity blade of air that prompted us to make the hand dryer could induce a lot of additional air behind it. This was the stuff of aerofoils and aircraft wings. It works according to Bernoulli's principle, a mathematical formulation set out by the eighteenth-century Swiss mathematician and physicist Daniel Bernoulli concerning fluid dynamics. The principle is well known in the aviation industry. What it means in practice is this: if a small blade of high-speed air exits, in our case from an annulus (a narrow slot for the air to exit that extends the whole way around the diameter of the hoop), it induces a region of low-pressure air, or suction, from behind it. Thus, from quite a small volume of airflow you can create one many times larger.

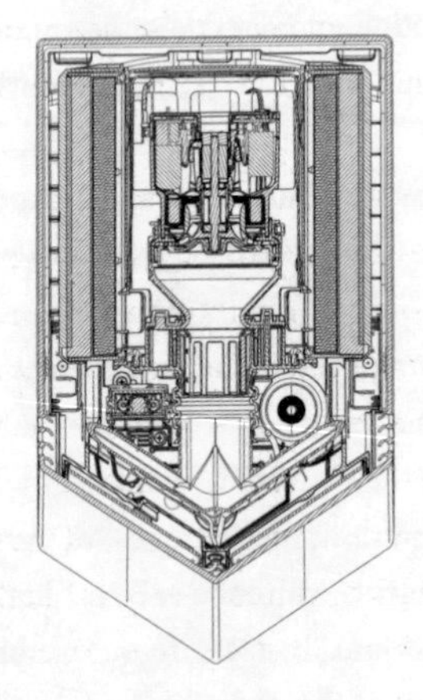

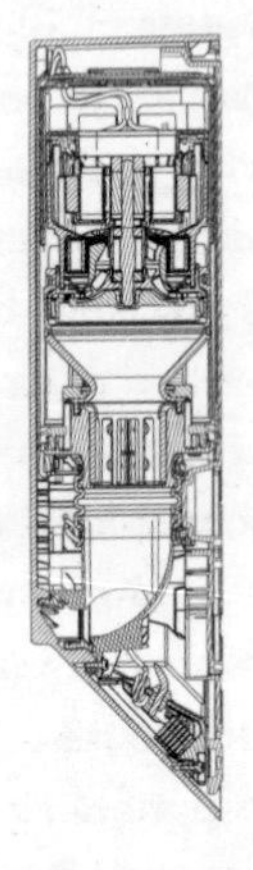

Airblade V

Air drawn into the base of the Dyson Airmultiplier fan at 27 litres a second by our motor passes over its aerofoil-shaped ring, creating low pressure, or suction, that pulls air from behind and multiplies its flow to 405 litres a second. Without the need for conventional fan blades to accelerate the airflow, the Dyson Airmultiplier fan is as quiet in operation as it is safe, and over subsequent iterations we have made it even more so. Fans with blades chop the airflow, which causes a buffeting effect for users in its airstream, which can be uncomfortable. The Dyson Airmultiplier fan was a reinvention of air movement. Electric fans were of a design that had remained basically the same since 1882, when Schuyler Wheeler, a 22-year-old American electrical engineer, whose real invention was the electric lift, made the first known version. Conventional fans have the heavy motor at the centre of the fan, which makes them top-heavy and continually slumping on their angle adjustment mechanism. On our fan the motor is in the base, with the centre of gravity around the middle of the base. We found that we could rotate the fan around an arc from its centre of gravity. This meant that at whatever angle you chose for a position, it would stay there perfectly balanced without the need for fixing or clamping.

With hindsight, perhaps we ought to have launched the Dyson Airmultiplier fan in Australia, Japan or somewhere else hot and humid in South East Asia rather than New York in the depths of a freezing winter. I drove out to New Jersey at four o'clock in the morning to be interviewed for CNBC TV's breakfast show. The presenters were intent on making this into a business story, but I wanted it to be about the invention itself – the technology. Desperate to turn the situation around in the midst of the interview, I stuck my head through the Airmultiplier fan's ring or annulus – you have to be careful how you say the word – and this changed the tone immediately. One of the presenters decided to blow-dry her hair with it, and then the other had a go. This slightly madcap TV appearance certainly got the Airmultiplier fan talked about. The following morning I was invited on *Good Morning America* for a more glamorous version of my head-through-the-ring routine.

After the show, I was in a yellow cab in Manhattan. I called Pete Gammack and said, 'It's bloody cold here. We've got to do one that puts out heat.' So, we started working on a heater. This wasn't so easy. I remembered the electric heaters of my youth, with that horrid smell of burning dust and the near-hysterical whirring of fans as hot air and dust was shovelled out in a single direction. We needed to come up with a heater that would create ambient heat in a room and regulate its temperature. We also wanted one that was cool to the touch, wouldn't burn dust and wouldn't smell. We settled on semiconducting positive temperature coefficient (PTC) ceramic stones as a heater, which, combined with the Airmultiplier technology, allowed us to heat an entire room and not just the area immediately around the fan. PTC ceramics increase their resistivity as they heat up, thus not getting hot enough to burn dust, so there's none of that acrid smell.

Having learned much about the science of HEPA filters from our development work on hand dryers and vacuum cleaners, we then created an air purifier that could also heat and provide cooling air while capturing harmful pollutants. Incidentally, a 30W Dyson Airmultiplier fan is far more energy-efficient than a 2,000W air conditioner. The blowing of atmospheric air over the skin causes the moisture to be released through evaporation, in turn causing a refrigerant effect of about 4°C. This is usually sufficient cooling, except in extreme conditions. Aside from a massive energy saving, this is also far more comfortable and certainly feels healthier than being in a room with a traditional air-conditioning unit.

Our purifiers use HEPA filters in combination with carbon filters. The HEPA filters trap particles down to a microscopic level, while activated carbon deals with harmful gasses like formaldehyde. However, the carbon quite quickly becomes loaded with the gases and starts to re-emit them into the room, in a process known as off-gassing. We developed a method of capturing formaldehyde for the lifetime of the product without off-gassing. This uses our Cryptomic technology, through which formaldehyde particles are captured by a catalyst and broken down into tiny amounts of harmless water and CO_2. Cryptomelane

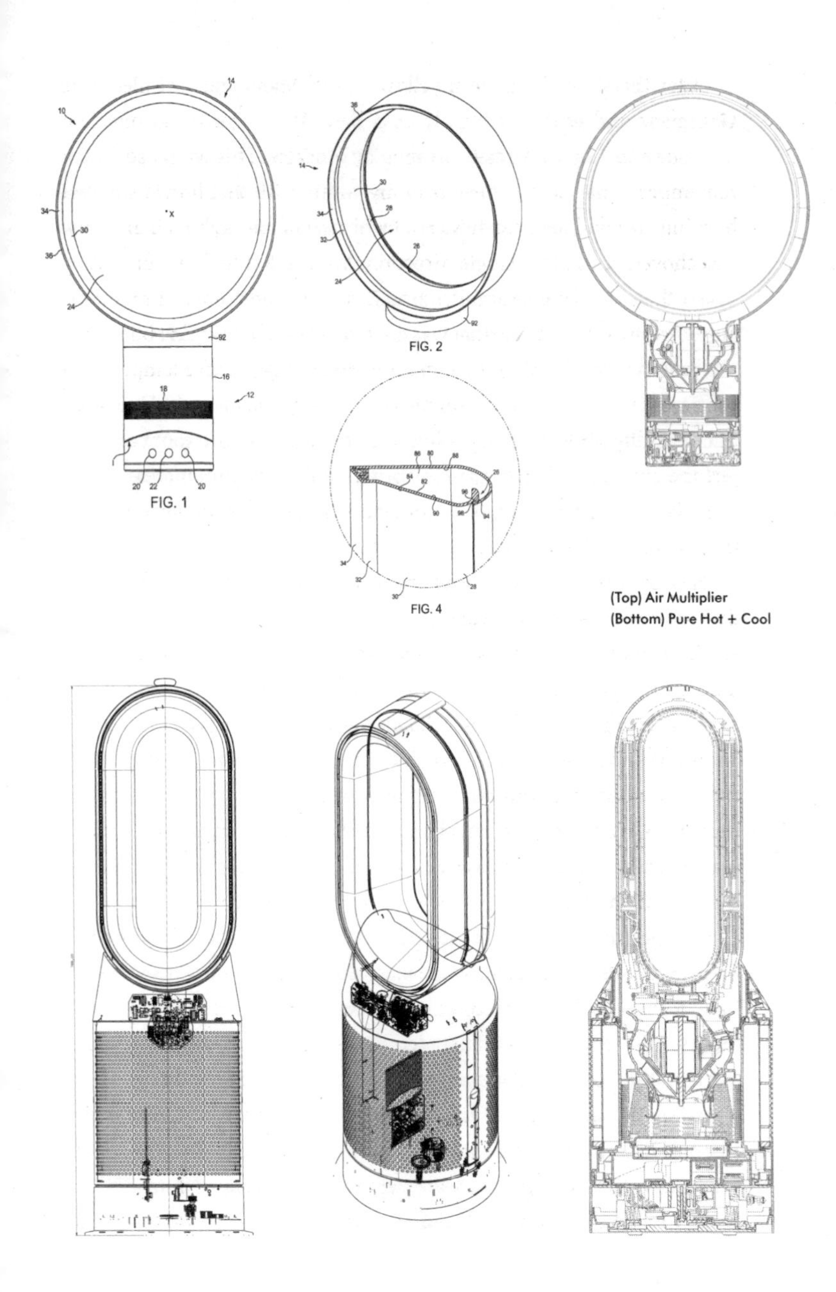

(Top) Air Multiplier
(Bottom) Pure Hot + Cool

is characterised by billions of atom-sized tunnels with the ability to destroy formaldehyde.

People spend the majority of their day inside, and indoor air pollution can be up to five times worse than the outdoor kind. The Dyson Pure Hot + Cool purifier cleanses the air of pollutants and even tells you what these are in a real-time display both on the machine and in app. Formaldehyde is released from many sources inside the home, including furniture, paint, wooden floors, scented candles, cooking, cleaning products, decorating products and plants, presenting considerable concern, particularly in China, where it is seen as a cause not just of skin irritation and conditions like asthma, but of certain cancers, too. Our purifiers work well and offer people an assurance other air filters cannot provide.

One of the ideas that Pete and I had discussed at the outset of the digital motor project was to make an even smaller motor than the one needed for vacuum cleaners to replace the bulky, heavy and inefficient motor inside hairdryers. Conventional hairdryer motors have 50mm or even bigger diameter fans. We proposed one just 28mm in diameter and a mere fraction of the weight. Because this would be very small, we would be able to put it in the handle of hairdryers, thus placing the centre of gravity – literally – in the palm of the hand. On other hairdryers, the heavy motor acts as an unhelpful lever, magnifying the effect of its weight. The head of our hairdryer need only contain the control electronics, the heater and a large hole or airway through the middle for the Airmultiplier effect.

For me, this was another of those products, used frequently by hundreds of millions of people, stuck in a technological time warp. Existing hairdryers were heavy and uncomfortable to use. Most seemed intent on drying hair brutally with excess heat, and, from personal experience, I knew how loud they could be, with someone else waking up the whole house in the early hours of the morning. Could we make a light, quiet hairdryer that could dry hair fast without damaging it?

To develop a better hairdryer, we wanted first to understand the science of hair, a complex field of study. I grew my hair so that I could learn by doing. Many of the male engineers also took to growing their hair long, keeping up with female colleagues, but couldn't tell their partners why. As usual with us, this was a secret project. Trichologists joined the team while some of our engineers went back to college to learn more about hair, including how to style it.

We learned that hair comprises two layers. The outer surface is covered with multiple layers of flat cells known as cuticula. Like roof tiles on a house, this protects the hair strand from damage. The inner and main structure of hair – the cortex – is full of keratinised hair cells. This gives hair its strength and elasticity and contains melanin pigments that give it its colour. Since hair is made up of dead cells, it is unable to repair itself. If damaged by heat, it will remain damaged until grown out from the root.

Excess heat can irreversibly damage hydrogen bonds within the cortex of the hair. These bonds are responsible for providing strength and elasticity as well as giving hair its ability to be styled into different shapes. At over 150°C, α-keratin slowly converts to β-keratin that, over time, makes hair weaker and less elastic. Over 230°C, hair begins to burn or melt, with strong disulphide bonds breaking down quickly. The surface of the hair becomes cratered and reflectiveness is lost, leading to a reduction in shine and gloss, as well as damage.

We needed to be able to control the powerful airflow and degree of heat precisely. We also learned that different types of hair around the world vary in strength and elasticity due to genetic differences, but healthy hair will withstand higher forces and stretch longer distances. Not only is hair much more interesting than we'd thought in terms of its make-up, it's also pretty much impossible to replicate. We had to buy in ethically sourced tresses representing different groups – straight, wavy, curly, kinky – and different thicknesses, and then find out how each type of hair responded to heat and other stresses. The hair had to have been free of all shampoos and styling products.

To deliver a high-speed and controlled airflow, we applied the airflow knowledge we had built up with our Dyson Airmultiplier fans.

This enabled us to amplify the airflow, producing a high-pressure, high-velocity jet of air. But fundamentally, the hairdryer demanded a new kind of motor. When positioned near the hair, other hairdryers with their flow-fans, suffer from restricted airflow, causing the temperature to rocket and inflicting irreversible hair damage. What was needed was a high-pressure turbine that could deliver the high flow even when the airflow was restricted.

The Dyson Supersonic hairdryer was put through its paces, as all Dyson products are, in our in-house test labs. We have sound-testing chambers, in which the pure sound of machines can be recorded and analysed, called semi-anechoic chambers. These are built to a standard format designed to prevent the echo of sound, in which the walls are not parallel and walls and ceilings are covered in elongated pyramid foam shapes. If the floors were covered in the same cones it would be a fully anechoic chamber but then you couldn't walk or work on it. If you spend any length of time inside one of these chambers – they are silent and echoless – you start to feel dizzy because you're not receiving any feedback of sound.

This process of testing and iterative improvement prompted us to increase the number of blades on the impeller to thirteen: two more than standard. This modification, late in the day, meant that we obliterated one tone of the high-speed motor with the frequency becoming supersonic – above, that is, the audible range for humans. This means we have a far quieter hairdryer than our rivals. It is also why the invention is called the Dyson Supersonic hairdryer.

We had this revelation only a few weeks before launch, and yet, due to the highly sophisticated automated production lines that milled the impeller from a single piece of aerospace-grade aluminium, we were able to apply the changes immediately. The few people who worked on the production in Singapore were unaware of the change made over our network from Malmesbury.

During four years in development and at a cost of £55 million, we made some 600 prototypes of the hairdryer with 103 engineers working

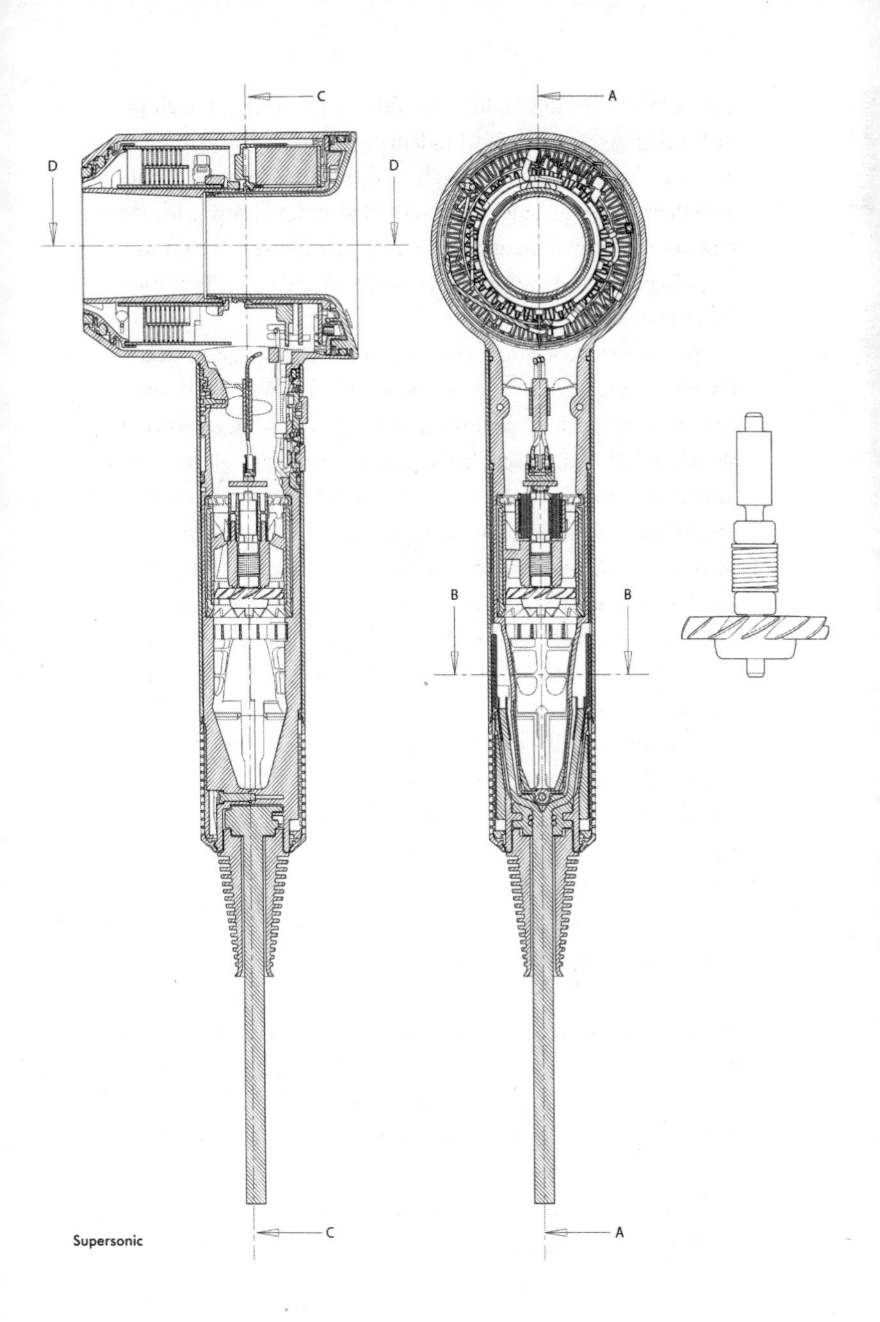

Supersonic

Airwrap

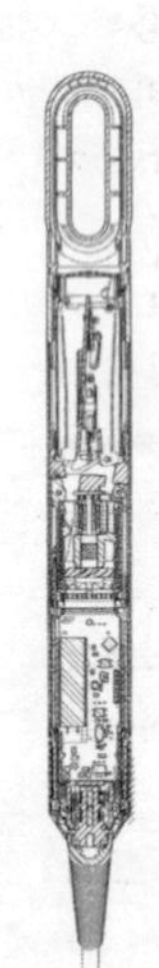

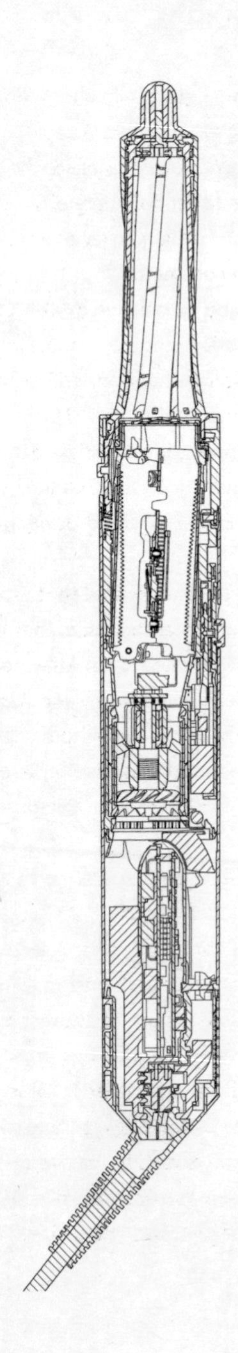

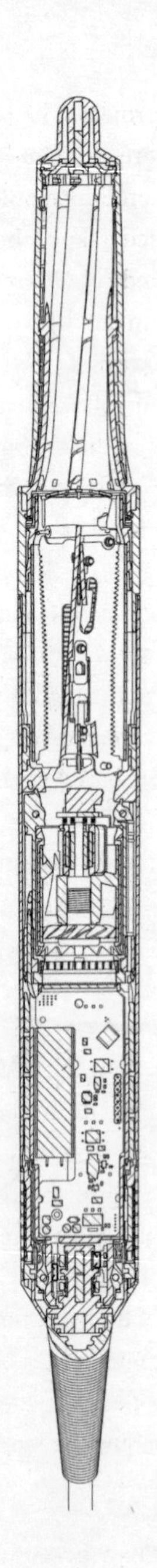

on the project. The product was quickly in short supply, since we had under-forecast demand, as we typically do. As soon as he heard about the Dyson Supersonic hairdryer, Karl Lagerfeld, the fashion designer, demanded to know how he could get hold of one, not for himself but for his adored cat Choupette, who had a quarter of a million followers on Instagram and her own Twitter account.

Meanwhile, we had become fascinated by the Coandă effect, by which air follows round a curved surface, wondering if we could use this to wrap hair around a barrel to shape curls. This became the Dyson Airwrap styler, which we launched in 2018. This styles hair with air rather than extreme heat, meaning it causes less damage. The effect is named after the Romanian inventor Henri Coandă, who identified the principle scientifically in the mid-1930s and developed it for applications in the aero industry.

The point about wrapping the hair around the barrel with air, unlike using a circular brush or mechanical means, is that the hair is free to fall off the barrel in perfect shape. The Dyson Airwrap styler enables all sorts of curls, shapes and indeed beach waves, or even straight hair, to be styled at a low heat, using the various tools and attachments we have invented to go with it. Jake and I did curls live on stage at the launches – curling a hair tress in New York, Paris and Tokyo – which shows just how easy it is to use.

We were also looking at hair straighteners and the damage they can do using temperatures of 210°C and even 230°C. Traditional straighteners have flat heated plates through which selected tresses of hair are passed. The idea is to apply heat and tension. The problem is that the plates grip the thick parts but not the thinner parts of the tress. The result is that the thick parts of the tresses are straightened while the non-gripped parts are left unstraightened. Thus it requires many passes of the tress through the plates, at ever-increasing temperatures, to straighten all the hair. The Dyson Corrale overcomes this problem by using flexible plates that retain the tress in a neat block without stray hairs missing the tension of the plates on either side. The tress is treated with even heat and even tension, with fewer passes through the plates and, therefore, less heat damage.

We also discovered that we could achieve the required styling more quickly while using twenty degrees less heat. Another interesting discovery was that maintaining a thick tress provides a cooling effect to its surface adjacent to the blades. This again prevented extreme heat.

The flexing plates were difficult to develop and even harder to manufacture. They were made from an interesting copper called beryllium, which can flex and behave like a spring. It can flex even more if you apply many cuts across the surface, reducing its thickness while increasing its ability to bend. We were able to machine the prototypes, but the process was far too slow for production. We make them with a process called wire erosion, where the copper is placed in a bath of oil and a wire charged with high voltages cuts through the material at high speed. The plates are then treated in an interesting way, which I can't share for reasons of intellectual property protection! The Corrale uses battery power, so users are free to style their hair without needing to be near a power socket. This is not only useful in the home but also if you happen to be in the car, at work, or perhaps glamping. Although batteries do add weight and cost, there is not the inconvenience, or tugging weight, of the power cord.

I was about to launch the Dyson Corrale straightener in New York in early March 2020. The venue was ready but then the coronavirus pandemic struck. Since I was in Paris at the time, the next day I did an online launch at our new Paris store facing the exuberant Paris Opéra. A camera followed me inside the shop while I explained how the Dyson Corrale straightener worked by gathering the tress and providing even tension, and how it worked using a much lower temperature than existing hair straighteners. A piece of showmanship maybe, but someone commented that it was good to see the owner of a business explaining his technology.

Throughout this period we became more deeply interested in robotics, AI (artificial intelligence) and vision systems. There was a lot of talk

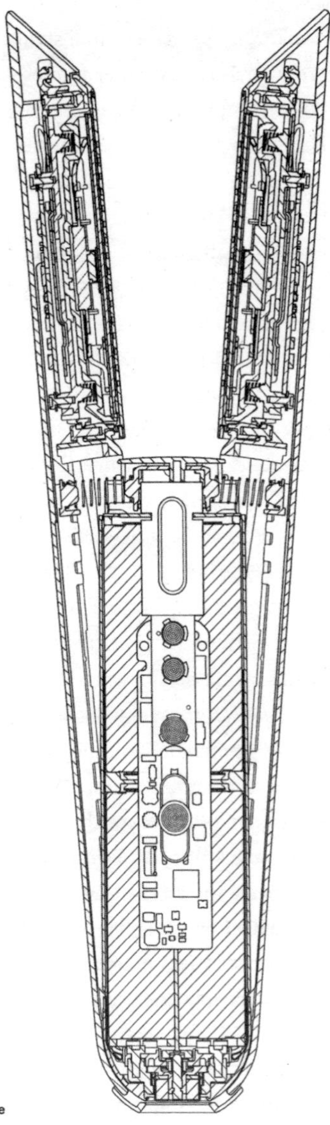

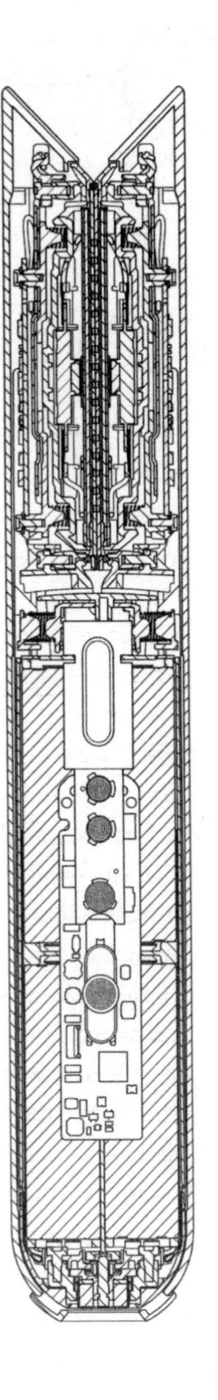

Corrale

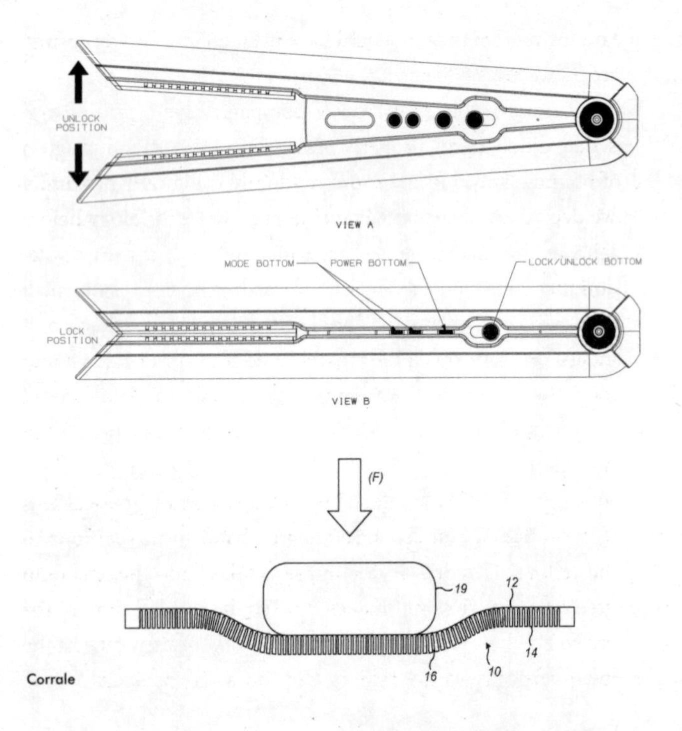

Corrale

in the 1990s of robotic domestic gadgets and machinery. While the idea of robots helping us in the home and garden goes back to science fiction comics of the 1920s and 1930s and to popular '50s films, for me the reality had never lived up to the hype surrounding them.

There was one further step to consider, however much we improved our vacuum cleaners. In the end people still had to do the cleaning. What if we could just let the vacuum do the job instead? This line of thinking took us deeper into the field of robotics. We came close to launching a robotic vacuum cleaner, the DC06, in 2006, but although it wasn't what I'd call a failure, it was terribly complicated. It cleaned the floor properly, but lithium-ion technology hadn't yet arrived, so it had eighty old-fashioned NiCad battery cells, eighty-four sensors of different types, an array of microchips and several motherboards. If it had gone on the market then, it would have been too expensive, costing somewhere between £1,800 and £2,000.

We had learned a lot. While other companies were putting out machines that didn't clean properly and navigated badly, through a process of bouncing around the room randomly using only proximity sensors, we decided to focus on advancing our own technology before trying again. In the meantime, we experimented with a mains-powered version that laid out a cord behind it, recovering it as it returned. Although it was very powerful, the cord limited its freedom of action. It could work just one room at a time and required the user to move it into another room. On the navigation side, we took a bold step and focused on a vision system with a Dyson-developed 360-degree camera that can see, film and interpret.

Our 360-degree camera is interesting. This is a hemispherical lens on top of the robot. It can see a 360-degree band from the floor to most of the ceiling. This 360-degree image is taken into the lens onto a 360-degree reflective ring and reflected up to the inside centre of the hemisphere where there is a round mirror. This mirror then reflects the images down vertically to the camera chip on a circuit board below.

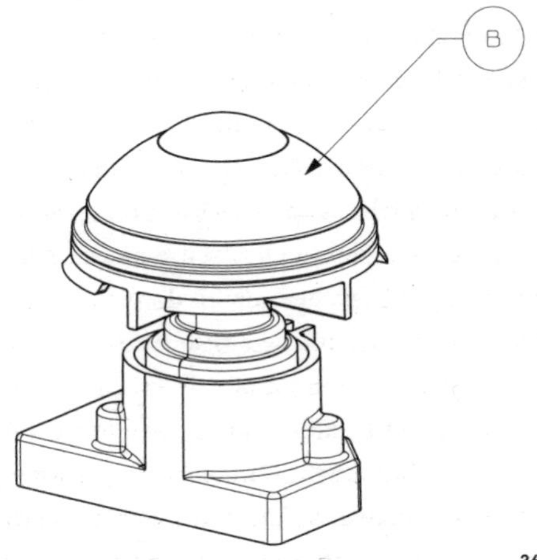

360 Eye camera

The camera comes to know waypoints, which might be the corners of a table, light fittings or the edges of doors, from which it can judge its distance and work out where it is. Actually, we humans work like this, avoiding table edges and judging where we are in a given space using waypoints in much the same way. From these waypoints, the robot knows where it is, what it has done and where it is going even more accurately than a human. Naturally, this vision system took time to develop.

Keen to find the world expert in the pioneering field of vision technology, we dispatched a member of the team to a conference in San Francisco. On arrival he was told that the world's leading brain on this topic, Professor Andrew Davidson, worked at Oxford, forty miles down the road from our campus. We have worked with Andrew ever since.

We eventually launched the Dyson 360 Eye robot vacuum cleaner in 2016 and four years later we updated it with the improved Dyson Heurist robot vacuum. These machines use vision-based mapping, Dyson-designed algorithms and our proprietary 360-degree camera. They are first and foremost highly effective vacuum cleaners, but they also have intelligence and can see, understand and learn the layout of homes so that they can clean thoroughly and efficiently using the Dyson Digital Motor and efficient lithium-ion batteries. They avoid obstacles – humans and pets, too – and return by themselves to the docking station when low on power. The Dyson 360 Heurist robot vacuum uses LED lights, combined with its 360-degree camera, to map in low light levels, for example under a sofa or bed. The Dyson 360 Eye robot vacuum returns back to its charging station and sets off again once recharged. We have received some lovely responses from customers. One Japanese customer built a charming little garage for his Dyson robot. Appearing to be real-life characters, they are often regarded fondly, a first, I think, for a vacuum cleaner.

Today, we work with more than twenty universities, the most significant of which is Imperial College London, to where Andrew moved from Oxford and where we have a dedicated Dyson lab. We have major programmes under way in Singapore and we've developed our own

360 Eye

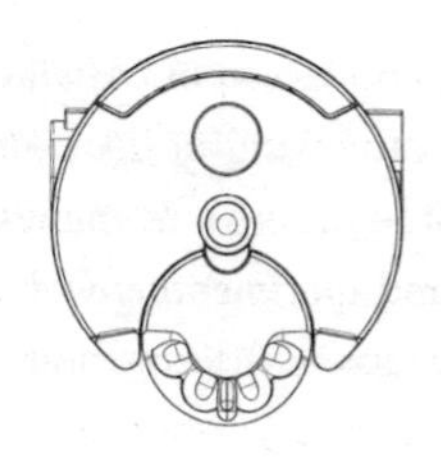

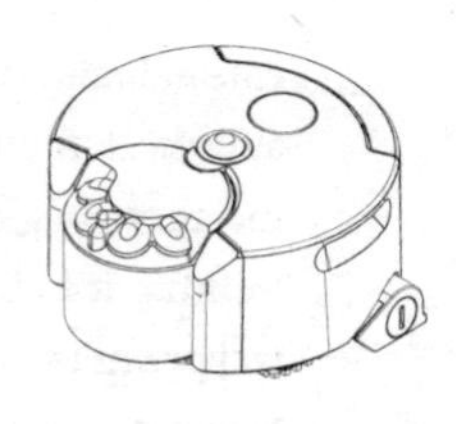

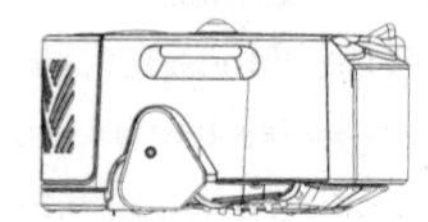

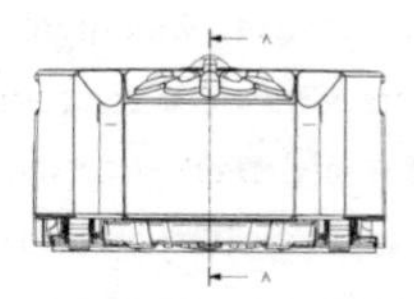

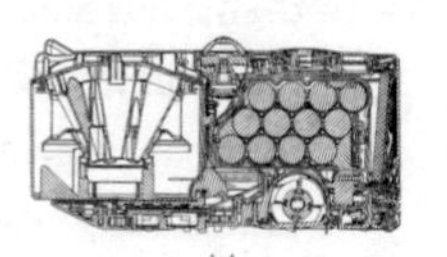

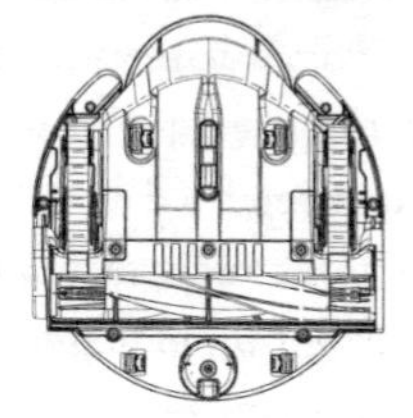

360 Eye camera

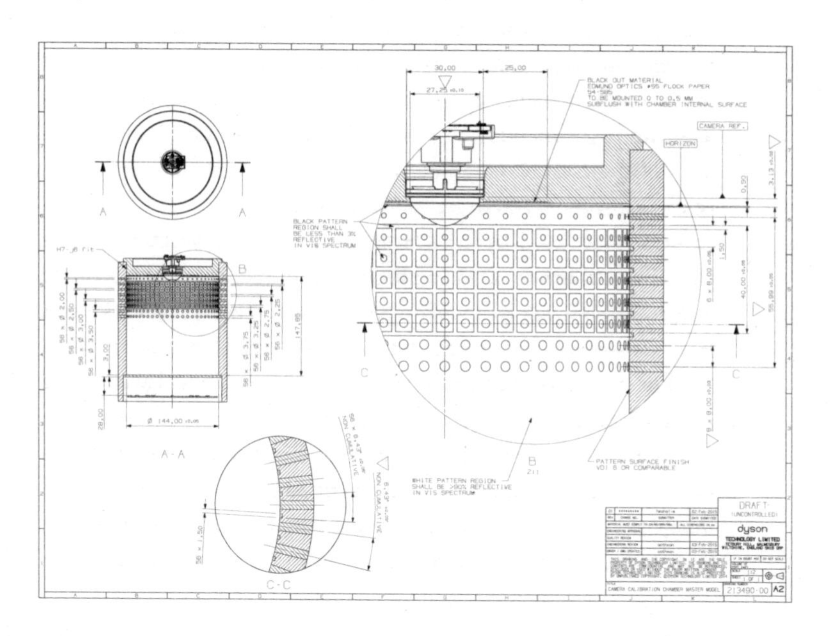

research and development hub at our Hullavington Airfield campus in Wiltshire, the biggest of its kind in the UK. These teams draw on an increasingly global group of software engineers, who develop the code to bring it to life. We're experimenting with various robots and all the technologies needed to make these far more useful in the home than they have been to date.

Meanwhile, there's still a long way to go with robotic cleaners. They can't climb stairs nor can they lift themselves off the floor. Home is about as difficult and unpredictable an environment for a robot as you can get. Most robots work in controlled and entirely rational conditions, like our digital motor factories, but when working at home, and while it maps and learns the domestic terrain, the robot never knows quite what to expect from one moment to the next – socks, rugs, Lego, stairs, phone-charging leads and pets.

We like our intelligent machines to do as much of the work for those who use them as possible, without intervention. Our purifiers, for example, continually 'sniff' the air and respond to the pollutants and volatile organic compounds they sense. They respond in real time to filter the air and provide an air-quality reading without interaction from the user. But, while half our engineers are now involved with software, there are practical, down-to-earth issues that remain critically important to the success of our products. These demand a willingness on our part to experiment and to get our hands dirty.

Some time ago, a Japanese magazine criticised all vacuum cleaners for leaving a fine film of dust on floors. Was it true? Was this something we had missed? The magazine report prompted Pete Gammack and I to get down to floor level and find out. We sprinkled powder over a sheet of black acrylic and ran one of our vacuum cleaners over it. The Japanese magazine was right. Although removing most of the powder, a fine film was indeed left on the sheet of acrylic. On our hands and knees we noticed that it required a damp finger or cloth to remove the film. It dawned on us that it might be sticking by static. We thought about the way we used to clean vinyl records with carbon-fibre bristles that conducted the static away. So we developed a manufacturing technique for inserting carbon-fibre bristles on our vacuum cleaner brush bar. The problem was solved and duly patented.

The depth of the research we're undertaking in programming, AI, machine learning, computational fluid dynamics, energy storage, acoustics, super-conductors and any number of pioneering fields has been scaled up significantly, yet there's still a lot more empirical testing to do in our labs. Ballistics, for example. We do a lot of work on ballistics, which sounds more like something from the First World War than the focus of scientists and engineers working for what has effectively become a global tech company.

As we developed our core technologies and created products that have sold in very many millions around the world, we have also focused on material science and the research and development of new forms of pseudo-plastics. This is all part of our quest to make products that use less energy and fewer resources. This is an area in which I see Dyson working increasingly closely with farms.

Of course, we can't do everything we'd like to do. We are, after all, just one – family-owned – company following its interests and passions. Yet with all the research, development and growth shown in this chapter, at a time when pretty much everyone thought us just a vacuum cleaner maker, it becomes easier, I hope, to understand why we had truly grown beyond the boundary of our rural Wiltshire idyll.

CHAPTER EIGHT

Going Global

I knew that if Dyson was to be a successful technology company, rather than just a British vacuum cleaner manufacturer, we couldn't be Little Englanders. We needed to become global, and quickly. England, and the rest of the United Kingdom, is simply not a big enough market on its own to sustain the constant and huge investment technology requires. This is in stark contrast to the United States, where the internal market is so huge, or Asia, where there is such rapid and exciting growth. An average-size American company is six times bigger than any large British company and can invest accordingly. This is what we were up against, and I knew it.

The difficult thing in the early days, however, was that we honestly didn't have time to go to Australia, the US, or even France, to try to set up companies. We were entirely consumed by the challenge of manufacturing enough vacuum cleaners for the UK, while simultaneously researching and developing new products, and coping with all the issues that emerge from being a fast-growing manufacturing company.

Production issues dominated the early days. Suppliers, for example, let us down with components, so we needed to find alternative and reliable, good-quality suppliers, especially as our volumes escalated. The dual sourcing of supplies became important in case of stoppages and to maintain a competitive edge between suppliers. Meanwhile, finding production space for assembly lines and for training people on new products was becoming difficult. Recruiting young engineers to work in Malmesbury rather than in London was also tricky. We needed more of them as we expanded and began new design projects. Recruitment at a time when the business was doubling in size every year was time-consuming, too. At this high rate of expansion, predicting sales volumes is critical. We usually underestimated what we needed, although this is a better problem to have than being overstocked. And yet, with a young and enthusiastic team, we learned to overcome these types of difficulties.

For our expansion overseas, I needed to find really good people who understood what I was doing and who were going to be able to excite other people about the technology. The problem, I discovered, is that it is nearly impossible to persuade really good people to leave an established company to join a start-up that absolutely nobody has heard of. It just doesn't happen like that. We may have been quite well known in England, but we were totally unknown everywhere else in the world. There was no reason for any retailer to take on Dyson, or for any household to aspire to buy our vacuum cleaner. Without proper explanation, ours was an odd-looking and expensive product. Other countries had their own established brands, and that was our problem.

Australia was the exception, for two reasons. Firstly, I landed on my feet and found a good person to run the business. This was Ross Cameron, the talented and energetic Australian in charge of commercial machinery at Johnson Wax, with whom we had a licence agreement. In 1995, he rang me late one night, completely out of the blue, and said, 'I'm leaving Johnson Wax because I'm fed up with globetrotting.' I said, 'Why don't you set up Dyson Australia? We won't be able to offer any help, because we are all too busy. But let's do it.' Ross said yes and did it.

Ross didn't take a salary in the first year, yet he committed himself entirely to setting up Dyson in Australia. Although he was an engineer, trained as an apprentice at the Austin Morris factory in Sydney, he was a brilliant natural marketeer and salesman, too. It was a slightly odd thing to be selling an upright cleaner in Australia in what is a cylinder – or 'barrel' as they call it – market, but Ross made a remarkable success of it.

The second reason, I believe, for our early success in Australia was the attitude of retailers there. There were two big retailers, one called Harvey Norman, a huge operation run by a wonderful and enterprising couple. It took a while to convince Katie Page of our credentials, but, once we did, she totally understood what we were about. Her enthusiasm for the product and the technology helped it take off to the extent that we captured 60 per cent of the market. The Pages are very good friends and very good retailers, with shops spread across the globe.

Now I had Ross Cameron doing Australia on the other side of the world, with retailers on board, but without much help from us. The time had come to employ a dedicated marketing person back in the UK. I put a notice up at the University of Bath. It read: 'Does any Modern Languages graduate want to do some marketing at Dyson?' Oddly, an Oxford graduate turned up at the office. The daughter of a schoolteacher, Rebecca Briggs knew nothing about business and even less about marketing. But she was bright and sparky. I took her on immediately and she joined our growing team of young graduates. Her husband, Alan, was to become our first in-house lawyer.

Rebecca's German-language skills proved to be a very handy asset indeed, since we'd been getting sales enquiries from British Army on the Rhine. I sent Rebecca off to Germany, with one of our English salesmen. Then, once I decided she'd had enough direct experience of selling, Rebecca came back to the office and got stuck deep into marketing, having never done anything of the kind before. She had a wonderful instinct for it and the nous and confidence to believe in her intuition. Because of our exciting and different technology, we didn't have to market our product in a traditional sense. I didn't want anyone to buy our vacuum cleaner through slick advertising. I wanted them to buy it because it performed. We could be straightforward in what we said, explaining things simply and clearly.

Rebecca ran our marketing very successfully for many years. She would go out, make store visits and do a bit of selling herself in stores, which is very important but for some reason it's very difficult to get conventional marketing people to do this. My only disappointment is that, whenever she took somebody on, it was someone with five or six years' experience. She wouldn't take on fresh graduates. We corrected that many years ago. I do understand how she felt as, apparently, it is safer and more responsible to employ someone with experience while, if you are very busy, it seems a bad idea to spend time training people.

In 1996 – three years after the UK launch – I decided it was time to set up properly in France with our own company. Germany, Benelux and Spain followed in 1998, and Italy in 2000. None of these ventures were runaway business-success stories. Getting the right team in place was a challenge. The first person we had running Dyson in France, for example, established the office in Créteil, in the suburbs of Paris, a soul-destroying place to visit, let alone work. I expressed my dissatisfaction to the team. 'We have to be more aspirational and optimistic than this!'

I spoke to Andy Garnett, who had been with me at Rotork, and had also left to start a successful company of his own. I had noticed that Andy was brilliant at finding the best offices and he knew Paris very well. I asked him if he could find us an office in Paris. Together we moved Dyson to rue La Boétie, a street leading off the Champs-Élysées and entering rue Saint-Honoré. It's not a shopping district, rather a place where smart people have offices. I asked Chris Wilkinson to design us an office with, at street level, a sculpture gallery.

We knocked through between the ground floor and the first floor and had a glass frontage to the showroom stretching up from pavement level to the top of the first floor. We installed two mezzanines connected by a stainless steel and glass bridge. The ground floor had nothing except a French limestone floor, white walls and plinths supporting a cylinder vacuum cleaner and an upright vacuum cleaner. Nothing was said about the products. There were no price tags. All we had on display were products on plinths. It resembled a minimalist sculpture gallery.

We had a limestone desk, behind which sat a sales assistant. And, rather like an art gallery, we had just a few select customers. We wanted to display our product how we thought it ought to be displayed. This is not a criticism of electrical retailers who know their business very well indeed and who have to display many companies' products and do so to maximise their sales. The result, though, can be a plethora of signs and discount tickets, offers and credit terms, with each manufacturer jostling for a good position. It was a relief in our Paris gallery to be able to have a sense of calmness with no signs and just one of each product on a pedestal. Admittedly this was a very expensive experiment, which could not be repeated elsewhere at that stage, although we longed to do so.

The Dyson Demo stores we have since opened around the world – including the large flagship store in the Opéra district of Paris – are descended from that first gallery shop in rue La Boétie. They are canvases on which to display and demonstrate our technology in exactly the way we want to. The first Apple Store, some years later, in SoHo, New York was exactly the same format, with a French limestone floor, mezzanines, stainless steel and glass bridges, and products on white plinths, with no writing.

I was keen to start our own company in Japan, although we couldn't do this until we were able to buy the licence back from Apex, which we did in 1997. At that point I flew on a Jumbo jet to Tokyo, on a trade mission with Tony Blair just after he became prime minister. I was looking for someone to run Dyson Japan. I used a recruitment agency, specifying that I wanted a young and inexperienced person to run the show, but the people they introduced me to were certainly not young. I tried another agency, with the same result. In those days, Japanese business was very, very traditional, and the older you were the wiser you were considered to be, and the more people listened to you. The agencies thought – on my behalf – that this was the type of person who should run Dyson.

In the end I said, 'The last thing I want is someone who knows what they are doing. I want a young Japanese person to run this business, because we're doing something different.' I knew these bright young

Japanese people existed, because I'd met some of them, but they couldn't find any of them to work for me. In the end I thought that a Scottish trade councillor from the British Embassy in Tokyo might be an unusual and intriguing enough bet to run the company and to win an audience with the big Japanese retailers. I took a chance and hired him. My hypothesis was correct. Dyson was finally established in Japan.

Japan has its own brilliant history of innovation. I loved my Honda 50, or Super Cub as it was properly known, and was inspired by the ingenious Sony Walkman. When in 2017 we sold our 100 millionth Dyson product, Honda announced sales of the Super Cub surpassing that figure and, in doing so, it had become the most-produced motor vehicle ever. In 2004, we took the DC12 cylinder vacuum cleaner to Japan, calling it the Dyson 'City'. It was engineered specifically for the tiny, perfectly formed homes of Japan. We were amazed by its success. Within three months it had captured 20 per cent of the Japanese market and went on to become the country's best-selling vacuum cleaner. As I had discovered for myself years before with the G-Force, the Japanese are truly fascinated by exquisitely crafted objects and new-technology products.

The DC12 was the 'bonsai' of vacuum cleaners, small enough to fit on a sheet of A4 paper and weighing just 1.5kg. Powered by the new Dyson Digital Motor and with a power-to-weight ratio five times greater than a contemporary Ferrari Formula 1 racing car, its performance was out of all proportion to its size. Given the size of so many Japanese homes, the DC12 all but specified itself, filling a niche in a market dominated by well-known Japanese electronics and household goods manufacturers like Sanyo, Sharp, Toshiba, Panasonic and Mitsubishi.

We decided on a grand launch in the piazza of Roppongi Hills, the vast urban complex constructed by Minoru Mori, the building tycoon, which had only just opened. I had tea with Mr Mori and Sir Terence Conran and received permission to build three pavilions for the occasion – a cube, a sphere and a triangle – made of DC12s. Pete Gammack and I, with Derek Phillips (who could make anything), also designed and built the world's first see-through velodrome for the occasion, with clear polycarbonate walls held together by tension wires. Everyone, at any level, could see the cyclists as they spun around a clear 'wall of death', demonstrating the very centrifugal forces by which dust is separated in the cyclone of our machines.

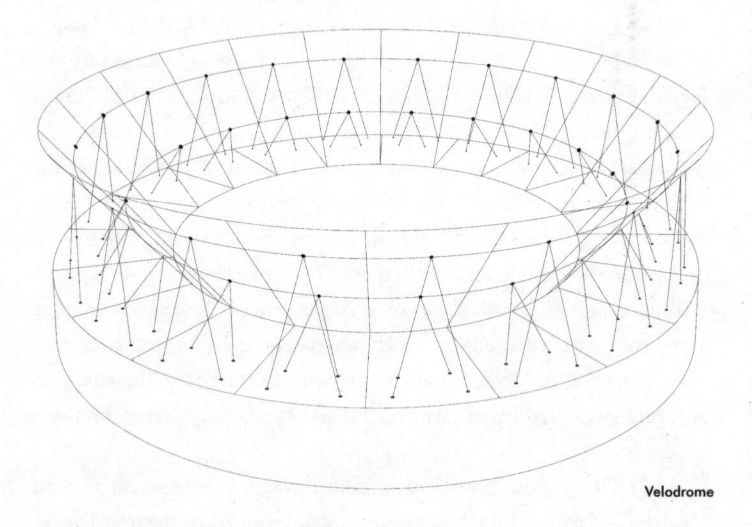

Velodrome

Issey Miyake, whose fashion show in Paris I would design a few years later, came along, as did the champion Japanese sumo wrestler, both fans of Dyson products. I don't know how much good launches do, but this one set us off on the right foot. We certainly got noticed. Retailers were quick to take on the DC12 and to make a success of it. Technology-savvy Japan has proved to be a fantastic market for Dyson and is very important to me personally, since it was the first to believe in my technology. As I have said, the Japanese company Apex licensed the pink G-Force back in 1985 and had sold it ever since, whereas other licensees fell by the wayside. Visits to Japan are some of my most keenly anticipated.

By 2002 we were ready for our next big adventure – the US. At the time the market in the States was dominated by big retailers like

(Above) Undulating roofscape of our first purpose-designed factory and now a research space at Malmesbury

Our see-through polycarbonate velodrome demonstrating centrifugal force at the launch of the Dyson DC12, Roppongi Hills, Tokyo, 2004

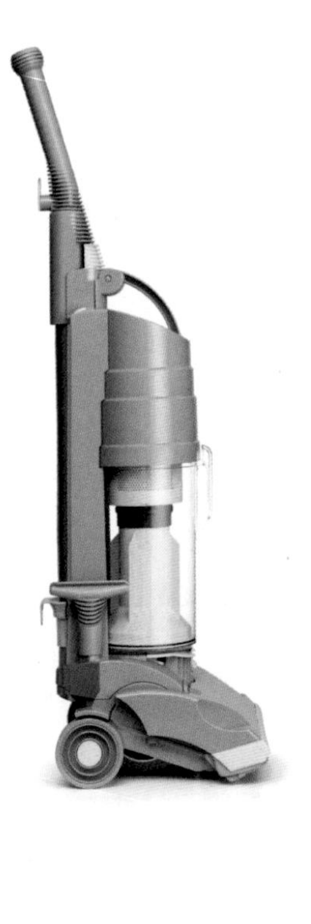

dyson
dyson

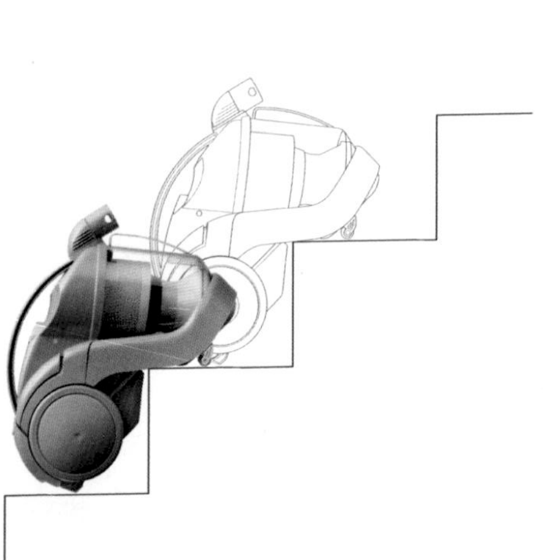

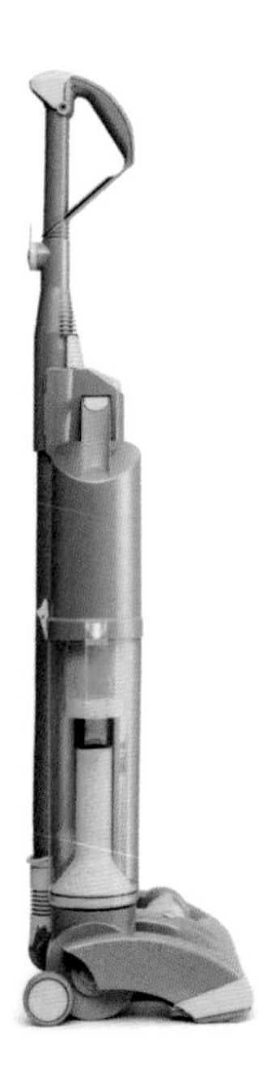

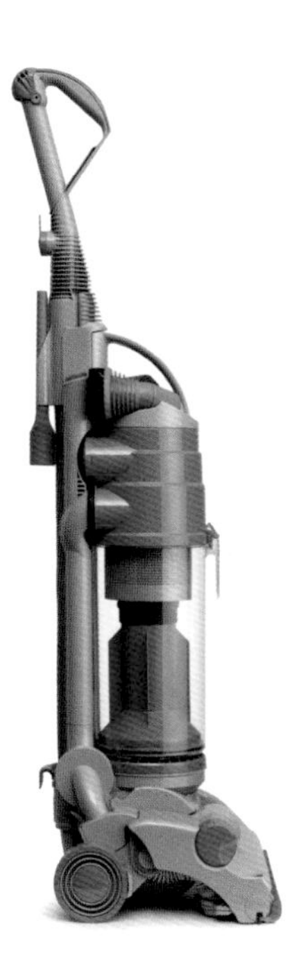

(Top row L-R)
DC01, Contrarotator, DC04, DC05
(Bottom row L-R)
DC02, DC03, DC06, DC07

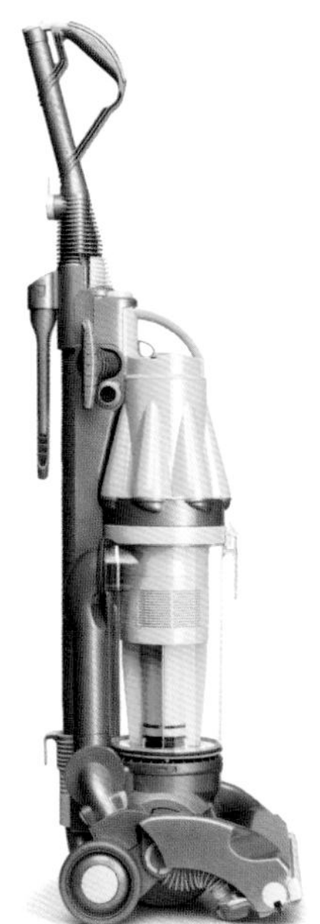

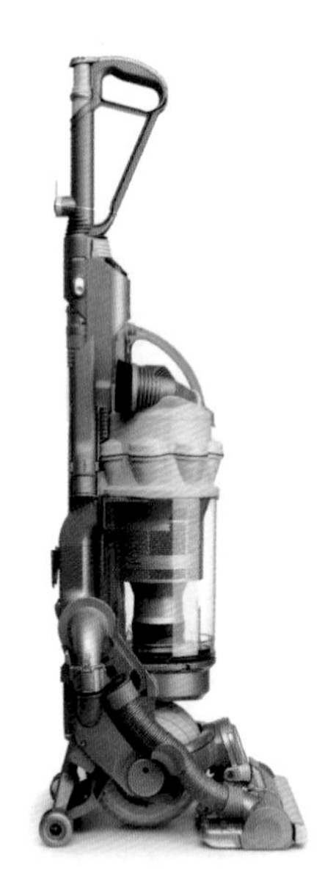

(Above, from top)
DC11, DC12, DC15
(Below)
DC16, DC35

The 'Dyson Symphony' in full swing at the Cadogan Hall, London in 2018 playing *Acoustical Anatomy*, an experimental meeting – with permission to fail – between the Orion Orchestra, Dyson engineers and machines like the Cyclophone made from Dyson components.
David Roche composed the piece, Toby Purser conducted.

'There goes James,' said Sir Terence Conran watching these 'ball' boats sail past the Design Museum at sunset.
They were part of the 2005 launch of the Dyson Ball vacuum cleaner at Oxo Tower Wharf, London, overlooking the River Thames and with views of St Paul's Cathedral.

Product image key: 1 Airblade db 2 Airblade Wash + Dry 3 Airmultiplier 4 Micro 1.5kg 5 Omniglide 6 DDM V9 7 CU Beam 8 Airwrap 9 360 Eye 10 Corrale

11 Dyson Pen 12 Dyson Hot + Cool 13 Supersonic 14 Pure Humidify + Cool 15 Pure Cool Tower 16 Airblade V 17 360 Heurist 18 Airblade 9kJ 19 Lightcycle Morph 20 Dyson EV

Demonstrating how to curl hair with a Dyson Airwrap at the product's launch in New York, 2017

Launch of the Dyson Supersonic hairdryer, with hair to match, on top of the Shibuya Hikarie skyscraper, Tokyo, 2016

Walmart, Target, Best Buy, Sears, Roebuck, Costco and Kohl's. These were national, not regional, retailers. Instead of starting gradually in a region, we had to cover the entire fifty states from the outset. Retailers have to be certain they can sell your product before taking it on, but we were an unknown company with a different-looking and expensive vacuum, and no track record.

We pointed out that we had been successful in the UK, but this cut no ice. To say the US is different is an understatement. No retailer would take us on – until, that is, a junior buyer at Best Buy took a Dyson home. This was the DC07, with multiple upside-down cyclones. The brave buyer went back to her bosses and said that the Dyson was everything we said it was. The bosses relented and agreed to put us in fifty stores. The trial was a success, and we were rolled out to most of their stores on the condition that we did national advertising on TV. This was no mean undertaking.

The US creatives we spoke to said, 'We think James should front the ad in person'. I was taken aback, recalling Victor Kiam of Remington fronting his ads with 'I liked the shaver so much I bought the company'. What they had in mind, thankfully, was something much gentler. The ad was directed by a wonderful British director, Nicholas Barker.

The ad was shot in one room at home in Wiltshire. I didn't say how brilliant my product was. I simply explained the technology. Deirdre and the children were horrified when I told them I would be fronting the advertisements. I found out why when *Saturday Night Live* parodied me with a fair-haired Englishman sitting on a lavatory following my script and explaining how he had to build a few thousand prototypes to get his lavatory invention to work. The thing, though, that really struck home with American consumers was that I said I'd made 5,127 prototypes of the vacuum cleaner. Americans like entrepreneurs and they particularly liked the fact that I had made and developed the vacuum cleaner myself. I wasn't a salesman–entrepreneur, I was an inventor–entrepreneur. We took off very quickly in America. Immigration officials in the US always say, 'You're the 5,000-prototype guy', so it must have worked. There is even a *Friends* episode featuring Monica's obsession with the Dyson vacuum.

Following our first success, Nicholas Barker shot many advertisements with me. I am not sure that personal appearances worked so well in the UK, although we did try. They certainly didn't work in Europe, where I think they wanted a product from a large and well-established company fronted by an executive in a laboratory dressed in a white coat. We did, though, try it quite successfully in Japan. The general idea was to show that Dyson was founded by Mr Dyson and that he was responsible for Dyson products. Big and long-established multinationals, most of them public companies, would not be able to put forward an individual in the same way. I believe that trustworthiness and loyalty come from striving to develop and make high-performing products and then looking after customers who have bought them. I am not a believer in the theory that great marketing campaigns can replace great products. What you say should be true to who you are.

Selling in a market as large as the US, however, required a significant increase in manufacturing capacity. In 2000, we made 800,000 vacuum cleaners and washing machines, but in doing so we had completely run out of assembly space to expand into and manufacturing staff at Malmesbury, where unemployment was very low and sometimes even zero. Desperately short of local labour, we had to bus people in from far afield. Even more importantly, all our suppliers of plastics, components and electronics were in South East Asia, while along with Australia and Japan, our fastest growing markets were there, too. Our overheads, meanwhile, had ballooned, so both logistically and practically we had to move production.

When, in 2002, I first announced the move of our vacuum cleaner production from England to South East Asia, I found myself in a quandary. While campaigning for more support for British manufacturing companies, we had been refused planning permission to expand our factory, but we needed to expand quickly. The solution was apparent. There were empty factories in Malaysia as electronics companies abandoned the country for cheaper production in China. Local workforces in both Malaysia and Singapore were highly skilled, enterprising and hardworking. We effectively had no choice, while the advantages were too great to ignore. The decision was a sad one for the 500 employees who lost their jobs with us, but it was the right one for the long-term future of the company. Dyson was trapped, with neither room to expand nor a sufficiently large supply of labour. Our suppliers were thousands of miles away.

Ever since, Dyson has been able to expand, both in the UK and around the world, on a scale that would have been impossible if we hadn't made the leap to South East Asia. We now employ 4,000 people in the UK, far more than before the move, mostly engineers on two large campuses in Malmesbury and Hullavington.

No one could have tried harder than I did to make things work in Britain. One logical move had been to try to buy land around us and to put up new buildings. But when we floated a design by Chris Wilkinson for a new factory sunk into the ground and invisible from the surrounding countryside, it was attacked by local interests, including the local Conservative MP. For us, it proved to be a case of checkmate. If Britain wants to retain manufacturing, it must begin reforming planning laws. A manufacturer will only commit capital to build an expensive new factory when he can see future sales, but then he needs it fast. He cannot wait the four years or more that our system takes to find out whether he has permission to build and wait a further two years for construction. He will go instead to a country where this happens in four months, as indeed it does with our factories in Asia.

I couldn't help thinking that although we were creating wealth in modern factories and laboratories, and even though Chris's design proposals were highly refined, factories and manufacturers in Britain were still looked down on as being somehow rough and grubby. The reaction to our proposal was certainly a very British case of 'not in my backyard'.

The attitude towards manufacturing, trade and industry could hardly be more different in Singapore, a city-state with three big sovereign funds investing in science and technology, and a culture in which this is celebrated. Singapore is an exciting place, a hub and heart of world trade. Each year, some 100,000 ships sail through the Strait of Malacca, the narrow 500-mile stretch of water between the Malay Peninsula and Sumatra, connecting the Indian and Pacific Oceans. This commercial armada carries a quarter of the world's goods, a fraction that will continue to increase as the energetic economies of Asia grow.

Independent since 1965, Singapore rose rapidly, initially under the premiership of Lee Kuan Yew, a dynamic Cambridge-trained lawyer, to become a highly successful multi-racial meritocracy, the very model in many ways of a free-trade economy at work. If we had gone ahead with the electric car, it would have been built in a brand-new factory of our own in Singapore. Dyson's presence in Asia has built up significantly since we started in Singapore, to the extent that we now have our global headquarters there along with campuses and research laboratories in Singapore, Malaysia and the Philippines.

Singapore was founded as a free port settlement in 1819 on behalf of the East India Company by the energetic globetrotter and former governor of Java, Sir Stamford Raffles. The population of the island was about a thousand at the time, but with its strategic location, deep natural harbour, reliable fresh water supply, readily available timber for the repair of ships and Raffles' enlightened leadership, it proved to be a good place to trade and live. Within five years, Singapore's population had risen tenfold. Within fifty it was 100,000.

The British, of course, went on to make a hash of Singapore when, in 1942, and with many more troops than the enemy, they failed to stop a brutal Japanese invasion. My father was one of the British soldiers who fought the Japanese back through Burma while Singapore, as valuable as ever, remained in enemy hands until September 1945, some weeks after atomic bombs were dropped on Hiroshima and Nagasaki. I don't think either Raffles, for all his flair, or even the industrious Japanese could have imagined just how Singapore would flower in the second half of the twentieth century.

Vacant factories aside, we met people in Singapore who were ambitious, wanted to make things, and wanted to grow their businesses. Their entrepreneurial spirit was second to none. By moving manufacturing to South East Asia, we would have the advantage of having all of our suppliers around us, local to us, which is important from a logistics point of view. It is also important from the point of view of improving quality and driving innovation. You can manufacture good-quality, pioneering technology much more readily when you sit side by side with your suppliers rather than 10,000 miles away in a different time zone.

We quickly found very co-operative suppliers who went out of their way to help us. And, despite what local and national press and media said back home, we were not going to South East Asia for cheap labour. If we had wanted that, we would have joined the rush that was under way back then to China. We were pushed at home by the inflexibility of the labour market, the twenty-one-year commitment to factory leases, problems with planning and the sheer time it took to build a new factory that was making life in England problematic. We had only one supplier left in the UK, for hoses, whereas all our other suppliers were in Asia. When our volume reached a certain level, the UK hose manufacturer did not want to make any more. He would have to have taken on more labour and a new factory, neither of which he was willing to do. I do not entirely blame him, as the risks he would run by committing to twenty-one-year factory leases and not being able to flex his labour force were too great. This prompted us to find an alternative supplier in Malaysia.

What we needed was highly skilled labour, and Singapore excels in this regard. Between here and Malaysia we were able to make things very quickly and to a very high quality, and we could employ brilliant, locally trained engineers. Since engineering is embraced and celebrated in the region, it was really exciting to go there to manufacture and to find ourselves working in real partnerships with local people who had passion for what they did. We also came across politicians who themselves were often industrialists with a knowledge of manufacturing, and thus far more sympathetic to what we were trying to do than were the vast majority of their distant cousins in Westminster, who have no experience of industry and little or no interest in making things.

By going to South East Asia we also learned a huge amount about what Asian consumers wanted. Asia is the fastest-growing world market and half our sales already come from there. Europe is also a fast-growing market for Dyson – very important and very enjoyable – but we must take an equitable view of world markets. Asia is growing at three times the rate of Western economies, and whereas the EU share of world trade is declining from 15 to 9 per cent, Asia's share is rising from 16 to 25 per cent. Furthermore, it would be arrogant to think that we can design for Asian consumers from Malmesbury. All countries have different attitudes and needs, so we needed engineers in Japan, Korea, China, Singapore and Malaysia developing products for those different markets.

In fact, since we have been in Singapore and Malaysia, we have balanced our research on products for the Asian market with those of other markets, feeding this back to Britain and so to the European market, too, because, aside from their appreciation of well-made products, people in this part of the world are fanatical about cleaning. Our manufacturing expansion was therefore in parallel with the rapid commercial growth we achieved as we began selling Dyson in Taiwan, the Philippines, Indonesia, Malaysia and South Korea. South Korea has become one of our biggest and fastest-growing markets.

In Korea, many people vacuum their homes twice a day, so they want small, light products that are readily to hand and easy to use. Along with cleanliness, there's also an extraordinary culture of beauty in Korea, influencing consumers elsewhere in Asia. This was something we didn't understand at first, yet when we launched the Supersonic hair dryer in Seoul, sales went through the roof. We followed up in 2019 with a Korean research lab dedicated to personal care products.

Here our engineering and technology teams meet and work with our customers face to face. Our Scanning Electron Microscope and Hair Mapping Analysis evaluate the condition of hair and, while making recommendations on treatments, the data contributes to future product development. This might seem odd to anyone not used to the idea of engineers and technologists at work in a beauty laboratory, but this direct human insight leads to entirely new ideas. We have been inventing new-technology products like the Dyson Corrale hair straightener as a result.

I admit that I was nervous about entering China. We had experienced much plagiarism there and I was unsure if it would be a good market for our designs. I shouldn't have been. The Chinese love new technology and the latest designs, while Chinese authorities are taking IP protection increasingly seriously and driving progress in ways other countries could learn from. The criminal prosecutions that have been handed out there in relation to counterfeits of Dyson products are some of the most significant seen in the country. In one raid in 2020, thirty-five counterfeiters were arrested and 277 hairdryers, along with accessories, packaging material and other items, were seized. The Shanghai courts subsequently delivered guilty verdicts, prison terms and fines for the four principal offenders and thirty-one accomplices.

We set up office in Shanghai and I did our launch there in an old steel factory. I like the contrast of disused industrial buildings with our new technology and design. We partnered with a delightful company called Jebsen, supplying the large department stores. Gradually we set up many of our own shops and started selling direct through the phenomena of WeChat, Alibaba and JD.com. We built up an engineering team there, too. China is one of our biggest markets as well as one of the most exciting and demanding.

As with other people in South East Asia, the Chinese are particularly keen on air purifiers. This, interestingly, is not, as many people would assume, because the pollution levels are necessarily higher than anywhere else in the world. Rather, it is because they are educated to be acutely aware of the dangers. They are much better informed about the dangers of indoor air pollution, with the effects of formaldehyde and other VOCs registering high in their consciousness. Quite right, too. Other countries have yet to catch up. When we developed air purifiers, these had to perform to a very high standard indeed to meet local expectations and we make the same specification for the rest of the world.

Where British companies used to create products for export markets dreamed up and shaped at home, today we are living and working in those markets, viscerally aware of what people might want and working in direct collaboration with them in terms of research, development and production. For example, we developed a Dyson Digital Slim vacuum cleaner that weighs less than 1kg and the Dyson Omni-glide vacuum cleaner that has a double Fluffy brush bar that can clean in all directions especially for Japan, China and South Korea. Launched in Asia first, they will now be supplied globally. As I have explained, responding to criticism we devised a carbon-fibre brush bar to break the static that makes dust stick to the floor. Customers in Korea and China, who often have highly polished floors, want to know when they have removed all the dust, and for them we have developed a laser that spotlights every speck of dust combined with a particle sizer and counter that shows just how many dust particles at each size are entering the vacuum.

Being in Singapore certainly gives you a very different take on the world. If you place Singapore at the centre of an atlas, you immediately understand how well it connects not just to China, Korea and Japan, or to Indonesia and the Philippines, but also to India, Australia and New Zealand. And to countries like Vietnam and Bangladesh, which are growing and developing rapidly.

On Singapore's doorstep is, in fact, the fastest-growing market area in the world, and countries with growing affluence and an appetite, as it happens, for Dyson products. The more time we spent in Singapore and Malaysia and then the Philippines, the more we became an Asian company that just happened to be owned by British people. While I'm proud to be English, started Dyson in rural England and continue to employ thousands of people there, I have never thought of trading on flag-waving. We sell our products on objective merit, on what they do, how they work and how energy-efficient they are. Dyson products sell purely on the basis of our latest technology, performance, design and quality. Our British research and development operations are hugely important, yet, as we expand in Asia, we will continue to invest heavily in research and development there, too. In any case, I think there is something exciting about bringing research based in Britain and research from Asia together. The more open the world is to ideas and trade, the more innovation there will be and potentially to everyone's benefit.

The move of manufacturing to Asia meant we could continue to grow our R&D in Malmesbury. Where we had formerly assembled vacuum cleaners and washing machines, we now built semi-anechoic chambers, an EMC (electro-magnetic compatibility) chamber, dust labs, microbiology labs, electric motor development labs, new-technology battery labs and small-scale production. The speed, flexibility and innate drive of those we work with in Singapore, Malaysia and the Philippines, meanwhile, means that we have been able to grow without the restrictions placed on us at home. Our product development process is now truly a twenty-four-hours a day process, and we can design and build a new factory or research lab in four months from start to finish. Standards are high, factories are often built on reclaimed land and Singapore is justly proud of its 'garden city' heritage.

At the outset we needed a few Dyson expats in Singapore and Malaysia as we built up our research and engineering base. All of our leadership team is based in Singapore, including the CEO, and nowadays we employ thousands of exceptional local people who know we're in it for the long term. We're there, and happily so.

We are growing so fast every year in Asia that it would be impossible for us to keep building factories large enough. Instead, we build close relationships with owners of factories so that we can build our machines in their premises. The tooling, assembly lines and test stations are ours and we control the purchasing and quality. We don't approach a sub-contractor and say, 'Make me a product of this or this design'. We tend to go to outfits which have never made vacuum cleaners before or hairdryers, robots, fans and heaters or purifiers or lights, and we teach their people to make these things using our production methods.

It's a heavily engaged and involved process of learning and improvement.

We need other factories because, expanding at the rate of 25 per cent each year, we simply couldn't cope with the planning and building of new factories even in Singapore, Malaysia and the Philippines. And we need two or three new factories every year. Today, we have a staff of 15,000 and a further 100,000 people working with us. We do, though, have our own dedicated Dyson-owned factories making digital electric motors, heaters and batteries. These are our core technologies which we have pioneered and even their manufacture is a major research and development project. By undertaking this manufacturing ourselves, further development and breakthroughs continue apace. They represent years of research and development. They are the inventions and technology that have enabled us to design and manufacture products that give us a quantum leap ahead of our competitors.

As we recruited ever more local graduate engineers, so the teams integrated and developed their own impetus. In 2009, we moved R&D in Singapore to a larger site at the city's Alexandra Technopark. This creative hub was close to the city's Central Business District and, by British standards, amazingly well equipped and serviced. A part of what is so special about Singapore is that everything, in terms of research, engineering, manufacturing and commerce, connects within very short distances, while, of course, the city is connected physically to a huge and growing regional market and, thanks to excellent communications, to the rest of world. The Singapore government understands the operational needs of companies developing and manufacturing technology. It has several sovereign wealth funds specifically there to invest in new-technology ideas and ventures. Intelligence from these wealth funds helps governments form their policies. Although a small island state, it has become the world's second-largest exporter of technology.

In 2012, we opened our first advanced and automated manufacturing facility in Singapore, known as SAM – Singapore Advance Manufacturing. Three hundred autonomous robots working on automated lines produced a Dyson Digital Motor every 2.6 seconds. We now have fourteen production lines spread across Singapore and the Philippines – the latter known as PAM. It's an amazing sight to watch the robots at work. They run twenty-four hours a day, seven days a week, with just an hour off for weekly maintenance. They glue the motors together with micro-dots of adhesive, sucking back the tiniest drop of excess glue. They spin the copper windings at amazing speeds and perform the most delicate of shaft-balancing procedures. They test the motors as they make them. There are no human operators here.

Our engineers, however, are always looking at how to improve the process. These lines are entirely bespoke and packed full of intellectual property, which we have to protect. They take over a year to build and are very expensive indeed, adding considerably to the cost of each motor. They represent the current state of the art and are the only way to achieve the precision that we need with quality and reliability. However, we are endlessly dissatisfied and iteratively improve the process version by version. We have now passed the 100 millionth Dyson Digital Motor manufactured on Dyson production lines at the Philippines Advanced Manufacturing facility.

Meanwhile, just as with every space we have occupied, it didn't take long before we outgrew our space at Alexandra Technopark. In 2017, a year after we launched the Dyson Supersonic hair dryer, we opened the Singapore Technology Centre at 2 Science Park Drive. Located in the heart of Singapore's start-up community and next to the National University of Singapore, there the centre's engineering teams are responsible for co-developing technologies for Dyson's growing personal care category. As in England, we work closely with a growing number of universities in Asia.

In January 2019, we announced our decision to establish Dyson's global headquarters in Singapore, reflecting the fact that Asia is the centre of our operations and that this had been the case for many years. We'd learned the importance of being close to and understanding our key markets, especially when it came to developing products that would suit Asian customers. This was the hub of our manufacturing and global trade and our fastest-growing region – it was also where our CEO and our executive team were already based. We had 1,200 employees in Singapore and the same number in Malaysia. Establishing the global headquarters in Singapore simply reflected these long-term commercial developments. But no matter how brilliant I think Singapore is, research will also continue in Malmesbury and at Hullavington.

We made it clear to the press at the time that, even as we were planning this, Dyson was also investing £200 million in new buildings and testing facilities at Hullavington, £44 million in renewing our Malmesbury offices and upgrading laboratories there and £31 million in the Dyson University to train the new generations of engineers, of all types, which Britain as well as Dyson need so very much. While in Singapore it has been relatively easy to find bright new graduate engineers as well as highly skilled workers, the same is definitely not true in Britain.

It was inevitable, I suppose, that I should be attacked in some quarters of the British media, who liked to suggest I was moving Dyson's head-quarters purely to lower outgoings, including tax. This just wasn't true. Both personally and corporately, and according to the *Sunday Times*, I am one of Britain's largest taxpayers, but the company also pays tax all over the world. Establishing the headquarters of Dyson in Singapore did not change where and how the revenue and profits of the company are taxed, and we still pay more tax in Britain than in any other country.

Cost was also not a factor; in fact, Singapore is one of the most expensive countries in the world. Besides, we continue to invest heavily in Dyson's operations in the UK, as well as our educational initiatives, both through loyalty and because I do believe Britain produces, while not nearly enough of them, some of the most inventive and enquiring minds anywhere in the world. With the addition of Hullavington and the Dyson Institute, our Wiltshire campus has grown significantly. It will continue to do so.

For me, a headquarters building should be a powerhouse of invention, experimentation, creativity, making and proving. The time had come to have a suitably inspiring space in Singapore, given its importance to the future of Dyson. That November, we agreed plans to turn Singapore's St James Power Station, an imposing building facing Keppel Harbour and the resort island of Sentosa, into our new headquarters. St James Power Station was very much an adventure when first built in the 1920s, as it will be again when its research and engineering spaces open.

It would have undoubtedly been much simpler to have moved into a brand new glass box – smart but soulless. But buildings have always been important to me and I believe that offices should inspire people. The coach house in Bath was, I suppose, my equivalent of a San Francisco garage. In Chicago our first offices were in the historic Montgomery Ward building, a former warehouse for America's oldest mail-order catalogue business and the largest reinforced concrete building in the world. In Sydney, Dyson people worked for a long time in a canal-side warehouse that was originally used to store wool for export during the Second World War. In Shanghai we started out in the French Concession in a space that became a car factory after the Communist Revolution. In Canada, Dyson people work in the aptly named 'Manufacturer's Building', designed by Toronto architects Baldwin and Greene. Hullavington airfield is also an inspiring canvas where, hidden beneath years of neglect, decay and dereliction, are stories that make this space an inspiration for all of us.

The St James Power Station, a steel-framed colossus sheathed in a garb of Beaux-Arts classical architecture, all red brick with high Roman windows, rises from what was swampland less than a century ago. It was designed, in collaboration with the engineering firm Preece, Cardew and Rider, by Alexander Gordon, a young Scot who came to Singapore in 1923 and was soon promoted Assistant Municipal Architect. For the St James project, Gordon shipped in a prefabricated steel frame from England, raising this on substantial concrete foundations and then wrapping the frame around with grand and civic architecture. Inside, the lofty spaces of the well-lit building are on the scale of a cathedral. It seemed exactly the right base for us in Singapore.

Opened in 1927, the coal-fired power station was decommissioned in 1976. At first it became an automated warehouse for the Port of Singapore, and then, from 2006, a nightclub complex run by Dennis Foo, known as Singapore's 'Nightlife King'. On a typical weekend, between 15,000 and 20,000 clubbers raved the nights away here. Again, the building was a powerhouse, with new Thai bands making their names here alongside clubs dedicated to disco. During Foo's tenure, St James was listed as a National Monument.

We took up the expired lease for the building in 2019 and, with Maple Tree and Aedas architects, worked up a scheme showing how we could shape building-like structures within the immense spaces of the building, animating these without affecting the existing fabric. For half a century – the occasional power failure aside – St James worked around the clock to provide Singapore with electricity. It then worked twenty-four hours a day for 'The Port that Never Sleeps', and all through the night for Dennis Foo's eleven clubs under one roof. Now, it will be a headquarters open all hours for Dyson engineers and scientists to focus on power electronics, energy storage, sensors, vision systems, embedded software, robotics, AI, machine learning and connected devices.

Construction work on the power station slowed during the 2020

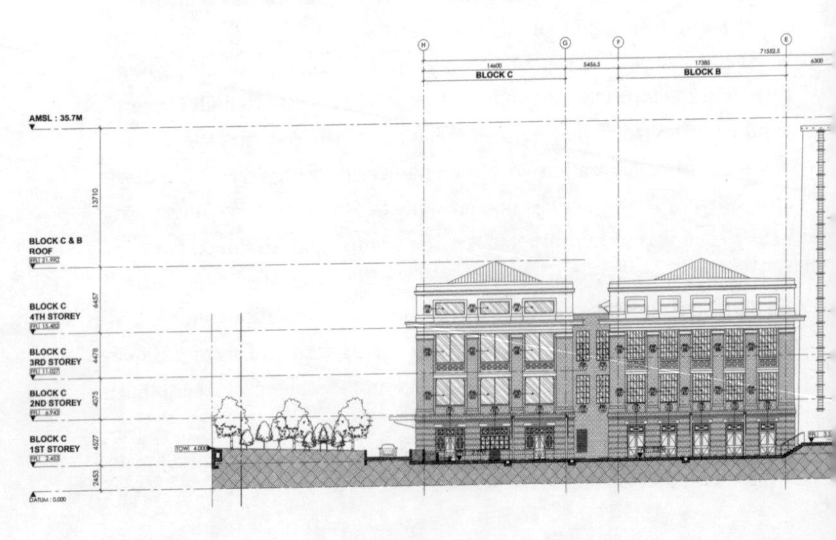

Covid-19 lockdowns, delaying our move. Our factories had to stop work, too, and our stock around the world ran down rapidly. We first saw the effects of lockdown in China as early as late January 2020 and could see the recession looming large on the horizon. We knew what we had to do: change everything.

Fortunately, over the past three years we had been striving to sell more products direct to our customers ourselves, either online or through Dyson Demo stores. By early 2021 we had 356 Dyson stores. We have been opening them around the world so that customers can try our Dyson products in the best possible way. There are two reasons for this. First, we like to have a direct relationship with our customers, who are buying our product for which we are responsible, and we want to know how we can help them.

Secondly, retailers round the world are declining in numbers and sales. They are nothing like the force they were, due of course to the decline of the high street and the rise of internet shopping. If you want to buy from a website, why not buy from the Dyson website! Why not deal directly with the manufacturer? Since we are now selling direct and are not dependent on the retail buyers, we can offer special products.

St James Power Station Architect's drawing, Maple Tree

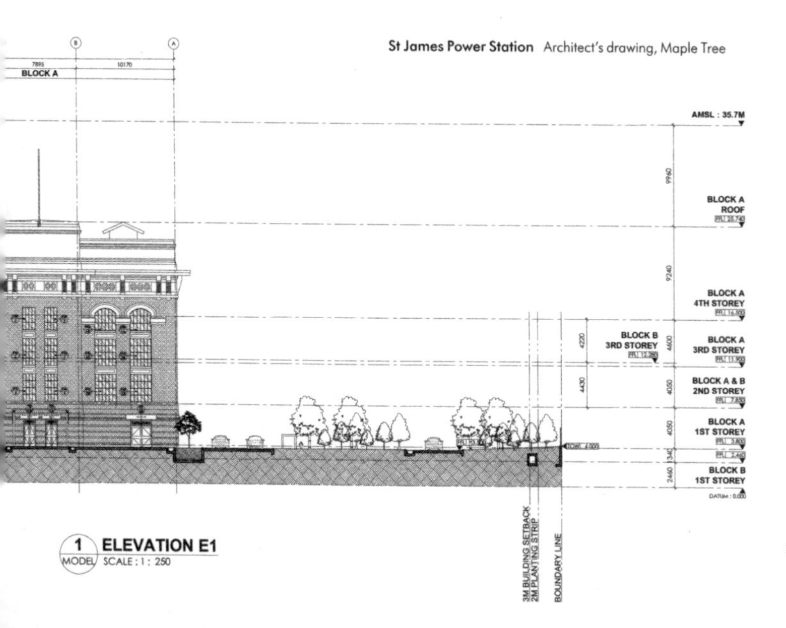

Online we can do one-on-one demonstrations. In a Dyson Demo store you can see special demonstrations, buy special versions and even have your hair styled for free. In early February 2020, therefore, when shops around the world were closing as a result of Covid-19, we accelerated our direct sales on our websites and continued to open Dyson Demo stores.

Early on with the Sea Truck I found it fascinating to explain and discuss products, their technology and their performance with potential customers, and, indeed, customers who were already using them. That is why I enjoy doing product launches so much, as does Jake. It is important to my family, as the owners, that we stand up to explain Dyson's inventions and why the technology will make a difference. We have a passion for developing new technology and bringing out different products, which we would like to share with each market. And I like to hear reactions first-hand. I am very lucky that I have excellent people to run the commercial side of Dyson, leaving me fairly free to concentrate on engineering.

I enjoyed the experience, too, of launching Dyson products in India, an Asian market we've entered only recently. Since 2017, we've sold Dyson products in India including the Dyson V11 cordless hand-held vacuum cleaners, with a run time of sixty minutes, our hairdryers and stylers, our cooling fans and purifiers, and the Dyson Lightcycle task lamp. The last of these, designed by Jake, is a revolution in lighting, providing light that adjusts automatically – colour, temperature and brightness – to suit local conditions and personal preferences. The ingenious cooling of its LED lamps will ensure these have a very long life. It took ninety engineers two years and 892 prototypes to get it as Jake wanted it. India is a country of engineers, so there is great appreciation of highly engineered products.

There's a whole new market in India of young professionals living in city apartments and without traditional domestic help. But while the market is sophisticated, living conditions in Indian cities are blighted as much, or even more, by pollution as they are by high humidity and blazing heat. We had the right products, but because there were no department stores or the equivalent of Carrefour or Costco to stock them, we went straight for direct sales from our website and our own Dyson shops.

I had been prepared to an extent for conditions in India. I visited the country in 2016 as part of a trade mission with Theresa May, the British prime minister at the time. To keep my Delhi hotel room cool and free of acrid smog, I had five air purifiers on the go at the same time. Mrs May sent one of her staff along the corridor to ask if she could borrow one for her room. I think we knew we had a potential winner with the air purifier. And, being located in Asia, we are constantly learning about other cultures and engaging with them. It's exciting. It keeps us on our toes.

Dyson has become as much an Asian business as a British one: our products are sold in eighty-three countries around the world, so we are arguably a truly global company. Having started in Britain and consistently grown in Britain, we, for some time now, sell over 95 per cent of our products in our other global markets. We will, I hope, continue to expand in countries like Indonesia, Vietnam and elsewhere. As we enjoy these markets, though, it will be with a sense of adventure, excitement and openness as we engage and learn from other cultures.

This global perspective encouraged me to back Britain's departure from the European Union. I do believe that Britain needs to be free to operate competitively around the world, a belief rooted in common sense and personal experience. In 2016, Dyson exported 19 per cent of its output to Europe and 81 per cent to the rest of the world. The fastest expanding markets lay overseas and even across 'uncharted waters'.

The EU's free movement of people, meanwhile, has been unable to bring Dyson the engineers it needs in Britain. We were not allowed to employ them unless they were from the EU. If we wanted to hire a foreign engineer, it took four and a half months, if we were lucky, to go through Home Office procedures. Hopefully, all this will now change, and we can recruit globally.

This local difficulty is compounded by the fact that 60 per cent of engineering postgraduates and 60 per cent of those engaged in engineering research in British universities are from outside the EU. After their studies, what does Britain do for or with them? We chuck them out, that's what. Why on earth would you send engineering graduates, researchers and doctorates packing when they have valuable technology at their fingertips that they take back to countries like China and use to compete against us?

I believe that Britain is not compatible with EU institutions; we do not understand their lobbying and their workings. The EU countries do not like our interference and our expressing different views – they are used to getting their own way. This is most obvious in their drive for ever greater integration and breakdown of each country's powers, except that they have a way of finding their way around any agreement or undertaking that they find awkward. While we value our legal independent sovereignty, they see the greater good of further integration if they can each avoid complying when it matters to them. In short, it is an incompatible marriage.

Dyson has been sitting on the International Electrotechnical Commission, the IEC (the industry body, a committee of electrical appliance manufacturers that meets to agree on standards and standard tests) for the past twenty-five years, so we have experience of the shenanigans and goings-on. For example, I have been fighting a legal battle with the EU itself since 2014 over their adoption of energy-label performance data that breaches their own EU laws and unfairly benefits one group (in itself illegal), in this case the German vacuum cleaner industry body. I'm afraid this requires a bit of explanation.

For a long time – indeed, I believe we were the first manufacturer to do so – we had supported a very sensible regulation that proposed placing a cap on the energy wattage of vacuum cleaners sold in Europe. This, it seemed to us, was a good way of incentivising manufacturers to make more efficient machines. It would compel companies to invest in research and development to use each joule of energy increasingly more efficiently. Other manufacturers – our competitors – did not like this approach. I can only assume this was because they did not have motors that were as efficient as ours. They therefore lobbied for an energy label that was based on a complicated matrix of lab-based tests, none of which included dust. The result was a regulation that stipulated that the energy performance of vacuum cleaners should be tested for efficiency when the machines were empty and unused. I don't need to tell you that this is nothing like the real world.

Dyson machines are engineered to have constant performance and we test them in real-life conditions using many different varieties of dust and other detritus. This is manifestly not the case with other products, particularly bagged machines where the energy performance drops as the bag fills with dust and the pores clog. Once the regulations came into effect, many vacuum cleaners sold around Europe quickly started sporting a green A-grade label for energy consumption and efficiency – as conferred by the European Commission test criteria. However, when you got them home and started using them, their bags and filters started clogging up with dust. Our subsequent testing showed that the efficiency of some of these would fall as low as a G-grade while 'in use'. In short, the performance you were promised on the label – and based on which you purchased the machine – was not what you got at home. It was conning the consumer and putting us at a commercial disadvantage.

Some manufacturers went even further and engineered their way around the cap on motor wattage, which came into effect at the same time. They used electronics to ensure their machines used a low motor wattage when empty and in a test state, but ramped up just as soon as the product was used in real life – thus appearing more efficient than they were. We were forced to bring a judicial review against the European Commission and multiple cases against our competitors.

The regulation, we argued, was unfairly discriminating against Dyson. We pointed out to the court that the performance of bag machines dropped off and continued to fall the moment the bags started to fill with dust. Thus the 'real use' or in-home performance was emphatically worse and not as stated on their energy labels. So the consumer was being grossly misled. Our plight revealed the complexity of European justice. Our case was initially rejected by the European General Court, in November 2015; the court accepted the Commission's rather odd argument that dust-loaded testing was not 'reliable or repeatable'. This was despite the fact that the International Electrotechnical Commission had long-devised just such a dust-loaded test, widely adopted by consumer test bodies and manufacturers worldwide.

We appealed and in May 2017 the European Court of Justice found that 'the General Court manifestly distorted the position taken by Dyson'. It 'failed to take into account certain items of evidence', 'infringed its obligation to state reasons', 'erred in law' and 'distorted the facts and failed to comply with its duty to give reasons'. Coming from a court, this was pretty strong stuff. What's more, in its ruling the European Court of Justice also directed that the energy test must adopt, where technically possible, 'a method of calculation which makes it possible to measure the energy performance of vacuum cleaners in conditions as close as possible to actual conditions in use'.

Eventually, on appeal, the General Court agreed with Dyson and annulled the regulation. To begin with, and despite having a strong case, the odds had been stacked against us. I am told that overturning an EU regulation has only occurred on a handful of occasions.

We stuck at it to defend our product and in the process shone a light on the murky practices that some companies will resort to and the dark arts of lobbying in Europe. As part of our necessary investigations, we were required to use Freedom of Information requests to obtain correspondence between German manufacturers and a commissioner in which they were lobbying the EU to further their interests at the expense of their rivals. Meanwhile, representations by European consumer groups were ignored.

The case cost Dyson hundreds of millions, not to mention the time and distraction when we would have far preferred to get on with developing new products. Now we are in another protracted battle for damages. Yes, we brought the legal action to protect our own position from discriminatory regulation, but fundamentally this was about protecting consumers from misleading regulations devised by EU bureaucrats and the European manufacturers who lobby them.

For me this was the tip of the iceberg. What was the point of the UK being in the EU if the EU was undermining UK manufacturers? Wouldn't UK manufacturers and consumers be better off making our own laws as a sovereign state, with free-trade deals around the world? The dynamic and open-handed Singapore model would surely be more in tune with the British character.

Much has been written about the infamous '*Non*' pugnaciously declared by General de Gaulle in the 1960s when Britain wanted to join the EEC (later the EU). It seemed unfriendly and ungracious at the time, yet he was just being honest. He understood the British, having spent 1940–1945 in London. The UK has a long history of global trading due to the Commonwealth and our pioneering spirit. This is a fortunate heritage and one based on free trade. The EU single market, however, is designed by Germany and France to be a single, closed market, protecting EU manufacturers and companies from foreign imports by imposing steep import tariffs. These are, for example, 30 per cent for roasted coffee and farm produce, 10 per cent on cars and 6 per cent on vacuum cleaners and hairdryers. This import duty, even on goods imported to the UK, goes straight to Brussels. Our entry into the EU in 1972 was a slap in the face for our former free-trading partners such as Australia, New Zealand, Singapore, India, Canada and the USA.

Since we became members, our trade balance with the EU has declined while our imports from the EU have rocketed. In other words, it was a good deal for the EU and a bad deal for the UK. Meanwhile, our exports to countries outside the EU have been rising during our membership. This trend in itself makes de Gaulle's point.

There are those in the UK who believe that by leaving the EU we are inflicting an act of self-harm. I disagree. Ever since we moved our production twenty years ago to South East Asia, Dyson has been importing its products into the EU, which includes the UK, from outside the EU and so paying the 6 per cent tariff. This has not been an act of self-harm for us; in fact, Dyson's sales in Europe have flourished during those twenty years. Indeed, currency fluctuations inflict far more damage than the paying of import duties.

As I write, Britain has already signed up sixty-three free-trade agreements around the world, putting the UK in a much better world trading position than the EU member states. In short, we have the opportunity to flourish. Furthermore, the EU would have been inflicting upon themselves an act of immense harm if they had not signed a free-trade agreement with the UK, putting its surplus balance of imports into the UK, of over £100 billion, at risk. The fact that, at the eleventh hour, the EU has agreed a free-trade deal that protects its trade with Britain is a good harbinger. The EU needs this deal, and on this basis it will flourish. I have a feeling that our more-than-arm's-length relationship, where we have to work things out diplomatically rather than being trapped inside, will lead to a new spirit of co-operation. I hope so. Like so many British people, I have always loved travelling through Europe and having businesses in European countries, not to mention the home I have had in France for many years.

CHAPTER NINE

The Car

I've always been horrified, even as a child, by the clouds of black smoke emerging from the back of vehicles, especially diesels. As a pedestrian, a cyclist or in a car following directly in the slipstream of a diesel vehicle, you breathe in a huge volume of this filth. In Britain alone, 34,000 people a year die from inhaling exhaust fumes. Exhaust is carcinogenic, affecting the lungs in particular. My father died prematurely from lung and throat cancer and, perhaps partly because of this, exhausts expelling dangerous particulates continue to repulse me.

I happened to make a trip to London at the time of the Great Smog of December 1952. Coal smoke was the villain of the piece then, but I remember not being able to see beyond my hand and, later, learning that at least 4,000 people died from lung conditions and something like 60,000 died from heart attacks after particulates had entered their bloodstream. The government hid the true statistics, and these remain hard to uncover. In recent years, poisonous smogs may be a memory in countries like Britain, largely because petrol and diesel engines have become less smoky. This is partly because particulates have been made smaller, yet while exhausts today may appear to be cleaner to the naked eye than they have been in the past, they are not free of dangerous gases or the same but very small particulates. Traditional automotive manufacturers and governments have long continued to ignore fundamental problems with internal combustion engines and diesel exhaust particulates.

Early in the development of my cyclonic technology, separating particulates from an air stream, I needed to count and size particles as small as 0.01 of a micron. Nobody else making vacuum cleaners at the time was interested in measuring particle sizes. It was a hard and very technical thing to do, but to make meaningful leaps in filtration efficiency it was something I had to understand.

In 1983, I visited a spin-out company from the University of Minnesota. It had developed a suitable aerodynamic particle counter. Sadly, it cost an enormous amount of money and I couldn't afford it at that time. While I was with them, however, the Americans showed me a report from the US Bureau of Mines that set me off in a new direction, exploring the dangers of diesel fumes. The report suggested that laboratory mice and rats were suffering heart attacks and developing cancer and other major health problems when exposed to diesel exhaust.

Back in the coach house, we began developing various particulate catchers, using cyclones and other novel technologies. This had led me to demonstrating one to Anthea Turner on *Blue Peter*. We ran this research project for around ten years, developing more sophisticated methods than the version we showed Anthea. We ended up with a rather different design, with the particles being captured on an electrically charged rod.

Once we had it working, we asked around within the automotive industry, showing it to both manufacturers and users. Nobody wanted to fit it. They felt that the difficulty of disposing of the collected soot was not something either they or their owners would want to deal with. Some claimed they would fit ceramic traps, but these would clog with particles like vacuum cleaner bags do.

Diesel pollution only got worse. During the 1990s, when diesel fuel was significantly cheaper than petrol in Britain and Europe, manufacturers with the support of the EU and supposedly scientific backing claimed that diesel exhaust was cleaner than the exhaust from petrol engines. While it's true that diesel exhaust produces less CO_2 than that of petrol engines, more significantly it emits dangerous levels of nitrogen oxide (NOx) and trace metals on dust particles. The European automotive industry was very successful in promoting the spurious case for diesel. It needed to do this because it had invested so very heavily in the production of diesel cars, unlike, say, the United States or Japan, where diesels were a rarity.

The European automotive lobby was so successful that David King, chief scientific adviser to the New Labour governments of Tony Blair and Gordon Brown, disgracefully fell into line with Brussels to promote diesel over petrol. In 2001, the British government cut fuel duty on diesel fuel, prompting a rush to buy diesel cars. The decision was clearly and dangerously wrong.

The problem of filthy car exhausts continued to haunt me. At Dyson, meanwhile, by 2014 we were developing ever more efficient batteries. We had been working on high-performance electric motors for some time. We also had research programmes on air purifiers and heaters. We were pursuing a range of new ideas for products, including hair care. It occurred to me that what we were developing was the very technology and knowhow that together spelled the development of an electric car.

At the time, the motor industry was still ignoring electric cars and industry projections for electric car adoption were tiny, a mere 5 per cent market-share by 2035. I thought they had it completely wrong. Surely people didn't want to go around polluting each other. They would vote with their feet and go electric, as indeed they are now doing. With hindsight, of course, it was a mistake to invest so much time, energy, emotion and money into what turned out to be a much bigger project than I had imagined. Nevertheless, the logic of using our core technologies in an electric car, a market that the incumbents had underestimated, was compelling. We wanted to build not just the best electric car on the market, but one of the best possible of all cars. We were very ambitious.

In 2014, we began putting together an exceptional team in our converted D4 building in Malmesbury. This is where the vacuum cleaner assembly lines used to be. WilkinsonEyre turned it into a secret office/studio away from all the other projects, well hidden so that nobody knew what we were doing. The project would soon grow. We needed ever more space, especially if we were to manufacture the car in Britain. Over the next five years, we developed a radical car loaded with technology. In doing so, we solved lots of problems traditionally associated with electric vehicles and made great progress, completing a very efficient and, I hope, original car ready for production. There was a huge amount to learn.

Typically, we wanted to do everything ourselves, to engineer a pure Dyson car. Of course, we had certain doubts. During the early years, for example, we tried to see if we could buy or develop a chassis from someone else, a bit like Tesla had done when it adopted a Lotus chassis even made by Lotus. For a moment this seemed a clever and streamlined way of going about things, yet, having explored the possibility and finding that no car company made one remotely suitable, we were free to start from scratch.

The cost of developing our own chassis was huge. It did, though, have the long-term advantage of allowing us to adopt a different chassis layout to create additional space for both passengers and batteries. As for range, or the distance it could travel on one charge, this was, perhaps, the key driving force of the project. Even the best electric cars on the market had a maximum range of 200–300 miles. Our research showed that people who were hooked on petrol and diesel cars would only seriously consider an electric car if it could do more than 600 miles before recharging. Interestingly, we also found that the distance of the long journeys people make by car tends to be the same length country by country, including even the United States.

We knew we needed to beat the competition in terms of how far the car would go before recharging. Two tiers of trays of lithium-ion batteries filling the long wheelbase gave the Dyson EV (Electric Vehicle, though our code name was N526) a range of 600 miles even on a cold, wet day with headlights on and aircon or heating working hard. I know people argue that many drivers only commute thirty miles a day, and this may be true, but we found that everyone does a 600-mile trip at least once a year. Having to stop to recharge on a trip to Newcastle, 300-miles from London, simply wasn't on. While it's true that recharging is speeding up, the aim was to give N526 the range and convenience of petrol cars; diesels, too, of course, until they go. How to achieve the 600-mile range was our key focus. This involved maximising the efficiency of motors, drive subsystem, heating and cooling, wheels, tyres and aerodynamics.

To get this range would mean a lot of batteries, which are heavy and take up a lot of space. As a result, the Dyson car would be big and expensive. Not only are batteries costly, but the battery management, electronics and cooling-system componentry associated with them are equally expensive. Reducing the cost of battery cells will not on its own reduce the additional costs of electric cars, unless the battery management electronics and cooling system are also simplified.

At Dyson our scientists are working hard on an entirely new generation of compact and highly efficient solid-state batteries. In lithium-ion batteries, the charge is transferred between the anode and the cathode through a liquid electrolyte, and this electrolyte becomes overheated, causing degradation, preventing fast charging and is prone to fierce fires. Modern batteries use cobalt, which, it is widely accepted, should not be used since it is a rare earth metal with a difficult supply chain. Solid-state batteries are constructed of layers of deposited material on an ultra-thin substrate that are then layered to form a solid multi-stack. Fires and overheating should be a thing of the past. However, the depositing of metals onto ultra-thin substrates in high volume, and forming multi-layers, is problematic. These would have replaced N526's lithium-ion battery pack, and – vastly increased efficiency aside – they would have made the car much lighter than its prototype kerb weight of 2,600kg.

Traditionally, batteries have been the Achilles heel of electric cars throughout their remarkably long history. Ideas for electric-powered vehicles date back to experiments conducted as long ago as the late 1820s. A revolution of sorts took place with the development of the lead-acid battery, with Gaston Planté, a French physicist, leading the way in 1859. Electric cars were popular from the 1890s. They were quiet, refined, fumeless and easy to start, drive and maintain. By 1900, of the 4,912 cars produced in the United States, 28 per cent were electric. Among them were Baker Electric Cars of Cleveland, Ohio. Aimed at women drivers, these highly crafted automobiles were in production from 1899 to 1915. An electric car was the first to break the 100km/h barrier, while some, as early as the 1890s, featured hydraulic brakes and four-wheel steering decades before either of these was offered on internal combustion cars.

The early electric car's fate, however, was effectively sealed for several generations by a combination of Henry Ford's mass-produced Model T of 1908, the invention by Charles Kettering in 1912 of the electric starter motor and, in the United States, by a glut of cheap Texas oil. From then on, and where they existed, electric cars were seen as dull if worthy machines little more exciting than a milk float.

When we came to designing our electric car, we knew it had to be special, but not a petrol head's car by any stretch of the imagination. We wanted a car that was a pleasure to own, drive and travel in. Every last detail mattered. The plug-in point for the battery recharger had to be as resolved, as refined and as elegant as the seats, controls and steering wheel. Heating and ventilation had to make the very best use of Dyson's knowledge of airflow and low-energy use. And batteries had to be the best we could offer, and then some. We looked forward to the day when we could replace lithium-ion batteries with solid-state batteries. This 'holy grail' of electrical engineering is currently being sleuthed by a number of companies worldwide.

The principles of solid-state batteries have been known since Michael Faraday's time – the 1830s – but the challenge to the automotive industry has been to find a battery with the strength and endurance to power cars easily and reliably over long distances. As recently as 2012, a report in the *American Ceramic Society Bulletin* noted that, given the current state of development, a high-range car would need between 800 and 1,000 solid-state battery cells, each costing $100,000. However, a solid-state battery offers high-energy density, fire resistance, faster charging, less heat, less weight and a longer life than lithium-ion batteries.

In 2012, we made a few investments in new-technology start-ups focused on technology areas that we were interested in for future products. One of these was Sakti3, a spin-out company from the University of Michigan, based in Ann Arbour. At the time, they were the most advanced team looking at solid-state battery developments. We wanted to get into this for the car and also for our battery-powered vacuum cleaners. Initially we became an investor and then, quite quickly seeing the great potential of what they were doing, we decided to buy Sakti3 outright, the first and only time we have made such a move.

We are not in the business of buying up other companies. It may be a quick way to acquire a technology or a business that would augment a company, but it can be difficult to assimilate the people and their ways of doing things. Usually, I feel, it's better to start your own research or your own business, which, although slower to begin with, develops organically and is stronger for it. The opposite occurred with our acquisition of Sakti3. The talented team have indeed supercharged our progress in batteries. We already had a team working on battery technology prior to this, but now we accelerated, with new teams in Japan and Singapore working alongside those in Britain and the States. The high-capacity battery pack of the car assembly was designed as an integral part of its body structure to optimise both weight and the space available for occupants in the cabin as well as providing the necessary rigidity and impact protection. Its aluminium casing was flexible in design to allow for a variety of possible sizes and types of batteries that we might develop and install in both existing and new Dyson cars without the need for any significant re-engineering.

Building on our years of experience with Dyson Digital Motor technology, we developed a bespoke, integrated and highly efficient electric drive unit (EDU) comprising Dyson digital electric motors, single-speed transmission and a state-of-the-art power inverter, meaning that electronic devices could be charged without compromising the performance of the car's batteries. These compact and lightweight units were mounted on sub-frames at the front and rear of the car.

N526 was designed not as a rigidly defined car but as a platform, so we could design other body styles to sit on it. The first model was a seven-seat SUV about the same size as a Range Rover although significantly lower and with a raked back windscreen. At 50mph the car would drop itself down on its suspension to take advantage of a lower centre of gravity. If driven through floodwater or across tricky terrain, it could be raised to give it extra ground clearance. It could 'wade' through water 920mm deep and, while I do not think many owners would willingly ford rivers, people do encounter flooded roads. It is comforting to know that in a flood emergency the Dyson car would perform and take you safely home. The car's independent air-spring suspension, designed in-house, has cross-connected anti-roll bars, doing away with conventional steel anti-roll bars and keeping it flat and stable around corners. This was always a big car that thought it was small. We used Tenneco adaptive damping with kinetic roll control, with a very long piston action, to achieve both comfort and stability.

The car is exactly 5m long, with tall 24-inch wheels giving huge ground clearance, helped additionally by the fact that it has a completely flat bottom. The wheels are actually one of the most interesting aspects of the car. The bigger the wheels, the less rolling resistance, and you can ride bumps and potholes more easily. Rolling resistance consumes vital battery power, reduces efficiency and range. As with Alec Issigonis's Mini, the wheels are right in the corners. I don't think you'll find any other car with the rear wheels as far back as those of the Dyson EV. The placement and size of the wheels gave us some unexpected advantages in terms of roadholding and comfort, especially over those potholes and bumps. When combined with the high ground-clearance, the entry and breakout angles were industry-leading. These are the angles of attack in front of the vehicle and behind it when attacking a steep slope from level ground or descending a steep slope onto level ground as, for example, when driving from a road up a steep bank to its side.

Placing the wheels on the four corners has one disadvantage: the turning circle is increased. This we overcame by having four-wheel steering, which reduces the turning circle when the rear wheels turn in the opposite direction to the front wheels at low speeds. If you are driving at high speed, the rear wheels can be turned marginally in the same direction as the front wheels, when changing lanes on a motorway, for example, which maintains stability.

The narrow wheels we designed were also less prone to aquaplaning and more competent in snow than wide-profile tyres. Our wheels were, in fact, bigger than any other on the market. There were, however, no tyres available for them. We went to Michelin. Their engineers were keen, so we researched the new tyres together even though they charged £500,000 for development work. Bridgestone approached us and developed a better tyre for free, and offered them at a lower price.

As well as giving good ground clearance, big wheels afford a commanding driving position. I had a thought that people were buying SUVs because they wanted a clear view of the road ahead as well as having a fear of getting stuck in a flood or in a field at a car boot sale. With our car we would have been able to claim all the things that Land Rover and Range Rover did without it being yet another 'Chelsea tractor'.

What's most striking when you get inside the car is the feeling of space. This is partly because of the position of the wheels at the extremities of its four corners, allowing a large, clear platform across the car, and partly because of the absence of an intrusive engine, gearbox and drive system. You get the internal space of a long-wheelbase SUV without the disadvantage of its massive external body.

We worked with virtual reality to imagine and show interiors, and to look at our car in comparison with other vehicles. Taking Issigonis's Mini design principles onwards, I aimed for an entirely flat floor. I wanted to have the same fully adjustable and ergonomic front seat also in Row Two, the rear seats. Why should the rear seats be compromised? We developed these with Faurecia, the big French automotive supplier. Initially we designed them to slide on four rails running the length of the cabin.

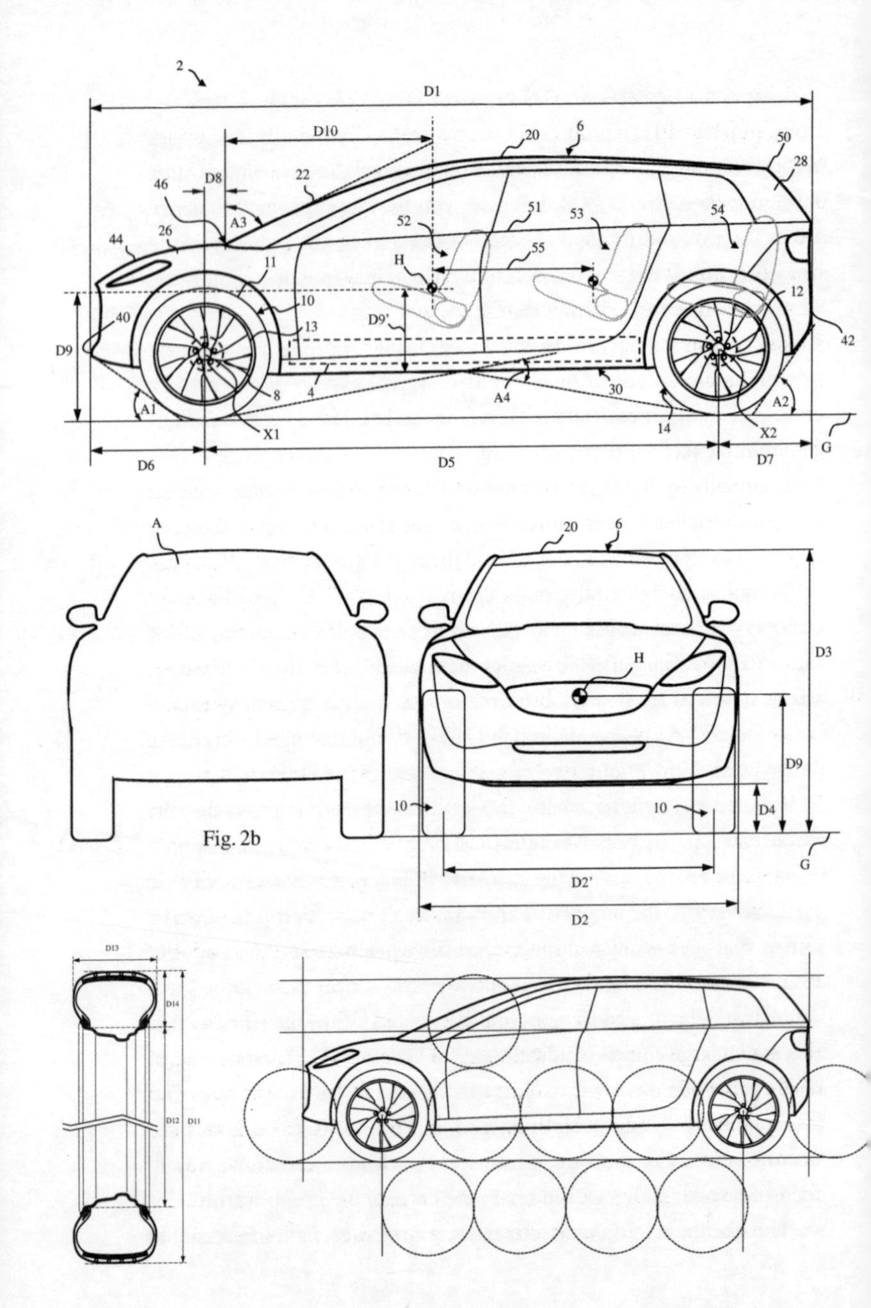
2
D1
D10
20
6
50
28
D8
46
22
A3
51
53
54
52
26
44
11
H
55
10
12
40
13
D9'
D9
42
8
4
A4
30
A2
A1
X1
14
X2
G
D6
D5
D7
A
20
6
H
D3
D9
10
10
D4
G
D2'
D2
D13
D14
D12
D11
Fig. 2b

I happen to hate the 1930s armchair look that car seats typically have, and I hadn't yet found a car seat with proper lumbar support. We wanted a more elegant, structural seat, with well-considered posture support. When you sit in this, it gives you that support in all the right areas. The car has three rows of seats, seating seven adults in comfort. I had never come across a comfortable car seat and I quickly discovered why. The legislation is highly restrictive, with rules stopping you doing almost anything you would want to do. In particular, the rules seem to prohibit any softness or yield in the seat. They also assume that in a crash you are shunted 100mm backwards, so in effect the headrest only comes into effect when you do crash.

Eventually we managed to comply with these rules while also having a comfortable seat. I wanted to be able to see the structure of the seats as you do in favourite chairs of mine like the Eames Soft Pad chair, an American design classic from the late '60s and still unbeatable in comfort. Our seat would have even more pads than the Eames, which also added to its breathability, important in hot weather. The exposed frame, which we anodised a bright colour, was made from magnesium, which is light and strong, with springing at the bottom and the back. We got its weight down to under 20kg, at least half of the industry standard. This meant a significant saving in weight. Weight saving is important for the handling, comfort, braking and range of the car.

We asked many customers in China, the biggest market for electric cars, and found that potential buyers wanted the car to be chauffeured during the week while serving as a family car at weekends. So a lot of focus went into having the front and rear seats of the same design and sliding on the same tracks running the length of the interior, so that they had a large choice of adjustment for spacing. It created no end of challenges when it came to avoiding design pitfalls like finger traps and losing coins. We tried everything and, in the end, decided it was too onerous and compromising, so we went back to a traditional way of fixing the seats though we did retain the entirely flat floor I wanted.

The car had a large windscreen, even shallower than a Ferrari's, to give lower wind resistance and good visibility. Although the car was necessarily quite tall to achieve the command drive position, the large, uncluttered ground clearance meant it still had quite a low frontal profile, excellent aerodynamics and a real sense of spaciousness about it. This was enhanced by keeping the dashboard as low as possible. Most cars have an imposing dash almost projecting into the car that you feel you have to peer over to see ahead. I wanted the opposite, partly to feel spacious but also to enhance the sense of control. Bearing in mind the amount of equipment housed under the dash, and the crash structures, this was difficult to achieve.

I wanted the car to be wholly uncluttered both inside and out. Indeed, the design of our dash and controls was engineered to keep the driver's eyes on the road at all times. All interacting controls, such as lights, indicators and audio adjustment were placed on the steering wheel. There were no controls on the dash, no need to look down there, distracting your eyes from the road, to see and operate the switches.

We then designed an enhanced head-up display, HUD. This would have the satellite navigation, speed, radio station, amount of range available, road warning signs such as speed limits, adaptive cruise control and other warnings on display on the windscreen. There were further enhancements coming through in 2026, with a digital micromirror device based on augmented reality and holographic capability. Our HUD meant that we had no need of a computer screen set on the dashboard in the middle of the car. Annoyingly, legislation demanded that we had to have a centrally mounted screen. I hope the outdated legislation catches up with technology, as those screens can distract drivers' attention from the road and are an unnecessary expense.

We used our own air-filtration technology in the car to control its environment, not just in terms of temperature but also to clean the air. A third of the power in electric cars is consumed by heating and air conditioning, so we went back to basics to find an efficient, lower-power system to save energy, using judiciously placed radiant heat and heat panels. Current heating and cooling systems rely on measuring and changing the temperature of the air inside the car. A car, however, is much like a glasshouse, hugely affected by the sun or the outside temperature, both of which are radiant temperatures affecting the structure of the car and the occupants. On most commuting journeys, the warm or cold air brought into the car's interior by the car's air systems has insufficient time to change the temperature of the car body and seats. This means that the occupant is significantly affected by the radiant cold or hot structure of the car. It takes a long time for the interior air to change the temperature of the interior bodywork. Even worse, the interior air must be ejected out of the back of the car in order retain a moderate CO2 level within it. All that expensively heated cold air is constantly being thrown away at the rate of up to 70 litres a second. We therefore concentrated on providing radiant heat and used less hot-air heating. This saved a great deal of battery energy.

We re-engineered the lights. The main beam was a matrix adaptive technology projector. For the sidelights and indicators we used a square area of acrylic with a live edge. From the front you could see a square of either white light or orange light when indicating. However, from the side you would see just one edge of the square as an upright, very bright line. From the side it was abundantly clear what the indicator was saying in an original way. The acrylic was fed its light by hidden LEDs; a feature of acrylic is that it conducts the light to where you need it, in effect a light pipe.

We established the basic principles of the car very much on our own. Pete Gammack worked on it full time along with a motors team and members of our aerodynamics team. We were very concerned with aerodynamics. These are critical in terms of the range of an electric car. We built carbon-fibre quarter-scale models to test at the Williams Racing wind tunnel near Wantage. We learned a huge amount from this. And then we tested a full-size model at the MIRA (Motor Industry Research Association) wind tunnel at Nuneaton. We then went to the former General Motors test track off the M1 at Bedford to put ten rival cars through their paces to analyse what was good and what was bad about them, which was interesting. We did a lot of analysis. As newcomers, we needed to.

The car was not styled. Its form emerged from engineering and aerodynamic requirements. I wouldn't have had it any other way. With the team, Pete and I made quarter-scale models in clay, something we both really enjoyed. The models were about massing and aerodynamics, and clay is a good material for creating designs quickly. This process involves heating up clay to 60°C in an oven. Once removed, you add layers of hot clay to a substrate – wood, for example – scraping back and adding bits as necessary as it hardens. For full-size clay models, we bought large milling machines running on tracks. These models would typically weigh 4 tonnes and would sometimes have real wheels. It is surprising how misleading a quarter-scale model can be! The full-size model reveals the car as it will be seen. Although the milling machine cuts the full form, modifications are often done by hand, adding clay and scraping to shape. When complete, the clay is scanned into a computer and the milling machine cuts a new full-size clay model.

The clay models were mostly about complying with legislation, crash structures and aerodynamics. Every part of the external shape, and the air intakes for cooling, affect the aerodynamics. The frontal shape needs to cut through the air, diverging as little of the air as possible, as smoothly as possible. Some of the air goes under the car, so that shape is important both for aerodynamics and lift. Some goes over the bonnet, windscreen and roof as smoothly as possible. Some goes around and along the sides of the car, where wheel arches create disturbance. Finally, at the rear, the design needs to prevent turbulent suction, which would slow the car down. All these factors contribute to the final form.

I did the body design with Pete. There was no outside studio involved, although we did ask for advice from experienced car people. There are all sorts of tricks involved in designing cars and to begin with we learned these largely by trial and error. We learned, for example, that when you try to make a long, straight line out of a car's body, it ends up looking as if it's sagging in the middle. You need just that little bit of curvature to balance the shape. I should have known this from my Classics background at home and school. Ancient Greek architects used entasis for the same effect. By giving columns of temples like the Parthenon a slight convex curve, they appear straight to the eye.

I am not a petrol head. I don't go to car races or rallies. I don't read car magazines in my spare time. I am only really interested in cars from an engineering and design point of view, and those I like best haven't really been styled at all, like early Citroën 2CVs, Land Rovers and the original 1970 Range Rover. The Range Rover was designed by the car's engineers, Spen King and Gordon Bashford. When the prototype was shown to Rover's chief stylist, David Bache, he said there was nothing much for him to do other than to improve the grille, lights, mirrors, door handles and badges.

I am, though, particularly fond of the Citroën SM, a grand tourer dating from 1970 and designed by Robert Opron. With its exciting profile, swivelling headlights, speed-sensitive steering, hydropneumatic suspension – sleeping 'gendarmes' didn't register even at 50mph – hospital-bed-type upholstered seats and compelling Maserati engine, I think of the SM as the best internal-combustion car ever. It might be half a century old, yet it wins hands down in terms of comfort and originality. I bought mine to show our design engineers at the beginning of our car project, to encourage them to be bold and original.

We had got a very long way down the line with the car by 2016 when Ian Minards joined us from Aston Martin and started to build up the team and the equipment we needed for production. We thought it best to mix traditional automotive people, with all their skills – many of them learned over decades of collective experience – with Dyson people, who approach problems from entirely new angles.

Ian had worked for Jaguar before Aston Martin, where he was involved with the final 1991 iteration of the XJS, designed originally by the aerodynamicist Malcolm Sayer, and its replacement, the XK8. With Aston Martin, Ian pursued the development of fast and powerful cars, including the V12 Vanquish – mechanical star of the 2002 Bond film *Die Another Day* – and the 200mph DB11, launched in 2016, by which time he felt it was time to move on to new challenges.

Ian joined us when the EV project had grown from just me and Peter Gammack to ten of us working in a tiny office to ninety engineers, half working on motors, Ian recalls, half on the air-con system and just two, without prior experience, on the chassis. The team grew to 500 and by mid-2019 we had a very convincing car based on an extremely strong and rigid aluminium monocoque body and chassis carrying a 150kWh lithium-ion battery in a two-layer stack of cylindrical cells located below the floor. These powered a pair of 264hp Dyson electric motors, one at the front, one at the rear of the car, driving all four wheels. Torque vectoring and traction control ensured that the car could be driven from rest, foot to the floor, in a smooth, straight line. With 480lb ft of torque available instantly, the car needed a sophisticated drive system. Although weighing 2.6 tonnes, the EV could have accelerated from 0 to 60mph in 4.6 seconds. Its top speed was estimated at 125mph.

It was quite a package. Highly original, beautifully engineered, subtly understated and unlike anything else on the road, our electric car was more than the machine itself. It was part of an impulse on our part to develop new technologies that we can put to work in a wide range of future products. When I first drove the car, though, I felt exactly the same as when I first used our hair dryer prototype, or indeed our very first vacuum cleaner. I enjoyed it, but immediately looked for improvements, as, instinctively, engineers always will.

By 2018, we had to make a decision about where to build the car. We could have done so on the Hullavington campus, but we needed government help. I went with key members of the car project to see Greg Clark, Secretary of State for Business, Energy and Industrial Strategy. We explained the plans and where we needed help. When we asked during the meeting if he could help us in any way, he flatly refused. I pointed out that he had just given JLR (Jaguar Land Rover) £250 million for a new diesel-engine plant, but he was adamant that he would not support us. I got absolutely nowhere. He wouldn't even think about it.

The following week I was in Singapore. I was invited to meet the long-serving prime minister, Lee Hsien Loong, who could not have been more helpful. He introduced me to members of his team who immediately set about finding land and development grants for building a factory. I came back from the trip and asked for a meeting with Theresa May. She refused to see me. Given the attitude of the British government and the fact that the main market for electric cars is China, it seemed a bit silly to think of trying to make cars in England and then shipping them thousands of miles. Even at Hullavington we would have needed to build a new factory that, however well designed, would have got us bogged down in the slow world of planning permissions. In addition, we were struggling to recruit sufficient engineers in the UK.

All signs pointed in the direction of Singapore, which also happens to be part of the ASEAN free-trade area group, and with China the main market. We started planning the factory in Singapore on reclaimed land in the new port area. We designed a big factory because we were doing so much of the car ourselves. Some suppliers, including manufacturers of specialist components like brakes, car glass, door fastenings, wiring looms, seat frames, seats and so on, were already there as if on tap.

Yes, we were new kids on the automotive block and so suppliers charged us more than they might an established car company because we couldn't guarantee the volume they wanted and because everything we needed was specific to us. Because of this, the bill for the parts we bought in for the car would be as much as 25 per cent higher than those sold to existing manufacturers. The car would be very expensive to make. We had huge overheads. The cost of materials was enormous. And, as we were planning to sell it direct rather than through dealers, we would need storage facilities and financing deals in every country we sold in. On top of this was the obvious fact that the fewer cars you make, the higher the cost per car. At a relatively low volume we would have to sell the car at £150,000. There are not many people who will buy a car at this price.

What was not immediately clear, though, was that those traditional automotive companies turning to the production of electric cars were making huge losses on them. This was because you simply can't make them for a reasonable price, although the reason existing car companies were willing to make them is that electric cars help to achieve specified exhaust emissions across their product range. So, if they make a loss on electric cars, they make a profit on polluting cars while appearing virtuous. Tesla, meanwhile, had been through $23 billion of shareholder money, grants and the rest of it, grants of which we wouldn't necessarily get even a small fraction. At the time of writing, Elon Musk is raising a further $6 billion.

Because of this shifting commercial sand, we made the decision to pull out of production at the very last minute. N526 was a brilliant car. Very efficient motors. Very aerodynamic. Wonderful to drive and to be driven in. We just couldn't ever have made money from it, and for all our enthusiasm for the project we were not prepared to risk the rest of Dyson.

The principal reason for the sudden and rapid acceleration in the development of electric cars by major automotive companies was 'Dieselgate', the exhaust-emissions scandal that changed everything almost overnight. In autumn 2015 the United States Environmental Protection Agency, prompted by the findings of an individual investigator, discovered that Volkswagen had programmed turbocharged direct-injection diesel engines in some 11 million cars to activate emission controls to meet regulations during laboratory tests, but not on the road. Diesels were producing unacceptable levels of nitrogen oxide and particulates by stealth. No amount of 'greenwash' by manufacturers and politicians, let alone evasive test reports, could cover up the fact that diesel engines are fundamentally dirty and a danger to health.

Volkswagen was not the only car company complicit in 'Dieselgate'. After the revelations of 2015, the game was up. From now on, automotive manufacturers had no choice but to shift to electric. Norway plans to be all-electric by 2025, China and Germany by 2030 and the United Kingdom has now brought the date forward to 2030. By December 2019, 4.8 million, or 1 in 250, of the world's cars were electric, half of them in

China. It's a start, but there is clearly a very long way to go in terms of creating electric cars and the infrastructure to support them, and how we produce the electricity needed to charge their batteries.

For us, 'Dieselgate' was both welcome and a commercial concern. Welcome, of course, in getting the automotive industry and car owners away from diesels, but a concern for Dyson because the major players were plunging headlong, in desperation, into making electric cars. These would undercut the price of our car by a significant margin. They have been able to do this not least because, by offsetting the zero emissions of the electric cars they produce against petrol and diesel cars, they can be shown to be meeting environmental targets. So, electric cars that are much more expensive to make are mostly sold at a loss. Tesla even sells its carbon offsets to internal-combustion engine manufacturers for many millions of dollars to boost its income. I am not sure about the ethics of that transaction. The extent of the incumbent car manufacturers' volte-face switch to electric cars, a delayed reaction to 'Dieselgate' on their part, had become apparent by 2019, meaning that it would be hard for us to compete at our elevated price and very risky for us to proceed.

When you stop a project it is a horrible, horrible thing to have to do. Everyone who had worked on it was thrilled with what they were doing, and we felt we were doing something really exciting and significant that had led to huge developments in batteries and motors. Suddenly all dreams were shattered by a decision that we made, and we realised that we were disappointing a lot of people – those who worked on it, but also those who wanted to buy it. There was a huge human cost and the great disappointment of not carrying out what we had spent the last period of our lives working on. As I said at the time, it tore all our hearts out.

We tried to make as few people on the team redundant as possible, while as soon as we announced the cessation of the project the predators started moving in. Car companies and recruitment agencies set up shop in local hotels to offer jobs to our people. I think everyone was placed in other roles very quickly, while those who wanted to stay on with Dyson are now working on other projects. We have been able to accelerate new projects at Dyson because of this influx of different talents into our fold and many of those who remained are now leading large teams.

Back in 2014 we realised that we had entered a competitive field but believed that competitors were mostly ignoring the demand for electric cars. What we did not predict was 'Dieselgate', nor that the landscape would change so much afterwards. Fortunately, we were able to stomach the £500 million cost and survive. We did, though, push ourselves to learn a great deal in areas including batteries, robotics, air treatment and lighting. We also learned more about virtual engineering as a tool in the design process and how, ultimately, we would be able to make products more quickly and less expensively. These were all valuable lessons for the future.

In the process we had acquired Hullavington. This is the extensive former RAF base close to Malmesbury bought as somewhere to develop, test and possibly to make the car. It is a special location, and it has given us a huge amount of space for current and future development work. There were, in fact, three redundant or little-used airfields not far from Malmesbury. At the time we were starting the car project, and given the space we knew we needed, it seemed a good idea to approach the Ministry of Defence (MoD) to ask if we could buy or lease one of them. Given that the MoD owns something like the entire land mass of Wales but only uses a fraction of it, I had imagined it would be easy enough to put in a bid for one of the three local airfields, especially as the MoD was short of money.

I had first set my eyes on RAF Lyneham, with its three long runways ideal for testing the car and now completely smothered with solar panels, but Hullavington proved to be a better bet with its two runways, multiple taxiways and spacious hangars. The MoD said no. No to all three airfields. I'm not sure why. In the end I went to see Prime Minister David Cameron and Chancellor of the Exchequer George Osborne. With their help, I was able to go ahead with my attempt to buy Hullavington.

I say attempt, because it proved to be a tortuous two-year process. The main part of the airfield occupied land that had been bought by compulsory purchase from five families of farmers in the 1930s. They had a vast number of descendants who we had to track down and negotiate with as part of what are known in Britain as Crichel Down Rules. This meant that previous owners, however long ago, had to be contacted and the land bought from them at market prices. They had already been paid the going rate for the land back in the 1930s, yet, incredibly, we still had to pay the descendants a premium. It was about the most complicated negotiation we had ever done.

Some of the families couldn't agree among themselves and didn't want to hurry. We, on the other hand, didn't have time on our side. In February 2017, we finally bought a freehold on various sections of the airfield. Other owners around RAF Hullavington have since offered us other land, which we have bought, taking the area we own up to 750 acres.

We then set about the painstaking restoration and renovation of the site and its buildings. Airfields are being dug up everywhere the length and breadth of Britain and I wanted to preserve Hullavington. It would become Dyson's new R&D hub. And while we developed the very latest technology here, I planned to bring working aircraft back to the site. When I said this in public there was local outrage, even one formal objection from Aberdeen, 500 miles away. I still have my heart set on at least one former Second World War fighter taking off from and landing here on a weekly basis and, indeed, hope to rehouse the historic Fighter Collection here from its current home at Duxford Aerodrome.

Hullavington has been as much a labour of love as a commercial venture. The hangars, for example, are lovely buildings but were horribly neglected. I don't think they saw any maintenance from the day they were built to the day we bought them. Chris Wilkinson has brought four key hangars back into action as elegant, light-filled, hi-tech spaces. Renovating them, and at the same time reducing their carbon footprint, has been more expensive than creating brand-new buildings here.

For me, a great part of Hullavington's attraction is its architectural, engineering and design story expressed in a number of innovative and spacious buildings that, special in the 1930s, have proved invaluable to us the best part of a century later. Significantly for us, the RAF airfields of the period were intended to be showcases of high-quality design and engineering.

Working under the auspices of the Air Ministry and with the Directorate of Works and Buildings, the RAF invested in dozens of new aerodromes in the five years leading up to the Second World War. They were meant to be very smart indeed. Ramsay MacDonald, prime minister of the National Government of 1931–5, called on the Royal Fine Art Commission to supervise their design and planning. As these were major, technologically-driven undertakings, there was concern about how they would fit into the British countryside and what image they should represent.

As a result, Hullavington boasts neo-Georgian domestic and administrative buildings arranged in a Beaux-Arts or Garden City plan designed by Scottish architect Archibald Bulloch in the style of those great twentieth-century English classicists Edwin Lutyens and Reginald Blomfield, and hangars and technical buildings that were at the forefront of new design and construction technology. At Hullavington, the design circle was squared: the new airfield would represent both modernity and tradition. The three-storey officers' mess was clad in local limestone, its lobby and main hall lined in polished oak panelling. The aircraft hangars, two of which have been converted into Dyson offices and workshops by WilkinsonEyre, were as up-to-the-minute as the design thinking of the most forward-looking European engineers and architects.

Around the perimeter of the airfield, for example, are late 1930s Type E hangars. With their earth-and-turf roofs, these curved concrete buildings are truly discreet. Seen from the road, or from the air, they might be mistaken for Neolithic barrows, for which Wiltshire, our county, is famous. Think of the ancient earthworks around Avebury, from the low-lying to the spectacular like Silbury Hill, England's echo of the pyramids of ancient Egypt.

But where Silbury Hill is, as far as we know, a solid earthen mass, the

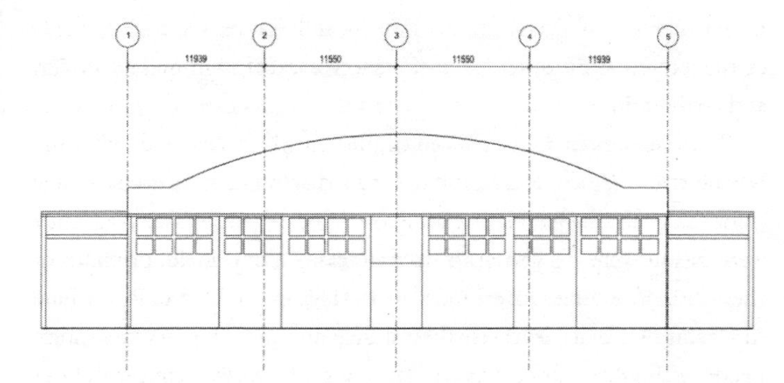

Hangar 85

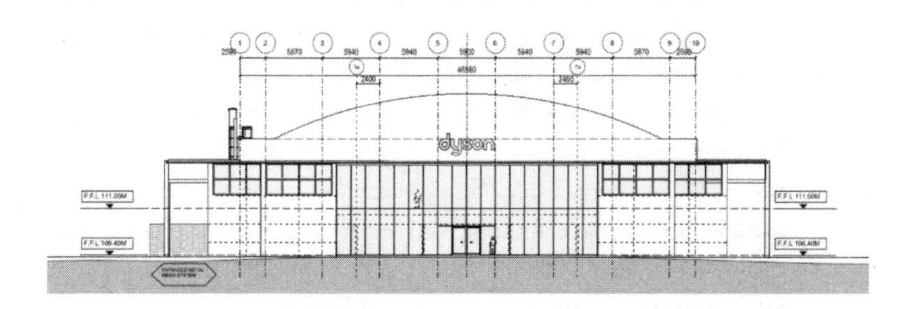

Hangar 86

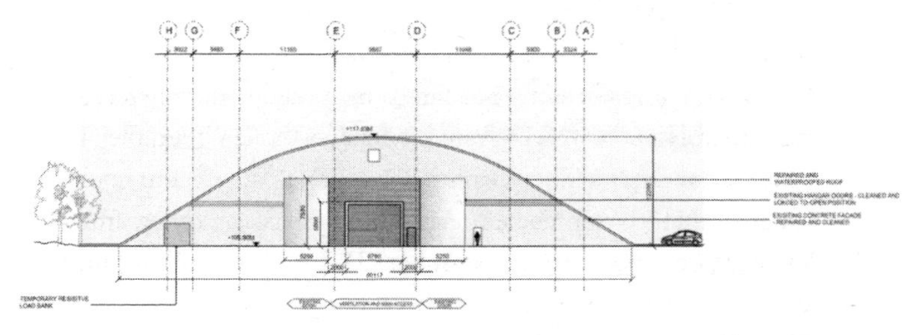

Hangar 181

gravel-filled bomb-proof doors of the Type E hangars at Hullavington open to reveal quite spectacular spaces, free of columns and shaped by ingenious concrete roof structures derived, in this case, from the work of Hugo Junkers, the German inventor and industrialist better known for the aircraft, including the Ju 87 Stuka dive bomber, mass-produced in his name at Dessau.

A left-leaning pacifist, Junkers was one of the principal patrons of the Bauhaus. He played a key role in bringing the new design school to Dessau. He had been a successful inventor and maker of gas water-heaters for bathrooms and fan heaters before turning to aircraft design. In 1925, Junkers patented a steel development of the timber 'Lamellendach', or segmental roof, prefabricated structural system. Junkers' net-like steel framework proved ideal for aircraft hangars. The 'Lamella' hangars were sold on the strength not only of their innovative structure and speed of construction, but on that of 'architectural beauty', too.

The concrete Type E hangars at Hullavington – the ones with a semi-circular roof covered in grass – with their roots in German design although employing a simplified construction system, stand testimony to both the evolving relationship between aviation technology and architecture and to the technology transfer between engineering concerns fated to represent opposing sides and irreconcilable political creeds in hugely destructive warfare. Kier & Co. built the Type E hangars in 1938. Eighty years later, the Kier Group was involved in the restoration of the two Dyson Type D hangars.

The Type Ds, erected primarily as aircraft storage units, are built of reinforced concrete columns supporting concrete bowstring ribs, or trusses, that form the roof. Their fifteen-bay walls are solid 14-inch-thick reinforced concrete with large steel windows at their upper level. Doors consist of six steel leaves, opening into concrete door gantries projecting from each side of the buildings. The construction of these hangars had been developed in France, where sophisticated concrete engineering informed not only new airfield buildings, notably at Montaudran airfield near Toulouse, but also the work of Auguste and Gustave Perret, whose church of Notre-Dame du Raincy (1922–3) remains a masterpiece of concrete construction. In certain lights, the Type Ds might be mistaken for temples.

When Hullavington airfield opened in June 1937, its runways were grass at first, later experimental tarmac surfaces, and its first aircraft biplanes. If you had been able to visit Hullavington in its 1940s heyday, you would have been impressed and perhaps even overwhelmed by the sheer number and variety of RAF aircraft gathered there, from Mosquitoes, Spitfires and Lancasters to Douglas Bostons, North American Mitchells and GAL Hotspur troop-carrying gliders. Although squadrons based at Hullavington took part in the defence of Bath and Bristol, the airfield was built primarily for aircrew training, as well as the training of flight instructors, and for the storage of aircraft.

Developing a car at a former air base has also been a reminder of the fascinating connection and collaboration between the automotive and aviation industries, from the work of Buckminster Fuller and his Dymaxion car, a kind of roadgoing airship in 1933, to the airship designer Paul Jaray's streamlined bodies for Hans Ledwinka's dynamic Tatra cars and the pioneering work of the engineer and former air force pilot Sixten Sason for Saab in Sweden.

All these inventor–designer–engineers were concerned, as the aviation industry has to be, with streamlining, wind-tunnel-testing, economy of means and, increasingly, with ergonomics, or ways in which we can design products, from vacuum cleaners to aircraft – if not, as yet, flying cars – in tune with human physiology and our cognitive abilities. When we developed our electric motor turbines in wind tunnels, we did so alongside Rolls-Royce aero engines. The latter might be very much bigger and more powerful than our tiny electric motors, yet the principles of airflow are equally important to the successful working of both.

As if by chance, Boris Johnson called me on 13 March 2020 saying he needed 50,000 ventilators in six weeks in response to the Covid-19 pandemic. The project was announced publicly four days later. We repurposed the very buildings at Hullavington that had been devoted to the development of the car. It was fortunate that we had a pristine new factory available to make a medical product such as this. We started as a relatively small team of engineers and designers, understanding the clinical need and the specification of the ventilator. Over the first weekend, we made lots of different prototypes, with varying degrees of sophistication.

We deliberately avoided traditional ventilator components due to potential supply issues, using instead those we knew were readily available and which we could rely on. The team grew quickly to about 450 people working across the UK and Singapore so we could achieve twenty-four-hour working cycles and draw on our global supply chains and knowledge. Our Singapore engineering team picked up while the UK team was sleeping and vice versa, but the days were still long, often stretching deep into the night.

It was an extraordinary project. The feedback we got from clinicians was good, and within two weeks we had a new ventilator ready for clinical testing. As the project progressed, we bought the rare components needed and reached the medical-grade manufacturing stage in just six weeks, turning an entire Hullavington hangar into a medical-grade production facility filled with Dyson people observing social distancing while assembling the machines.

The teams were highly driven. They were diverted from developing commercial projects, which were delayed as a result. We only ever thought the ventilator would cost us money – rather than make it – but we did it for the right reasons as part of the national effort at this time of crisis. I wasn't surprised by the way the teams responded: this is how we work and what we do. The teams did all of this at the height of the pandemic, working all hours, away from their families. It was selfless and showed just what engineers can achieve together in a short time. Many amazing people across the company contributed to the project, not just the engineers but also those who came in to make us meals, the cleaners who kept us safe from the virus, the people taking our temperatures every few hours, the security team, the team who prepared the hangars for manufacture – the list goes on. The speed with which we were able to deliver the device is testament to their hard work.

One element of this project that made it very hard to manage was the ever-changing specification. We were designing and making the ventilator at the request of, and under an order from, the Cabinet Office at 10 Downing Street. A civil servant there provided the specification. As far as I am aware, he was neither a clinician nor was he connected with the medical equipment approvals arm of the health service department. He initially told us to do a self-contained portable ventilator. It should have its own air pump powered by batteries with HEPA filters. It would support or replace the breathing of the patient in tune with their breathing rates.

At this point, the first of many outside consultants were brought in by the Cabinet Office. They condemned the idea of an air pump providing the air for the ventilator on the grounds of toxicology while ignoring the five HEPA filters that we had incorporated for that very purpose. We had to change our design so that it was powered by the mains electricity supply and fed with the compressed air supply in hospitals (the toxicology consultants were not, I noticed, asked to comment on the toxicology of the compressed air supply in hospitals). We completed this design and were then told that we had to incorporate a vacuum mechanism to suck accumulated gunk out of the lungs. We incorporated this in the design. We were then told that the ventilator had to provide assisted breathing as well as taking over the breathing. We incorporated that addition as well. The difference is that a ventilator has a pipe which is put into the patient's throat, under anaesthetic, and takes over the breathing. An assisted breathing device works through a face mask, like an oxygen mask, and allows the patient to stay in the habit of doing their own breathing, making recovery much more likely. Our device had to do both.

All this development, building and testing of prototypes and component-sourcing took place over six weeks. In parallel we set up medical-grade production facilities at both Hullavington and Singapore.

The Singapore facility would provide ventilators for South East Asia as well as fulfilling the desperate enquiries we had received from around the world.

Our design was much smaller than a traditional ventilator, about the size of a briefcase, and would sit on a bedside table or trolley. It was an aluminium box with a one-piece polycarbonate fascia. This was similar to the control fascia that we had developed for our washing machine, itself a first at the time. A 'clear' polycarbonate sheet is formed with the printing on the underside. This means that there is no wear to the print and the polycarbonate fascia forms a hygienic, wipeable and waterproof surface. Formed into the polycarbonate sheet were small, raised domes that, when pressed, provided snap-control switches, in turn actuating tact switches (tactile switches) on the PC board underneath. The graphics are printed on the back of the polycarbonate, leaving unprinted portions in the shape of clear circles or strips in the clear polycarbonate so that lights on the PC board underneath can shine through the clear shapes as illuminated indicators.

By mid-April 2020 it became apparent to clinicians around the world that it was better not to put patients onto ventilators if it could be avoided. They were to be used only as a last resort. The requirement for additional ventilators besides those already in stock in hospitals vanished overnight. It took the Cabinet Office several weeks to admit to us that our efforts were not needed. We decided to voluntarily bear all the £20m of costs.

The Cabinet Office had come to us asking for the all-but-impossible: to design, develop and produce a huge order of ventilators from scratch in a matter of weeks. They kept making radical changes to what they wanted and then, because of the change in medical thinking, didn't need or want it after all. They brought in an army of consultants who, in my view, earned their expensive keep by putting their oar in and creating unnecessary obstacles and delays.

What I have observed in Britain is that those in the civil service have a deep distrust, bordering on hatred, for those in the private sector. This is strange since their salaries are paid for by the wealth-creating private-sector taxpayers, the very people they despise. Their salaries are guaranteed whether they perform well or badly, and their jobs are secure in recessions. This sense of security may account for their contemptuous attitude towards the private sector, whereas in private enterprise we only get paid, and have a job, when delivering something of value that someone else is prepared to pay for. We have a measure for what we do and we are servants to the customer, whereas in stark contrast, civil servants need no customer and are servants to nobody.

Unlike government hierarchies, we have a very flat management structure at Dyson. This helped to empower individuals to take decisions quickly and to get things done. We had all the right people around the table, and everyone knew their role. Given the possibility of people getting ill due to the virus, we also had a lot of crossovers, with people understanding each other's roles and responsibilities in detail. Having such a clear purpose, understanding and direction allowed our

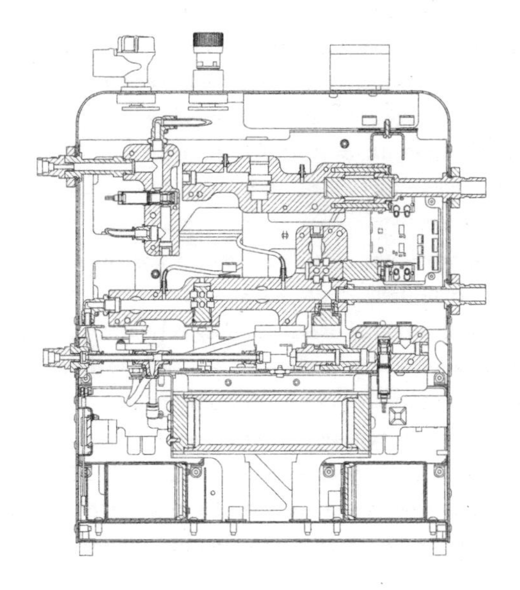

Ventilator

people to use their expertise to maximum effect.

In many ways, this is how we operate in normal times. Ours is a flat organisational structure, with everyone in a product team dedicated to delivering their product on time. All the necessary specialists, such as software and electronics engineers, are in each team. We also have many other specialists who can parachute into a team to help resolve a problem. We are always doing things we have never done before. We solve problems every day and we move quickly, failing and learning as we go. We answered the government's direct request for help with the ventilator because I believed it was the right thing to do.

Sadly, our good intentions were questioned at the end of April 2021 by the BBC. They twisted the truth and as a result the prime minister and myself were implicated in 'sleaze'. The story, by the BBC's Political Editor, Laura Kuenssberg, was at a time of important regional elections – a test of the Government's majority. It focused on my text messages to the prime minister following a request for clarification that had been made to the Chancellor of the Exchequer about how UK tax rules would be applied to non-UK employees working on the project.

From the BBC's perspective, I was 'lobbying' the prime minister – despite him having contacted me first, rather than the other way around – and because lobbying is conflated with sleaze in Britain, I was apparently seeking some form of wrongful preferential treatment. To support their narrative, I was described as a 'prominent Conservative supporter'.

Precisely what favours I was trying to exact from the prime minister were, however, anyone's guess. Far from any gain, the ventilator project cost Dyson £20 million. Far from being private, I had sent copies of the messages to officials in Number 10 and the Treasury as a record of the conversation and this is precisely how they came to the BBC's attention. Besides, text messages are an established method of communication, which I am sure BBC reporters themselves use – there is nothing underhand about it.

Finally, they were just plain wrong to portray me as a 'prominent supporter'. I have never attended a Conservative Party social event. I did not give as much as a penny to the Vote Leave campaign. I don't like or make grand political gestures. Laura Kuenssberg said I had spoken at a Conservative Party conference, which indeed I had, although she failed to consider that I have also spoken at a Labour Party conference, speaking on both occasions for a better understanding by British politicians of engineering and of our need for more engineers.

The BBC later admitted that its claim I was a 'prominent supporter' was in fact based on charitable donations, shown on the Electoral Commission register, of £11,450 from The James Dyson Foundation to the Wiltshire Engineering Festival. This was a charitable gift to an event encouraging schoolchildren into engineering careers – it was not political and not for the Conservative Party. It is well known that registered charities are not allowed to donate to political causes. In my view, the story was simply a smear on me and the prime minister.

I am proud of Dyson's contribution to the ventilator project. It was, then, deeply disappointing for me, and the hundreds of Dyson people who gave the project their all in the midst of the crisis, to have our efforts so mischaracterised and used for the purpose not of saving lives, but for political mudslinging. To their great credit, sections of the media did investigate the BBC's claims and found them entirely false. This attempt to mire Dyson and myself in a political sleaze narrative failed, because it wasn't true. It took a while, but eventually the BBC did apologise for its inaccuracies:

> Sir James Dyson apology – Various outlets, Wednesday 21 April 2021
> *We accept that Sir James Dyson is not a prominent Conservative supporter as was stated in some of our coverage of his text messages with the Prime Minister. The James Dyson Foundation made a charitable gift to support the Wiltshire Engineering Festival for school children. We accept that this does not signal affiliation to any political party and we would like to put the record straight. Sir James also raised concerns about the accuracy of other aspects of our reporting. We wish to make clear that Sir James contacted Number 10 in response to the Prime*

Minister's direct request to him for assistance in relation to the urgent need for ventilators and incurred costs of £20m which his company voluntarily absorbed in trying to assist in the national emergency. His text messages to the Prime Minister were also later sent to officials. We are sorry that these facts were not always reflected in our coverage, and we apologise for not doing so.

Covid-19 created the most difficult circumstances facing business and the public in my lifetime. Many economies face a grave future as the world begins to recover, although the difference between the performance of Western and emerging and developing Asian economies is significant. In December 2019, national debt in Western countries was 103 per cent of GDP, rising to 124 per cent a year later. Asian debt rose from 53 to 63 per cent of GDP, showing how these economies, although affected by the pandemic, were running much tighter ships than the West. Where Western economies contracted by 5.8 per cent between December 2019 and December 2020, the figure for Asia was just 1.7 per cent and, according to the IMF (International Monetary Fund), these economies are expected to grow by an average 5.9 per cent by mid-decade, well above what can be expected of Western countries.

It was not an easy time to be in business. Some businesses seem to have resigned themselves to remote working. We have found working remotely deeply unsatisfactory as we make physical things that exist in a matter-of-fact world. These require physical interaction and specialist equipment, so lockdown has brought significant challenges and delays to some of our projects. The years of training, research, design and testing behind all our products just cannot be done from home; they require labs and equipment, and progress slows without access to these or without face-to-face discussion and interaction between engineers looking at prototypes. Nevertheless, our UK campus remained open throughout the crisis for people to work on the ventilator and we have been trying as best we can to keep projects on track, relying on our support staff, thousands of new desks to help with social distancing and, unlike government advice, mandatory wearing of masks from the start.

We developed an entirely new ventilator in thirty days. Mercifully, it was not required in the UK, but we don't for one moment regret our contribution to the national effort. Ventilators are not a product we would choose to make and so the venture ends there. Dyson spent around £20 million on this project. While governments in the countries in which we operate have offered money during the crisis, we have not accepted any public money from any of them, neither have we furloughed any staff.

When the ventilator project ended, we realised that we needed to move some development projects from Malmesbury to Hullavington to give them more space to grow. The campus is thankfully being put to good use and has afforded us lots of space for new research and development into robotics, environmental care and lighting. I am confident that we will make many exciting future discoveries here, reaching, if not perhaps literally, for the sky.

CHAPTER TEN

Farming

Farming is not exactly in my blood and yet I have developed a passion for it. Today, I own 36,000 acres of high-yielding, good-quality farmland in Lincolnshire, Oxfordshire, Gloucestershire and Somerset. Dyson farms have evolved into an increasingly seamless part of our family business, employing the very latest science and technology to prove how agriculture can be profitable, how it can reduce our reliance on imported foodstuffs and how it can be truly sustainable in the very long term.

It might seem strange to think we can bring our farming and technology operations together – peas and vacuum cleaners, potatoes and hairdryers – in a symbiotic relationship, with the hope of finding novel approaches to driving performance and sustainability in both Dyson products and Dyson foodstuffs, and yet farming is not so different from manufacturing. It's about producing something – food in this case – and, just as with a factory, the set-up should be well-designed, well-built and work as efficiently and sustainably as the best machines, using advanced technology. Initially our farming operations were entirely separate from Dyson, quietly getting on and improving the ways they worked. Now we see the mutual benefit of bringing our farming and technology businesses slightly closer together.

Dyson has always focused on making long-lasting machines that use fewer resources while achieving higher performance. Lighter machines, resulting from developing new technology and reinventing the format, consume less energy and are not only better for the planet but also more pleasurable to use. Our cord-free vacuum cleaners, for instance, are a fraction of the weight and use a fraction of the electricity than their predecessors did. This has come about by taking an entirely different approach and developing new technology, motors and batteries, from the ground up.

Material science, energy creation and energy storage are at the core of this and farming has much to give, growing materials and creating energy that can be used in our machines. Meanwhile, Dyson technologies, among them robotics, vision-based sensing, AI and energy storage, will increasingly drive further technology innovation on our farms, leading to greater efficiency and superior products.

The parallels between the two businesses – Dyson Ltd and Dyson Farming – are greater than might be expected, with the future for both dependent on investment in research and development and on continual improvement. I'm hopeful that Dyson Farming will contribute towards the very necessary transformation of agriculture, while protecting the countryside, and that it will drive meaningful advances in sustainability.

Sustainable food production and food security are vital to the nation's health and economy. Britain currently imports 30 per cent of its food. This is unnecessary since we can grow almost everything and the inwards importing transport results in a poor carbon footprint. The same eco warriors who are so keen on the re-wilding of farms think nothing of consuming avocados flown in by air and other imported foods. Within our glasshouses, and with our fertile soil and more-than-ample rainfall, we can grow the food we need. At the same time there is also a real opportunity for agriculture to drive a revolution in technology and vice versa. As mentioned, we are already developing ways to use agricultural materials in our products. This, as yet, is a company secret, but the research is fascinating and we're getting there.

Farming sustainably, and without recourse to 'greenwash', is very much to do with a cycle and recycle of energy and materials. The same ought to be true of manufacturing. We must move ever closer to a culture whereby we minimise the use of materials through lean engineering along with the recycling of products at the end of their lives. It's not just okay to politely offset our carbon footprint. We have to deal with it at source. This has certainly always been our intent, but there is always room not just for improvement but also for radical change through invention and innovation. Engineers and scientists have made, are making, and will make the difference, rather than the endless grandstanding by commentators.

Farming is a visceral form of manufacturing with much to teach us. As in making anything on a large and repetitive scale, however, all the basics have to be right. Whether it's the drainage, ditches, tracks, hedgerows, walls, soil quality, woodlands, the control of weeds, the nurturing of wildflowers, insects, birds and other wildlife, the quality of farm buildings, machines and yards, the well-being of farmers and their staff, everything must be of the highest quality, with everyone and everything working together harmoniously. There is no reason why a farm, its equipment, tracks, ditches and yards should be messy, muddy, or untidy. We should always maintain high standards.

All but one of the farms we've bought to date had suffered from underinvestment and lack of maintenance; this is why they had become run-down and inefficient. On every farm there were blocked drains, overgrown ditches, worn-out tracks, a lack of grain stores, black grass competing adversely with wheat and barley, improper crop rotation, missing hedgerows and collapsed walls. This, though, hasn't been for some innate lack of interest on the part of most farmers, who were passionate about their farms, but because of a lack of capital and profit from the sector more generally. Not one of the farms had a modern grain store, so they would contract out the storing and drying of grain, meaning a further cut in their tight margins while losing the added value of maintaining a high-quality grain store. For the first years, we had to spend heavily on infrastructure before we could make our farms work efficiently and indeed profitably.

All this, though, is jumping the gun. I need to go back several steps to my childhood and upbringing in north Norfolk. At the time, this was a deeply rural region projecting out into the North Sea and cut off from the rest of the country. Second homes here were a thing of the future, while, if anything, communications through north Norfolk declined as rural railways were abandoned. This was even before Dr Beeching wielded his axe in the 1960s, closing thousands of miles of cross-country lines that left many towns and villages isolated.

I was at Gresham's School in Holt with a number of farmers' sons. Farms surrounded our small market town. In my teens, I ran across farmland every day. During school holidays I worked on local farms. Deirdre joined me in college days. Although we went on to lead a very different life, the coach house at Bathford where I developed the cyclonic vacuum cleaner and our factory in Malmesbury were both in rural settings. I think I had always wanted to own a farm, although only recently have I been able to see what I might be able to contribute to farming.

Certainly, I have never had an interest in speculating in farmland or owning land for the sake of it, as if for a matter of prestige. Nor did I think we might make money from farming. Things changed quickly from 2013, when we bought 8,000 acres of the old Nocton estate stretching across heathland and fens a few miles south of Lincoln. Rich in nutrients, the black soil here has long been ideal for growing crops like peas and potatoes.

The estate itself, once owned by the Earl of Ripon, had changed hands many times and was broken into three separate farms by the time we bought all three and brought them back together again. A part of the estate's history, though, and what caught my attention, was its attempt to mechanise using technology from outside farming. In 1919, Major 'Jock' Webber, the Nocton estate manager, hit on the idea of buying railway track and rolling stock from the Army surplus store at Arras in Pas-de-Calais. Webber had served there, on the Western Front, during the First World War. He saw how efficient light narrow-gauge railways, operating across sodden fields, were. These had been the invention of a French army colonel, Prosper Péchot, working, from the late 1880s, with the railway and locomotive engineers Charles Bourdon and Paul Decauville. While these light railways were used to transport troops, equipment, supplies, weapons, ammunitions and the wounded in wartime France, they would surely make good sense in the peacetime fields of Lincolnshire.

They did. Eventually, the Nocton railway extended to twenty-three miles, its former ration and ambulance wagons loaded with crops collected from every field on the farms. The railway also took post, supplies and fresh water to remote farm cottages. It fetched and carried gangs of potato pickers, food and water for livestock, muck for spreading and coal to the fen pumping station. Most importantly, it carried thousands of tons of potatoes direct from the fields to a mainline railway station on the Lincoln to Sleaford line, from where they could be distributed to anywhere in the country.

The trains were hauled across the fields by tiny Simplex petrol locomotives, converted to diesel, although until the mid-1930s the heavy trains to the mainline station ran behind a pair of steam locomotives. Their tanks were replenished during the day from standpipes, one of which – in the guise of a handsome fluted classical column – we have restored and put back on its original site. The locomotives shared this standpipe with steam traction engines used for ploughing, threshing and pulling wagons.

In 1936, Smith's Crisps, a hugely successful concern, bought the Nocton estate. The nation's hunger for potato crisps kept the railway going at full stretch until new farm roads and lorries led to its closure from 1960. The railway's engine shed was what is now The Hive, our meeting and conference centre at Wasp's Nest, a wonderfully named hamlet on the estate. We've laid a new section of narrow-gauge track inside the renovated building, not simply for nostalgia's sake, but to remind us of how technological innovation has played a key role in Nocton's history.

As for those millions of packets of crisps, in the 1950s, when we were sailing with my father at Blakeney Point on the north Norfolk coast, we used to stop off afterwards at the Anchor Inn at Morston for ginger beer shandies, beer and, yes, Smith's Crisps with those blue twists of salt hidden somewhere inside the greaseproof packets. Crisps were invented by Mr Smith in London, and, as well as buying Nocton, he set up a factory in Sydney, where they are still called Smith's Crisps. Unlike in the UK, where they are now, disappointingly, Walkers.

Our farms at Nocton and Carrington, about twenty miles east towards the sea, benefit from the rich soil they rise from. These are the Fenlands. They contain about half of England's Grade 1 agricultural land. But, until the seventeenth century, when Dutch engineers from whom I am descended were invited to help drain them, the Fens were water-sodden and of little use to farmers. Following pioneering if not always successful drainage by those Dutch engineers, real advances were made in transforming the Fens in the late eighteenth and early nineteenth centuries by landowners and engineers of the stature and invention of John Rennie, a Scottish farmer's son. Man-made rivers, canals, ditches and embankments served by pumping stations – steam from the 1820s, diesel a century later and electric today – made this an engineered landscape, with a haunting beauty of its own, that works in all our interests in producing prodigious quantities of food. The vital pumping stations, all 286 of them, are fully automated and work around the clock to prevent flooding. In a twenty-four-hour period they can pump water from the equivalent of 16,000 Olympic-size swimming pools.

In the 1950s grants were paid to farmers to improve drainage by means of perforated clay pipes across fields, and ditches around the fields, to carry away the water to larger ditches and dykes. Over the past fifty years the pipes have become clogged and the ditches filled in with soil, bushes and trees. Poor drainage leads to lower yields or ruined crops. We have cleared the ditches and excavated deep so that the cross-field pipes could drain into the ditches, and at the same time cleared out the pipes across the field or replaced them. We have also criss-crossed the fields with drainage channels made by towing a stainless-steel mole. These provide very effective drainage channels.

We've been doing our own bit. We've built a 50-million-gallon reservoir at Nocton to maintain the balance of water levels across the estate and throughout the year, while on a micro scale we drip-feed water through irrigation tapes to growing crops like potatoes. The Fens are a waterland. They need constant surveillance and maintenance, including the annual clearing of reeds, privately funded by farmers, to prevent clogging of the waterways, yet nothing – not a drop of water – need be wasted if the technology and infrastructure is right and evolves to become ever more efficient and courteous towards the environment.

We've incorporated a specially designed lip around the edge of our reservoir, so water is always available for wildlife to drink from, and birds, among them oystercatchers, kingfishers and egrets, are flourishing here. The reservoir is surrounded by 15 hectares of wildflowers, so it's a haven, too, for pollinating insects like bees and butterflies. Engineering interventions in the landscape can be wholly in tune with it rather than impositions upon it.

Inventions, and their innovation, revolutionised English agriculture from the late seventeenth century. The impact of machinery, like Viscount Jethro Tull's horse-drawn seed drill of 1701 or the Scottish engineer Andrew Meikle's threshing machine of 1786, raised the productivity of British farms many times over. Norfolk witnessed revolutionary improvements in soil health as its farmers – notably Charles 'Turnip' Townshend at Raynham Hall – adopted a four-year crop rotation cycle with fields given over successively to wheat, barley, clover and turnips. Each of these crops acted on the soil in different ways, each benefiting the others.

English farms became extremely productive. The population grew rapidly. With a large supply of food and labour, a new generation of factory proprietors was able to think big. If unwittingly, the revolution in British agriculture during the seventeenth and eighteenth centuries, which also witnessed major improvements in the quality of sheep and cattle, did much to support the Industrial Revolution.

In turn, the Industrial Revolution made farms ever more productive.

In the 1850s and while engaged in the repair of agricultural machinery in a Kentish workshop at Rochester, Thomas Aveling had watched teams of six heavy horses dragging stationary steam engines to and from farms. He likened this to 'six sailing vessels towing a steamer', a thought, he said, that was 'an insult to mechanical science'. In 1859, the 35-year-old farmer and mechanic invented the steam traction engine that, in turn, led to the internal-combustion-engine tractor and, ultimately, when combined with digital technology, the new generation of machines Dyson Farming uses today.

My buying of farms began with a conversation with Nick Worboys, a farmer's son, who came to work with me as estate manager of Dodington Park when I bought this rather run-down Gloucestershire estate as a home ten years before one part of Nocton came up for sale. Nick was very enthusiastic, saying the best farmland in England was in Lincolnshire and Norfolk. By the time Nick had bought Nocton on our behalf, we started to think how we might make a success of farming.

Our first task on buying the three farms to reunite at Nocton was to spend capital and much effort modernising and rebuilding the infrastructure. As neighbouring landowners saw our restoration activities, they wondered if we would like to buy their farms. We never approached a farmer. It happened the other way round. Sometimes the price was too high. One of the best-maintained farms belonged to Michael Cornish. I knew him and his father, who owned Linpac during my Ballbarrow days. He wanted to retire and thought that we would nurture and develop his beloved farm.

If we could revive these and other farms and increase their yield, we could perhaps make an impact on the country's over-reliance on imported food. In the great scheme of things, this would reduce the energy-cost and carbon footprint of the food we buy. If, at the same time, we could generate our own energy from waste on our farms, we might – given the latest technology – become net exporters of electricity to the National Grid.

It got more interesting. Although this is not a new idea, I thought that if we were able to sell food direct from our farms, we could cut out the middleman – supermarket buyers – and so earn significantly more for our produce than English farms have been doing in recent decades. If you buy a packet of peas for £1 in Waitrose, for example, the farmer gets about 20p, one-fifth. I know this because we supply thousands of tons of peas to supermarkets. If this sounds a bit harsh for farmers, there's more to it. Supermarkets only pay for peas when they need them. We plant the peas well ahead of this. We harvest them, bear all the risk and act as bank for the supermarket. Alone, our three pea harvesters cost £1.8 million. We might be good at growing peas – I think we're the largest pea grower in England – but we earn very little from it. Over the last ten years we have lost an average of £100,000 each year. Nature can never be taken for granted.

The food market is fundamentally skewed, so we're working out how to change the commercial model. If we can make a success of selling direct to the consumer, and so cutting out the middleman, we can also reduce our reliance on subsidy. As Britain leaves the European Union, powerful voices object to farm payments, which means that subsidies will decline and big farmers, the main target, will be hit the most. The same people want farmers to be paid to turn their farms over to re-wilding and to be used as leisure spaces for everyone. They neglect to mention that if we did this we would have to import all our food. Some imagine that big farmers are pocketing farm payments. Some of these, however, are for proven greening practices for the climate and environment, on which most farmers lose money. What they also neglect to mention is that it is very hard not to lose money growing food, whether you are a large or small producer.

This will be the second major reform of subsidies. Prior to 1992, throughout the EU farmers received 'price support' for their produce. When the flat rate for land owned was introduced, it was seen as the saviour of farming! I'm not sure that farmers are seeing the latest proposals in anything other than a negative light, putting them at a distinct commercial disadvantage to well-subsidised EU farmers. As I know from experience, supermarkets buy where they can find the cheapest source, often switching to EU farmers as currencies fluctuate. Managing without subsidies and without a level playing field for British farmers may be a spur to fresh thinking, invention and innovation. However, farmers may yet give up growing food and Britain will need to import more and more food to feed the nation.

Food production is, of course, an old story, with the retailer taking the lion's share of margins. It's much like when I started out with the vacuum cleaner business. Wholesalers and retailers made most of the money then, which is why today a lot of our sales at Dyson are direct. Quite how we get to sell all our farm products direct is not an easy problem to solve, but we're working on it. We supply restaurants direct. We are looking at food boxes and markets. You can now buy Dyson Peas, Dyson Potatoes, Dyson Beef and Lamb and Dyson Strawberries, direct from us.

Along with grain and vegetables, we also make and sell electricity. Our Lincolnshire farms provide electricity through anaerobic digesters to more than 10,000 homes. If we could sell electricity to households at a lower price than they pay today, we would be only too delighted. But electricity bills don't come direct from the National Grid. The electricity 'retailers' have a monopoly of distribution, courtesy of Ofgem, the supposed regulator, their demands sent out from an array of middlemen who do absolutely nothing except to play with computer programs and prepare and post bills. They neither generate electricity, nor take risk, nor maintain the infrastructure. Cutting out the middleman, and those who add no value, ought to be a popular national campaign. It would mean a possibility of profit for risk takers and producers, and lower prices for consumers. And why is the regulator, Ofgem, preventing producers directly supplying users?

The glory days of Nocton, meanwhile, were long behind it when we set to work. When visitors ask where the great house is that ought to be at the centre of an estate this size, we can only point to the ruins of Nocton Hall. The house mysteriously went up in flames at around the time we were talking to the local council about restoring it.

While its style was Tudor, the house, now in ruins, actually dates from 1841. It rose from the ashes of an earlier Jacobean and Carolean house, highly impressive from all accounts, destroyed by fire in 1837. During the First World War, Nocton Hall served as a convalescent home for American soldiers. A military hospital again in the Second World War, it was run by the RAF until 1983 and then by the USAAF until 1995. It had a brief life as a residential home before being bought by developers in 2000. One day, perhaps, we'll be able to buy the ruins, restore it and make Nocton Hall special again with some new and interesting purpose. We are in farming and the stewardship of this and other estates for the very long haul, so we can bide our time to make things right.

The ruins of Nocton Hall stand close to ancient woodlands that had been invaded over the years by domineering rhododendrons. They are also home to glorious oak trees, some more than 600 years old. This was a wonderful place when we first tramped through it, yet there was clearly a great deal to do. Fortunately, we were in the position of being able to invest heavily in the estate. A part of this investment was to bring together a brilliant team of enthusiastic graduate farmers and specialists in any number of related, if new, disciplines, among them a dedicated woodland manager.

Along with shepherds and farm managers, we employ agronomists, researchers, engineers, drone pilots and technical data analysts. Someone you see knee-deep in sillion in the middle of a field here, come rain, shine and Lincolnshire wind, might well have a doctorate in soil research. While we're investing in sophisticated modern machinery at Nocton, we're also developing the best natural ways to nurture soil and so weaning farms off their dependence on chemical fertilisers.

Many people, unconnected to agriculture, may well be suspicious of farming and science working together, yet today's innovations are increasingly born from nature itself and are beneficial rather than destructive to the land and wildlife.

The reason for this is important. Here at Nocton we can improve the soil we use for growing crops with what's known as 'green manure', or plants that encourage the fertilisation of the soil. These are called 'cover crops' and are planted typically between summer harvest and spring planting to improve the soil by retaining nitrogen, reducing soil erosion and controlling weeds, thus avoiding the use of weedkiller. We have done considerable research and trials to come up with the most effective recipe for cover crops. Without chemicals or fertilisers, flora and fauna flourish. We aim to establish a healthy balance so we can grow large quantities of healthy and delicious food while allowing nature to thrive.

Walk through Nocton and you'll see brown hares racing across fields, buzzards, and kestrels in pursuit of rabbits, owls emerging silently at sunset from the many hundreds of boxes we've made for them, countless bats and dragonflies in our woodlands and so many more shy animals living as in harmony with us as is possible.

Instead of spraying potato fields with pesticides to keep unwanted insects – aphids in particular – at bay, we intersperse the rows of potato plants with 'tramlines' of meadow flowers. This might mean the loss of a certain percentage of the potato crop, but the wildflowers attract ladybirds that feed on the aphids and so save very many more potatoes. The flowers also attract bees, other essential pollinators and birds that feed on insects.

At the same time, we plant hedgerows to protect topsoil from the challenging winds that blow across the flatlands of Lincolnshire. Hedgerows are also havens for birds. They are attractive, too. I take pleasure in the fact that good-looking farmland is also the most productive and most sustainable farmland.

At our Gloucestershire and Somerset farms, we're rebuilding 15 miles of Cotswold drystone walls with the help of the Dry Stone Walling Association. I like the fact that we have the only man who can build drystone walls staying on one side of the wall. This may sound pointless until you imagine the bother of having to keep climbing over the wall to the other side as well as having to maintain a sufficient pile of stones on each side waiting to be used. Our man has just one pile on his side.

I admire the skill of such craftsmen. Built without mortar, drystone walls are costly to build, not least because those who know how to make them are in short supply.

When we lived in the Cotswolds, Deirdre and I tried to build a 6-foot-high stone wall. It collapsed in a rainstorm before we rebuilt it properly. We hadn't understood the subtle structure of drystone walling. We learned and rebuilt our wall. These walls serve a purpose, of course, as they have since the Bronze Age. They act as field boundaries. They serve as shelters for sheep in blazing sun, fierce winds and snow. They are home to many species of moss, lichen, small birds, mammals, invertebrates, lizards, insects and wildflowers. You'll find them in areas where it has always been hard and even impossible to establish hedgerows. Drystone walls are a beautiful, hardworking and enduring part of the English landscape. Some of those you see in the Cotswolds and Yorkshire Dales are hundreds of years old.

Protecting and encouraging wildlife, while maintaining a natural balance that allows us to farm efficiently, is a part of good stewardship of land, and a code of being that we aim to live up to. It's exciting, for example, to watch our giant combine harvesters at work. These employ the latest digital technology that allows them to work fields with a degree of accuracy farmers could only have dreamed of until recently. We use a combination of GPS and a beacon system, and it is the beacons that give us millimetre accuracy for each row of planting and working. Drones survey and map every last square millimetre of our fields. With this data, the combine harvesters are programmed to cut around the ground nests of curlews, lapwings and rare marsh harriers, a breed of bird that until recently was very nearly extinct. We're seeing a significant rise in the number of marsh harriers at Nocton and Carrington, and this has been achieved through the latest technology. We can farm with increasing efficiency while the birds thrive.

At the same time, between them our tractors and harvesters sow, fertilise and reap crops with a precision that means there is very little waste. We can farm with increasing efficiency while nature thrives. To me, all this is common sense. I am not by any means a hair-shirted conservationist or ardent re-wilder. I don't plan to release wolves into our estates. Rather, I aim to be a highly productive and profitable farmer growing as much good food as possible. But, with the aid of invention, innovation and technology, I can help to shape farms with my colleagues where human demands are in balance with those of nature.

It's hard not be impressed by the latest farm machinery. Harvesters, for example, are huge and highly complex machines marrying mechanical and digital technology. We have several Claas combines. These are German machines. Claas, in fact, invented and built Europe's first combine harvesters. In the early 1930s, though, the Claas brothers – there were four of them in business together – were unable to get support for their idea from German farmers.

They were faced with that Henry Ford conundrum: 'If I had asked people what they wanted,' he said, 'they would have said faster horses.' Instead, he offered them a reliable and affordable automobile that farmers took to like ducks to water. Taking the lead in the face of the faster-horse brigade in rural Germany, August Claas said, 'Then we'll just do it on our own.' The brothers designed and made their combine harvesters down to every last nut and bolt. When farmers saw what Claas combine harvesters could do, of course they wanted them.

This reminds me of my own experience when existing vacuum cleaner manufacturers turned down my cyclonic and bagless prototype. When you're sure of your invention, you have to pursue it as much for the sake of proving to yourself that it's a good idea as in the hope of revolutionising the market for it. The Claas brothers revolutionised European farming. They had to give up production in 1943 when Nazi Germany was up against the wall and fighter planes, tanks and other weapons were valued more highly than the latest farm equipment. As they did with Ferdinand Porsche's radical air-cooled Volkswagen, it was the British who recognised the significance of the Claas brothers' combine harvesters. Orders from Britain lifted the company back on its tracks soon after the fall of the Nazi regime.

Claas farm machinery is highly efficient and reliable, but not cheap. Our machinery is a part of the £110 million we've invested so far beyond land-purchase costs. It was, then, a little galling to find an article in *The Times* in August 2020 – August is the height of the silly season in national journalism – trotting out the lazy claim that I had bought farming land for tax avoidance reasons.

'The billionaires,' wrote Sir Max Hastings, the well-known military historian, 'who merely speculate with the countryside should stick to oil futures and vacuum cleaners.' I wrote to the paper saying that I found this 'insulting to the 169 intelligent and dedicated people I employ on my farms ... these people are completely dedicated to advancing the state-of-the-art of their profession and should be celebrated for the huge progress they are making.'

I wrote about the initiatives we have been taking and the ways in which we are working hard to change English farming for the better for consumers, the countryside and farmers themselves.

'If this is tax avoidance,' I wrote, 'why would I bother with all of this? There would be far easier ways!' I happen to pay a very large amount of tax to the British exchequer, but the point is that I would have to be seriously mad to go to all the bother of creating a hugely expensive and productive new commercial adventure, to employ so many people. To his credit, John Witherow, the editor of *The Times*, sent Alice Thomson, a feature writer, to see what we're doing, and the result was a lively and accurate picture of Dyson Farming.

Dyson Farming has grown very quickly to become one of the biggest such operations in the country. We haven't wanted to be big for the sake of it, but to benefit from certain economies of scale, although perhaps not so very deep down the romantic in me wants to save, protect and nurture as much English farmland as possible.

Scale comes with both increasing possibilities and responsibilities. I had no intention, for example, of becoming a landlord. In fact, when we began to buy farms, I really wanted land with no residential property. Now, though, we own some three hundred properties on the farms, of which 180 are homes to rent. Restored barns or other brick and flint buildings around redundant farmyards, each one has been renovated to a high standard. I remember with fondness the flint, brick and pantiled farm buildings in north Norfolk. The farm buildings in Lincolnshire are similar and it is delightful to be able to restore them and turn them into homes for families in our old farmyards.

On our Lincolnshire farms we've built two anaerobic-digester power plants. These produce biogas – methane mostly – from maize, other crops and waste products from the farms, that feed Austrian-built GEC-Jenbacher 1,500hp engines. In turn, these generate heat, which we use to dry grain and heat glasshouses, and electricity, which we feed into the National Grid. The engines run around the clock though they need constant nursing and regular servicing. The anaerobic digesters also produce organic fertiliser for the farms. Combined with run-off water from our yards, excess heat and gas from the digester on the Carrington estate is used to grow strawberries out of season in our big new greenhouse and to dry grain in our grain stores.

Built by Dutch contractors, the greenhouse's filigree structure resembles a lightweight, contemporary version of Joseph Paxton's Crystal Palace. It's been designed to produce 750 tonnes of strawberries a year from 700,000 strawberry plants. What this means is that there is now less need for retailers to buy strawberries picked abroad in winter and to fly these thousands of miles to reach British homes, restaurants and hotels. At Dyson, we are trying at every turn to touch the ground lightly in everything we do, to make more from less and to create a circular system of production through which we aim to recycle everything we use, whether in our farms or our factories.

Of course, there's a long way to go. For all our research into vision systems and robotics, Dyson farms – for now – need skilled manual labour to sort potatoes and select strawberries. Human eyes and hands are, to date, quicker at spotting unripe strawberries or green potatoes than machines.

As to the big question of what our farms can teach the rest of Dyson operations, one answer is the possibility of new materials for the millions of products we make every year. Our scientists and research engineers are working hard on long-term solutions. At least one answer lies in the soil. This is corn starch. Corn starch comes from plants grown in soil rather than from the soil itself. For a number of years now it's been possible to extract polylactic acid (PLA) from corn starch; we are currently producing it from sugar beet. This is a carbon-neutral and biodegradable alternative to petroleum-based plastics. The idea of being able to grow organic plastic on farms sounds very exciting. These, though, are early days. While PLA is used to make an increasing number of what are known as commodity plastics for food containers, packaging and even tea bags, as well as for 3D printing, it is not suitable for engineering plastics used in the manufacture of, for example, vacuum cleaners and hairdryers.

What I can say is that if you came back to see what Dyson's up to in five, ten, twenty or a hundred years from now, whether with our products or through our farms, things will be very different indeed. It's all tremendously exciting and we should have cause for optimism.

Equally, I do have reservations about some of the things we're doing now. It seems to me that there is a question about whether the production of energy trumps the production of food. How much of our effort and investment, for example, should go into the production of electricity? On the one hand, it's satisfying to think that we can generate 'bio-electricity'. On the other, it's questionable as to whether we should be growing maize to feed anaerobic digesters rather than using the land to grow crops to feed people. Although, that said, we are making good use of the heat and waste by-products.

And, of course, I have my own particular likes and dislikes. I happen to hate wind turbines – they disfigure the landscape, kill birds and massacre insects, too – but vast wind farms off the Lincolnshire coast are connected to the National Grid underneath tracts of compulsorily purchased land from our farms and there is nothing we can do about it.

I feel the same about solar panels. You see them disfiguring buildings

Launching the Dyson Corrale straightener at our Paris Demo store as the world closed down when COVID-19 struck in 2020

Conversion under way of the former British-built St James's Power Station on the waterfront in Singapore into a new, open all hours, Dyson HQ with R&D laboratories. The windows are the tallest in Singapore

Computer rendition of the Dyson EV (electric vehicle) as if about to set off from Hullavington on a 600-mile drive without the need to recharge

With a wheel at each corner, like the Mini, passengers were given as much space as possible within the envelope of the car.

Seats with generous lumbar support are a homage to Charles and Ray Eames Soft Pad chair, a favourite design

How the *Sunday Times Magazine* of London reported my decision to pull the plug on the car project. I'm standing with a clay model in one of the hangars at Hullavington

RICH LIST | INTERVIEW

"I BLEW HALF A BILLION QUID... ...ON A CAR"

Sir James Dyson spent £500m developing an electric car to rival Tesla's. Then he scrapped it before the first prototype took to the road. He tells *John Arlidge* why

Cutaway of compact and energy efficient Electric Drive Unit incorporating Dyson digital motor, power electronics and transmission to front and rear wheels

(Right) Working on the clay model of the car

Cultivating, ridging, bed-forming and planting potat at Dyson Farm at Nocton, Lincolnshire

Three pea-vine harvesters at work at Noctor

The peas – billio of them – go fror field to freezer in under 120 minut

Producing 750 tonnes of Englis strawberries at Carrington, Lincolnshire, ou of season using electricity and heat from our anaerobic diges

side one of the
e-assembled
ds, with cross-
minated timber
alls and ceiling,
at, together,
m the student
commodation at
e Dyson Institute
Technology
d Engineering
our campus
Malmesbury,
ltshire.

esigned the desk
d bench, made
us by British
mpany, Centrium

ight and below)
part of the
escent of student
ods. Delivered
truck, lowered
to place by crane
nd connected
p. Each slotted
recisely into
lace alongside,
bove or below
s neighbour.

Carvey Maigue, a student at Magúa University in the Philippines won the Sustainability category prize in the 2020 James Dyson Awards for a revolutionary system that converts fruit waste into ultra-violet sequestering windows that can then produce electricity, with buildings becoming solar farms

(Below) Always the optimist: in 2019, and just as the COVID-19 pandemic struck, I unveiled our six-year plan for Dyson in Hangar 85 at our Hullavington campus

(Opposite, top)
The first intake of Dyson Institute undergraduates at Malmesbury, September 2017.
These talented young engineers graduated in September 2021

(Opposite, second from top)
The Roundhouse, by Chris Wilkinson, is a café, library, cinema and lecture theatre

(Opposite, second from bottom)
Opening in September 2021, the Dyson Building is a new centre for STEAM (Science, Technology, Engineering, Arts and Mathematics) cross-disciplinary education at the heart of my old school, Gresham's in Norfolk, its lightweight steel frame adorned by green plants

(Opposite, bottom)
The top-lit central space is centred on a giant auditorium stair. We first did this in our Development Centre in Malaysia

My son, Jake, and I meet in half a Mini at Malmesbury when he joined Dyson in 2015. We share the same tailor, my son-in-law, Ian Paley

Deirdre continues to paint as well as creating rugs. Here she is, in 1996, at work in Provence

and stretched across fields that would be better used to grow food. I am mystified, too, by the fashionable argument that the English countryside should be turned into a kind of outdoor rural leisure centre for people to wander around and soak up nature, and with flowers replacing crops. Who is going to maintain millions of acres for this purpose? Surely the challenge of feeding our own people with our own foods is of prime importance? It's like the argument that Britain doesn't need or want to manufacture. We do make really good food in this country when we choose to, but not on a scale large enough to stop us importing huge and what should be unnecessary quantities of staple foods, let alone exotic fruit and vegetables, from around the world.

And why should we expect people to grow food for us in poor countries when they should be growing it for themselves, especially where water is scarce or where export of food drives deforestation. There is, of course, the argument that British people neither want to work in factories nor on the land. This, though, is when invention, innovation and technology can really help us. We will rely increasingly on digital and robotic technology to farm for us, but we can and will do this while being in tune with nature and learning from it. I simply can't bear to think of English agriculture going the same way as British manufacturing.

CHAPTER ELEVEN

Education

Of the 5,127 prototypes I made in the coach house of the cyclone technology for my first vacuum cleaner, all but the very last one were failures. And yet, as well as painstakingly solving a problem, I was also going through a process of self-education and learning. Each failure taught me something and was a step towards a working model. I have been questioning things and learning every day ever since. This, of course, is what engineers do, beginning as often as not as those children whose curiosity leads them to take toys and clocks and radios and, indeed, the very latest technology, apart while not necessarily being able to put them back together again.

Having studied and practised engineering, there is nothing quite as instructive as seeing a test fail before your eyes. This is why at Dyson we don't have technicians. Our engineers build their own prototypes and then test them rigorously so we properly understand how and why they might fail. The action of making the parts to do a test is also really important. It allows you to spot opportunities to do things in different ways, hopefully for the better.

Learning by doing. Learning by trial and error. Learning by failing. These are all effective forms of education. My view, though – especially since Dyson became a global technology company – is that the education of budding and future engineers in all fields, starting at school, is essential as the pace of change and competition intensifies. I think we miss a huge trick in not catering for young people for whom academia is of no real interest if purely cerebral and not applied.

Young children love making things and yet, all too often, this innate curiosity and experimentation realised and expressed through our hands is stamped out by educational systems that see no virtue in such natural creativity. Because parents and schools are driven to get children through school and into university, and because the educational system is geared – notably in Britain, the nation that triggered the Industrial Revolution – to cerebral subjects, the making of things is considered a waste of time as pupils face the challenge of public exams.

And, because of this attitude, design and technology are very poor relations in the school curriculum family. Schools tend to be poorly equipped for the subject and qualified teachers few and far between. The result, aside from a woeful lack of engineers, is the commonly held view that engineers and technicians are somehow lower down the professional and social pecking order than bankers, YouTube influencers and brand managers.

The global economy, however, has moved on apace at an ever-accelerating rate. Driven by new technologies, it needs huge numbers of highly skilled and inventive engineers working in, for example, the fields of robotics, vision systems, signal processing, machine learning, computer architecture and systems along with aerodynamics, acoustics, thermodynamics and structural analysis. We could, of course, decide not to engage with this new world and to tend our gardens instead. Yet, however attractive a proposition this might be, the fourth Industrial Revolution is not going to dissipate any time soon.

At the time of the first Industrial Revolution, there was an equally demanding challenge in finding artisans and engineers with the right skills to shape the railways that spread the gospel and goods of this new world order. In terms of technical education, British governments were slow off the mark. It was left to railway companies themselves and their employees to remedy the situation. From the 1830s, Mechanics'

Institutes appeared in Scotland. The following decade, they emerged south of the border. In 1855, the Great Western Railway opened its Mechanics' Institute in Swindon. Perhaps it was significant that the Institute's Gothic Revival building resembled a church much more than it did an industrial workshop, for by this time railways were a kind of fervent religion and the Great Western the highest church of all.

These institutes were places of learning and of entertainment and social welfare, too. The Mechanics' Institute at Swindon boasted Britain's first lending library while its comprehensive system of health-care was one of the models for the country's National Health Service, founded after the Second World War. My godfather ran the Great Western complex for a while during the war. Such institutes were also funded by industrial philanthropists including Robert Stephenson, already a world-famous railway engineer, James Nasmyth, inventor of the steam hammer and pile driver, and Joseph Whitworth, the machine toolist, whose Whitworth Scholarships for the advancement of mechanical engineering are still awarded to outstanding students, the money drawn from a donation of £128,000 (the equivalent of about £7 million in 2020) he made to the government in 1868.

The British public at large were made aware of enormous strides in new engineering skills not just by the arrival of the railways but also by the Great Exhibition of 1851 – the first Expo or World's Fair – visited by a third of the population. Held in a truly revolutionary building – Paxton's Crystal Palace – the exhibition included manufacturing processes and machinery in action along with a wealth of products made in Britain for export to the world. Altogether there were 100,000 exhibits, while millions of people, for whom the malodorous earth closet was a staple of life in yards outside their homes, were introduced to penny-in-the-slot flushing-lavatory cubicles.

The exhibition was highly profitable, allowing the creation of what became the Victoria & Albert, Natural History and Science museums along with Imperial College, the Royal College of Art, the Royal College of Music and the Royal Albert Hall. This was highly sophisticated public and professional education on an unprecedented scale. To date, postgraduate scholarships awarded by the Commission for the Exhibition of 1851, on which I sat for a few years, have funded thirteen Nobel Prize winners.

In shaming contrast, there was absolutely nothing challenging in the Millennium Experience of 2000. This was a colossally expensive show of nothing particularly new or worthwhile housed inside Richard Rogers's capacious and likeable Greenwich 'Dome'. It was meant to be the Great Exhibition revisited for the much-hyped new millennium. In the event it displayed no depth, no industry and no magic whatsoever. Hardly like to inspire, much less fund, future Nobel Prize winners, the Millennium Experience was a vanity project spearheaded by ambitious politicians. In all, it cost £1 billion, had nothing to teach us or inspire us and left no legacy. Whereas the Great Exhibition was visited by a third of Britain's population, the Dome hosted less than a tenth. Imagine how much better spent that money would have been on scholarships for young people working to invent and shape innovative ideas, better products and happier lives for millions of people around the world. I had expectations that Tony Blair's New Labour government, which had inherited the Millennium Experience project when it came to power in 1997, would cancel it and build much-needed new hospitals instead.

In a sense, the Millennium Experience was a reflection of Britain's lack of education in science, engineering and technology on the scale needed by the turn of the third millennium. The problems dated back to the end of the Second World War, when a new Education Act spelled out the manner in which new generations of children were to be taught. There were selective grammar schools aimed at preparing young people for university, secondary modern schools for those less academically gifted and technical schools for those who, theoretically, would replace the very many thousands of engineers and skilled technical workers killed between 1939 and 1945.

In practice, few technical schools were built. Some spent too much time trying to ape grammar schools. Others taught girls how to cook and keep home. In any case, there was a lack of suitable teachers while the schools and their curriculum faced hostility from trade unions that, powerful at the time, saw 'techs' as unnecessary rivals to industrial apprenticeships. No more than 3 per cent of British pupils attended technical schools. As these ebbed, a new generation of comprehensive schools had, and still have for the most part, little to offer in terms of design and technology courses. Under-resourced and lacking status, 'design and technology' in schools needs all the help it can get from outside sources.

Of course our schools and universities need to teach subjects like philosophy, ethics, pure maths; music and art, too, should be a given at school. Yet the shortfall in the number of the hands-on engineers society relies on ever increasingly, to make knowledge-based and connected economies work, has to be made up somehow. Unfortunately, education has become, in recent years, more and more about how much a young person can spew out in an exam setting. The idea that we should learn parrot-fashion, repeat facts back in an exam and then forget them, is misplaced. This approach may suit some professions – though I can't think which – but for disciplines engaged in problem-solving, like engineering, it simply doesn't work. People who do well at exams at school are not necessarily those who do well at work. Students are rewarded for following the train of thought set in textbooks. If they think for themselves or question textbook knowledge, examiners can't give them the marks their original minds may well deserve.

I love the story of Concorde engineers, at the early design stage of the supersonic airliner, making paper planes and throwing them around in their drawing offices to test ideas for an ideal wing. Some of these models are in the safekeeping of the Science Museum in South Kensington, a happy reminder of how things teachers and examiners might well disapprove of – throwing paper planes in class – might just lead some young people towards some of the greatest designs of all time.

Children appear to be born with the desire and ability to make things, to experiment and play. This is a gift lost by all too many adults, but, especially in a robotic world of the future, humans need to be more creative than ever, to dream up, invent and make things that a machine, for all its algorithms, might find hard to imagine and give shape to. Frank Whittle is a good example of how you don't need to be good at exams at school to change the world. Having left school at fifteen, he became a test pilot. His was a life of learning, and tenaciousness, as he went on to get a double first at Cambridge and then to invent the jet engine, changing the course of aviation, and our lives, with it.

Education should be about problem-solving rather than retaining knowledge simply to pass exams. Problem-solving is something young people are naturally good at. Because there is a huge shortage of engineers, we need a way of educating young people that makes the subject as wonderful and as exciting as it is. In Britain alone we lack something like 60,000 engineers and yet, partly because of the way the subject is not taught at school, all too many young people feel it can't possibly be for them. Surely it's just about bridges, tunnels and perhaps aircraft. No. Although these are exciting things to work on, today engineering covers an enormous and eclectic territory embracing software, robotics, vision systems and virtual reality. It can be about helping the elderly, infirm and ill. It can be about cleaning our oceans and the environment, developing sustainable materials, inventing products that use little energy, conceiving energy-generating materials or, of course, inventing things that don't yet exist.

Sadly, there's very little in the media for aspiring young engineers to get excited about and, certainly in Britain, there's still a snobbery and dislike of manufacturing. To be unable to change a plug, repair a lawn mower or hang a picture is all too often seen as a mark of cultural refinement and social superiority. And yet our humanoid ancestors set out on the story of invention and making things some 3 million years ago when they discovered tools, and, in doing so, found a new way to feed and clothe themselves and to create new forms of shelter. Progress appears to have been slow between then and the Bronze Age. Through the great civilisations of ancient Greece and Rome it gathered speed, and, with a few hiccups on the way, burst into a frenzy of invention in the period leading up to the Industrial Revolution. Ever since, inventions have tended to compound inventions, although the attractive idea of a wholly benevolent technological progress has been rocked time and again by the demands of war, terrorism and retrogressive ideologies.

Without that first discovery leading to the invention of hand tools millions of years ago, there would be no electric light, no phones, no bicycles, buses, trains, cars, no aircraft, let alone computers, artificial gravity and spaceships. Invention, then, is a human imperative. And, while the story of invention is littered with failures, the determination to make a success of something new – whether the printing press, the steam railway locomotive, the telephone, television, the jet engine, the World Wide Web and even the bagless cyclonic vacuum cleaner – is not only its own reward, it also brings previously unknown benefits to potentially billions of people.

The drive to invent cannot be resisted by those in its thrall. Inventors rarely set out to make money per se, and if they do theirs is more often than not a pipe dream. As for children, they need to experiment and play as they learn. Sometimes they might make something useful at school or handy at home. Equally they might make something that seems very interesting but doesn't quite make sense, or at least not for now.

Young people today have their own agenda of sustainability and caring for the planet. We know, through the James Dyson Award, that young engineers are capable of inventing the solutions: we need more young people to study science and engineering to solve these problems rather than the daily stream of the same grandstanding campaigners. But we should surely want young people to think imaginatively and cut across the grain of perceived wisdom. These are certainly the young people we want at Dyson. Although there is no guarantee of success, disruptive ideas can revolutionise a company and its finances through intuition, imagination and risk-taking as opposed to market research, business plans and strategic investment.

During my time at the Royal College of Art, manufacturing was all too often treated by government as a grubby political shuttlecock rather than as a crucible of creativity. Factories were a useful tool in terms of sopping up unemployment in regions of the country where this was unacceptably high. Central government gave generous grants for new manufacturing plants in the employment trouble spots. The real trouble, though, was that while grants made it an attractive proposition for companies to build new factories far away from the sites of their core business, factories could be, and were, built in places where the local workforce was either unskilled or used to working in very different ways and where no appropriate investment had been made in technical education.

When, for example, the Rootes Group built the new Linwood plant near Glasgow in 1963 to manufacture the Hillman Imp, a credible rival to Alec Issigonis's Mini, and to create 6,000 jobs, things were to go very wrong. The Linwood experience caused Rootes to stumble financially. The long-established British car conglomerate was taken over by the American giant Chrysler. The Linwood factory closed in 1981 with the loss of many thousands of jobs. Rushed into production, the Hillman Imp had been underdeveloped in engineering terms and badly built by a workforce drawn from redundant shipyards on the River Clyde. Industrial relations were bad from pretty much the day the Duke of Edinburgh drove one of the first cars out of the factory. The Linwood story shows the danger in both commercial and human terms of governments with little care for, and no understanding of, engineering and manufacturing, and playing politics with industry. Very few MPs today have worked in factories. Remarkably few have ever had anything like a recognisable job. Their entire careers have been as full-time politicians.

Partly to help raise the profile and understanding of engineering, and partly because of my interest in education, in 2002 we formed the James Dyson Foundation with funding from the company. One of the first things we did was to go back to schools, primary and secondary, to encourage young people to share our excitement with the world of engineering.

We started by visiting local schools, our young Dyson engineers giving 'master classes'. The first thing we discovered when visiting the local Westonbirt School was that they had no products to experiment with in their design and technology classes. We started sending vacuum cleaners to schools and running classes ourselves to make the school curriculum more exciting and relevant. Asking young people to solve a problem, we learned, really fires them up. Discovering this was a big turning point; this is the key connection between how young people think and engineering.

The one thousand or so 'Roadie' boxes we sent out were labelled 'First Identify the Problem'. They also included 'Challenge Cards' asking students to respond to popular challenges like marble runs, balloon car races and underwater volcanoes. At the same time, we began introducing schoolchildren to the worlds of ingenious inventors, engineers and inventions, among them cat's eyes – the ones headlamps catch in the dark to help keep us safe as we drive by night – the small-wheeled Moulton bicycle, Kevlar, the Tesla coil, geodesic domes, computer programs, memory foam, Thomas Edison, Sugru (a mouldable glue), windscreen wipers, Isambard Kingdom Brunel, Velcro, zip fasteners and the Smart Fortwo car

Not long after Tony Blair announced his new 'academy' schools in 2000, he asked me if I would consider getting involved. These were state schools independent of the Department of Education designed to raise standards in city centres and managed and part funded by business. Given the challenges we could see and the interest we had, I said 'Yes, I would love to' and committed £12 million to the project.

We wanted to establish what would be a sixth-form college, feeding into local universities – Bath and Bristol universities have excellent engineering faculties – and to teach young people design, technology and engineering. Bath City Council suggested we bought the old Stothert and Pitt quayside crane factory and redevelop it. Innovative steam and electric cranes made here in Bath had been sold successfully around the world.

The site was very close to where Jeremy Fry had built his new Rotork HQ in the '60s, facing the River Avon along the Lower Bristol Road. The City of Bath said it was keen to demolish the Victorian factory. We asked

Chris Wilkinson to design a scheme to replace it. Its architect, however, Thomas Fuller, had moved from Bath to Canada, where he designed the Canadian houses of parliament in Ottowa. We received several letters of protest from Canadians and others.

I went to see Sir Neil Cossens at the body in charge of 'listing' at English Heritage to try to persuade them not to list the building, as a listing would mean that the building would have to be painstakingly restored to its original state and could not be demolished. He told me that Bath Council, the very people telling us to demolish it, had applied for listed status two years earlier.

Chris Wilkinson designed another scheme incorporating Fuller's factory. Time ticked by and costs mounted as, one by one, we complied with all the City's stipulations, including a new river wall, an arcade by the road for pedestrians, a footpath and, not least, a bridge across the River Avon for fire engines. At this point, we had a great design, but strangely Bath Council backtracked and told us that we would have to share the site with Bath Spa University. I said this site wasn't big enough. The council then asked for sealed bids for the whole site. Who would offer to pay the most? We lost to Bath Spa University, which then dropped out six months later, leaving us back in the frame.

We really were being given the run-around. But it only got worse. Before I could speak – limited by law to just two minutes – at the Council's public planning-permission meeting on the project, we heard long speeches from the Environment Agency condemning the site and the City architect, urging people to vote against our building. The vote did go our way, but the scheme was referred to the Secretary of State, Estelle Morris, at the insistence of the hostile City architect. The City architect's advice, as the Council subsequently told us, was entirely false. I was in China when I got a call from Estelle Morris to say that unfortunately the referral could not be reversed.

After we had spent £4 million on the project and invested a great deal of time and emotional energy, the end came for me when I went to see the Blair government's education secretary, Ed Balls, at the House of

Commons. I wanted to know if the government was still serious about the Bath academy. He said 'no'. At least we knew.

When in 2010 the Labour government was replaced by a Conservative and Liberal Democrat coalition, the new Secretary of State for Education, Michael Gove, said he wanted to do the Bath school. This was soon after the publication of a report I'd written at the invitation of David Cameron, the new prime minister, 'Ingenious Britain: Making the UK the leading high tech exporter in Europe', looking at ways in which Britain might reawaken its 'innate inventiveness and creativity'.

The report was an opportunity to put forward my own views, and those of some of Britain's leading industrialists, scientists, engineers and academics, in a coherent form. Britain could, as the prime minister hoped to hear, become Europe's leading generator of new technology, with industry, science and technology creating jobs and wealth. Britain was in the depths of a recession at the time and evidently overdependent on financial services, and yet this was a country that, to date, had produced 116 Nobel Prize winners, second only to the United States – 320 at the time in 2010 with five times the population of Britain.

The biggest challenge for the government was to bring about a culture in which science, technology and engineering were held in high esteem. This came down to education. We needed, as we do even more ten years down the line, to nurture young, creative minds with students pursuing science, technology, engineering and mathematics – the STEM subjects in the national curriculum – in further and higher education.

I said I would employ 3,000 extra engineers overnight if I could find them. They were not available. In surveys, schoolchildren were saying they wanted to be fashion models, YouTubers, celebrities or simply rich, and yet if more had the opportunity to study enthusiastically and creatively taught science subjects, they might well enjoy working in engineering, science and research-based companies.

Beyond this, the government needed to commercialise the knowledge we have in our world-renowned universities, encouraging the practical application of 'blue skies' research in order to create world-beating products. There must be new ways, I wrote, of financing investment in high tech and other companies with fresh and committed support for research and development. With long-term government vision, focus and support, the nation's instinctive talent could propel Britain forward out of recession. The global economy had taken a nosedive in 2008, but we had brilliant, brilliant minds, I said, and a good dose of obstinacy, which was ideal really.

Cameron and the Chancellor of the Exchequer, George Osborne, subsequently enshrined in law the fiscal recommendations of the report, centred on high tax relief for research and development spending, up to 220 per cent of the money spent. This was designed to avoid the government having to pick winners, at which they are notoriously bad, but instead to help those who had already invested in themselves. Technology entrepreneurs who had spent their own money on research would be able, at the end of the tax year, to recover 100 per cent of the money they had spent on research even though they might not yet be paying tax. Tax relief was also given to people investing in technology start-ups. Start-ups are inevitably risky, and these measures were developed to encourage investors to support them as well as match-funding the entrepreneurs themselves. I am happy to report that between 2010, when the tax incentives were introduced, and 2018, British companies had doubled their expenditure on research and development.

Anyway, it was nice to have Michael Gove's support for the school in Bath, so we went back to the City Council saying that we had the go-ahead from the Secretary of State for Education, even though it was an unusually expensive scheme due to its location and the demands of the Council for a fire engine bridge, a riverbank cycle path and many other add-ons. They said no. They had enough schools. Since then, nothing has happened on the River Avon site. But the James Dyson Foundation refused, in a quiet if determined way, to abandon Bath. While our experience with local and national politicians had been, to put it politely, discouraging, I felt we really had to do something to prove the value of design and technology in schools.

In 2012, we decided to focus the Foundation's attention on five Bath schools, Ralph Allen, Chew Valley, Hayesfield Girls, Wellsway and Writhlington. With the schools' support – they put up a quarter of the money – we brought real-life engineering into classrooms along with hi-tech equipment like 3D printers and laser cutters. The project lasted six years, during which time the uptake of the new design and technology GCSE courses in the Bath schools rose from 23 to 32 per cent. While just 16 per cent of girls chose design and technology in 2012, this was up to 38 per cent in 2018.

The project introduced the students to what engineering is really about and in so doing opened up the possibility of engineering as a career for girls as well as boys. Among the female uptake, three times the number of girls expressed an interest in becoming an engineer. The success of the curriculum lay partly in the fact that students were excited about the idea that the jobs they might take up in the future had yet to be invented. By the time they left college, technology would have moved on in yet further leaps and bounds. There was also the real-world problem-solving nature of the projects that were set, and their relevance to the students' lives. The teachers, who enjoyed the courses, found that they helped improve students' ability in maths and physics.

By the third year, we really felt we were achieving something. The schools, their students and teachers were defying the national odds. It all seemed very exciting and an adventure for both us and the schools. But when we spoke to the government in 2015, Michael Gove, as Secretary of State for Education, was busy downgrading design and technology in schools across the country. This caused a temporary drop in design and technology student numbers that year, more, I think, because parents were worried that this might not be the best course to help their children into university, rather than the attitude of pupils themselves. Numbers were back up again in 2017–18. Over the six years of the Foundation's engagement with the five Bath schools, uptake in design and technology in Britain as a whole fell by 54 per cent. This was very upsetting, and I let it be known.

Forty per cent of British employers were reporting a shortage of science, tech, engineering and maths students and there was clearly an urgent need for more engineering graduates, yet Michael Gove was continuing to push focus on the English Baccalaureate and the core school subjects, at the expense of design and technology. The situation was made worse by funding cuts. Because it needs tools and other equipment, design and technology is an expensive course compared to others based on textbooks. It is easy for schools to make significant cuts in their budgets by abandoning design and technology. Partly because of this policy, A levels in design and technology were no longer needed for students applying for engineering courses at British universities. What this meant was a further separation of academic from practical research and manufacturing as if design and technology was just old-fashioned woodwork for boys and neither academically rigorous nor challenging.

In 2010, David Cameron had asked me to sit on the Prime Minister's Business Advisory Group, something I did for five years. We met at 10 Downing Street every three months to discuss the issues that faced businesses. Naturally, I represented the cause of engineering, technology and manufacturing, constantly stressing the need for more engineers and how important technology was to create exports. I was also the only representative for private businesses, which are always overlooked in government circles, where public companies and the lobby organisation, the Confederation of British Industry, are powerful voices.

On one occasion, Michael Gove sat in on the Prime Minister's Business Advisory Group meeting in Westminster. I can't say much, but I didn't hold back in expressing my dismay about the downgrading of design and technology. What could I do? What could the Foundation do? In 2016, I went to Westminster to see Jo Johnson, Minister of State for Universities and Science. The meeting started rather badly because I went to the loo and discovered that he, in his British government building, had a foreign hand dryer. It wasn't a Dyson, and my hands were wet, so I began by voicing another complaint. I don't know if he ever changed his hand dryer, but to his great credit Johnson threw down a challenge on the shortage of engineers. If I wasn't getting anywhere with the education system in my quest to raise the number and quality of engineering graduates, why didn't I start a university of my own?

At the time, Jo Johnson was nurturing his controversial Higher Education and Research Act through parliament. If this became law, it would give degree-awarding powers to new education providers. And this could mean Dyson. While very excited by the idea of starting a university from scratch, I was fully aware of the dangers of Johnson's Bill. The National Union of Students took against it, calling it the 'marketisation of education'. There was significant opposition in the House of Lords led by Lord Patten, Chancellor of Oxford University, espousing the same reason. The universities of Bath, Cambridge, Oxford and Imperial College London also took umbrage at Jo Johnson's Bill.

I was also well aware of the dangers of turning universities into big businesses with vice-chancellors acting like CEOs and many of their management staff earning salaries to match their new ambition. I don't think we should be bringing foreign students into ever-expanding British universities for their money rather than for their culture and knowledge. I feel they are short-changed, and this is wrong.

I decided to be enthusiastic despite residual doubts because of my largely negative run-ins with politicians. At Dyson we certainly had good and really important reasons to take up Johnson's challenge. We had been taking on a high proportion of graduates for years and when you walk around the Malmesbury campus it does feel much more like a university than a company. The average age of Dyson staff is young and, though we work very hard, it's an enjoyable place to be.

If we had operated in a narrow field of engineering, the university idea would not have worked, but since we work across and research a broad range of disciplines, I thought we could give the right range of experience to potential undergraduates, everything from fluid dynamics, new-technology batteries, software, new electric-motor technology, turbine development, AI, algorithms, electronics, robotics and aerodynamics. We also had some land to spare where we could build an undergraduate village.

We approached a number of universities to partner with us, as in the first few years we would not be allowed to call ourselves a university or award officially recognised degrees. They said no. There was clearly some concern among Russell Group universities – a collective of those concerned with research – about us entering the sector. This seemed odd given that we are such a small fish. The WMG (Warwick Manufacturing Group) department at the University of Warwick, however, had the imagination to work with us. This large and influential department collaborates closely with research and education programmes in industry. One of its key founders, Kumar, later Professor Lord, Bhattacharyya, the British-Indian engineer, educator and government adviser, turned the situation around for us and we have had nothing but praise for Warwick ever since.

It all happened very quickly. Within eighteen months of my meeting with Jo Johnson and for an initial outlay by the Foundation of £35 million, the Dyson Institute of Engineering and Technology received its first cohort of forty-four undergraduates. These were top A-level candidates drawn from across the country. They had chosen us, a brand-new venture, over established universities including Imperial, Oxford and Cambridge. It was humbling and thrilling at the same time to see these immensely bright young faces at Malmesbury. We had no idea in 2017 that just one slot featuring the launch of the Institute on BBC Radio 4 would prompt 900 applications for the forty-four available places.

We did offer certain attractions. There were to be no tuition fees. Our undergraduates would work three days a week with Dyson on real research projects alongside young Dyson engineers, earning a proper salary, and we would teach them for the remaining two days. When the first undergraduates complete their four-year course in 2021, they will, or should be, debt-free. I find it extraordinary and truly sad that most British graduates are saddled with huge debt at the very moment they go out into the world. Our graduates will have studied intensely, worked for a living and gained immense experience. They are not tied to Dyson.

(Top) **DIET Pods** WilkinsonEyre
(Middle) **Roundhouse** WilkinsonEyre
(Bottom) **Desk Bench**

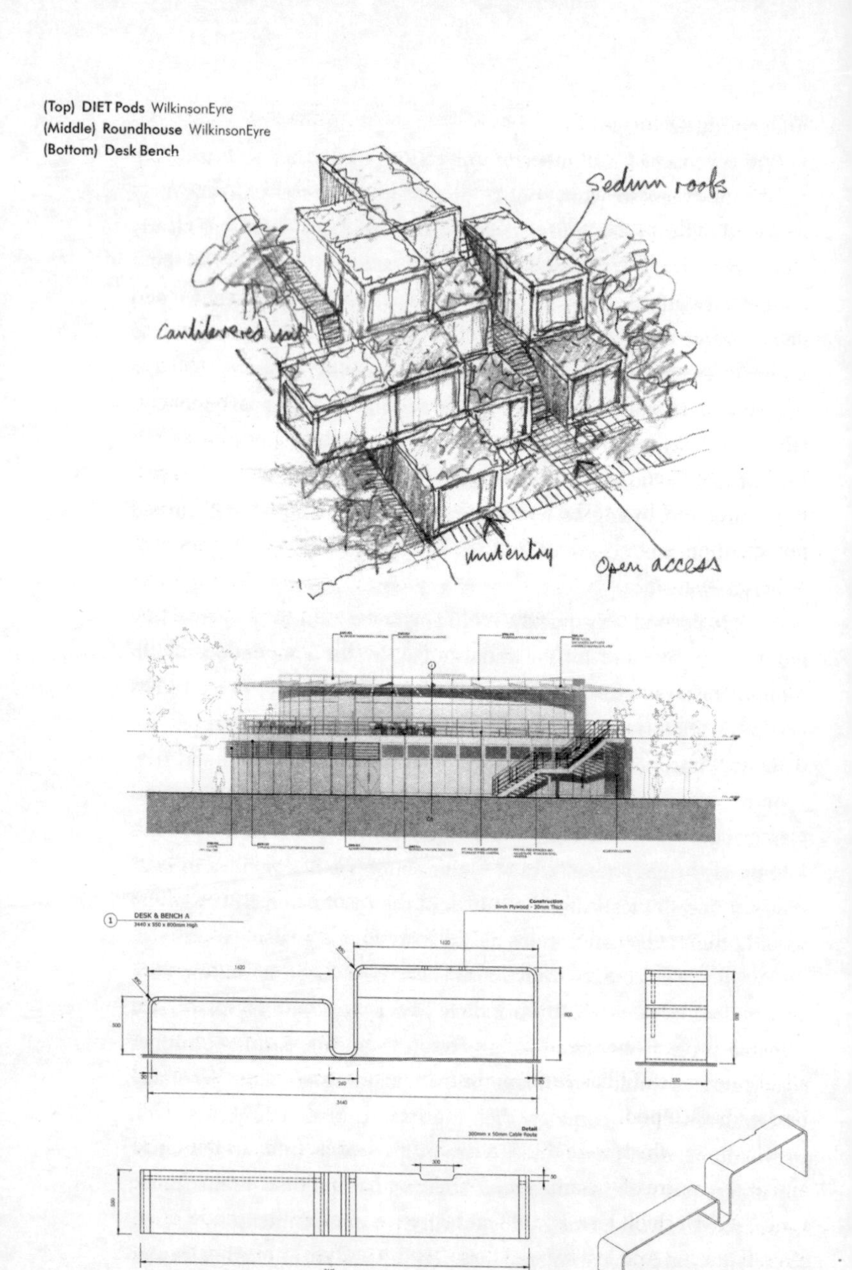

They can go where they like. They will owe nothing financially, although I hope as many as possible will want to stay with us.

The idea of starting a university from scratch raised a number of issues, from the practical to the pastoral. Where would the undergraduates live? I asked Chris Wilkinson to come up with a scheme for student accommodation on site at Malmesbury. His first version was for a student block. However, I was keen that students should feel like independent individuals when they came to join us at Malmesbury and live in houses of a kind that would feel their own. I wanted these to be made in factories, a symbolic reference to our world of manufacturing products.

I remembered the Canadian exhibit called Habitat at the Montreal Expo, designed by Moshe Safdie, with concrete box apartments juxtaposed at right angles and cantilevered on top of one another. Ours were to be rectangular boxes built from CLT (cross-laminated timber, a sustainable material) for the floors, ceilings and walls. Insulation would be put on the outside of the box and the entire structure clad in aluminium. The timber sequesters carbon from the atmosphere and, as we would need virtually no heating because of the wonderful insulating properties of the material, our student houses would be sustainable.

So, Chris and I agreed on something new, a cluster of factory-built pods, each a home of its own that could be slotted, fully fitted out, into place on a site overlooking the Wiltshire countryside. A number of the pods spaced regularly through the scheme are kitchens where undergraduates can cook and eat together (Joe Croan, our wonderful three-star Michelin head chef, teaches them to cook) and launderettes so they can easily wash their clothes. One pod is allocated solely for students to receive their online orders. I really enjoyed designing an all-in-one piece of thick bentwood ply to form a desk, folder store and bench for each pod.

The pods, which were lifted into place by cranes, form a semi-circle and at the centre of the arc Chris Wilkinson has built our Roundhouse Café, library, clubhouse and a separate sports hall in the shape of an aircraft hangar. And just beyond these, there are Dyson's other cafés and laboratories. We knew we would be taking on the care of young people, many of them away from home for the first time. What's more, theirs was to be no easy ride. Undergraduates at the Institute, a third of whom are women, work forty-seven weeks a year, much longer than the twenty-two weeks at other universities, but they have certainly been up for the challenge.

In 2020, by which time we had 150 students, we were the first institution in the UK to receive new degree-awarding powers (DAPs). It was a challenging journey to getting these powers, but this is quite right since we are being trusted with a most vital ingredient of a young person's future. We, for our part, are putting our faith in these young engineers and in the revolutionary technologies that they will pioneer – investing in them and backing them with a supportive environment that encourages risk-taking and new ideas. The aspiration is that they choose to remain here long after graduation, and quickly become the leaders of the business. In this way I believe the Dyson Institute represents a modern, forward-thinking, twenty-first-century education relevant to the world of work.

The undergraduates have a unique education at Dyson. They are working with the best engineers and scientists in the world – real practitioners. They learn and invent alongside them. They experience a very wide range of disciplines on products and research that go into production quickly. It is not for the faint-hearted. As well as having a forty-seven-week year, they are here for four years rather than three. I estimate that this represents two and a half times the teaching offered by traditional universities. Plus, the students are working on real, live products. We positively encourage them to think against the grain, to come up with entirely new ideas. The aim is to give students wings as well as degree certificates, allowing them to take creative flight. With their training, they are also professionals from day one.

I am often asked what we will do next. Expand? Take on many more undergraduates? Build a bigger campus? In fact, I'm keen to keep the Institute, or University as it will soon be known, small. We certainly

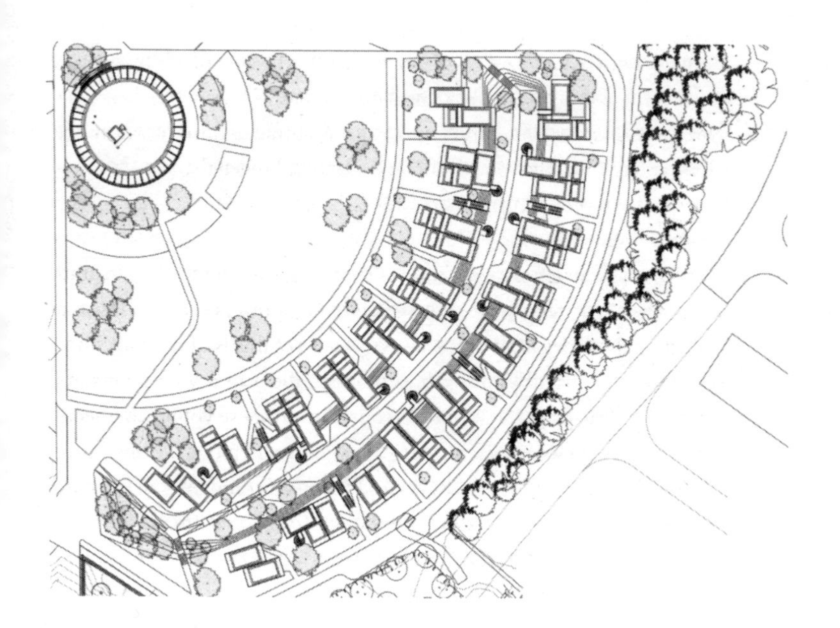

DIET Village WilkinsonEyre

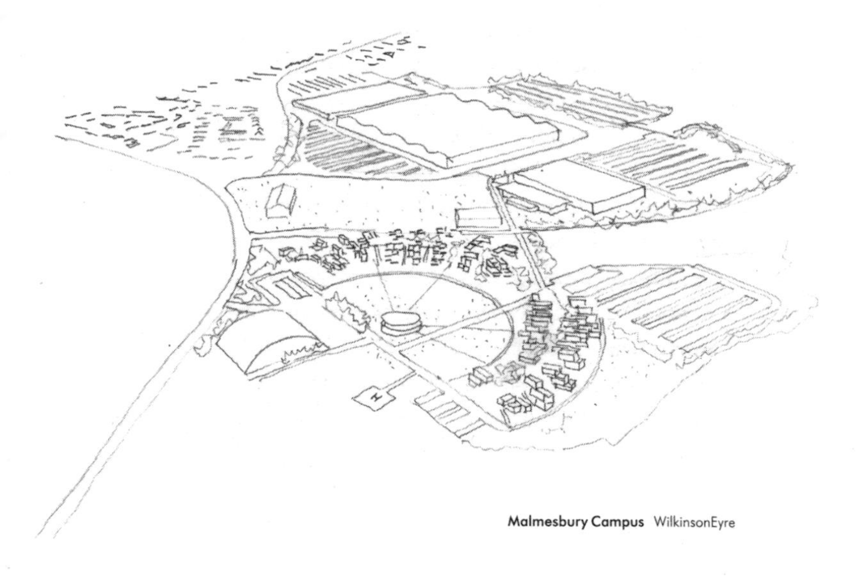

Malmesbury Campus WilkinsonEyre

have all the space we need to expand at Hullavington, but I think being bigger will spoil the special experience undergraduates have here. For one thing, there would be the danger of the ratio of undergraduates to Dyson engineers getting out of kilter and I don't want to lose the sense of excitement we enjoy here.

However, we would like to offer masters degrees and doctorates soon if not straight away and perhaps in time we might want to start something similar in Singapore. We certainly want our undergraduates to have the experience of working in our overseas centres. They have been very excited by this, but in 2020, conditions surrounding the Covid-19 pandemic, both political and practical, meant this was, temporarily at least, out of the question.

It has certainly been more than worthwhile. One day in October 2020, I received an email out of the blue. I share it here for obvious reasons. Education can be so very exciting and rewarding.

Good morning James,

Thank you for your praise and thank you for starting the Institute in 2017 – I'm having a fantastic time!

Here is a brief summary of my achievements this year:

Averaged a 1st in my first-year studies for a University of Warwick Engineering degree.
Designed and built an automated system for the Dyson Motors Advanced Test Systems and Validation team.
Completed LabView Core 1 and 2 certifications.
Became an AWS certified Cloud Practitioner, whilst deploying a cost saving improvement to the Dyson cloud infrastructure – at Dyson's Bristol office.
Learnt about rotor dynamics and used CAD to construct a visualisation rig whilst working in the Dyson Motors' Mechanical team.

Pitched and led a summer series project on capturing microplastics from our waterways which won the 'Most Linked Project to the UN Sustainable Development Goals' award.

All the best

William Thoburn
Undergraduate Engineer
The Dyson Institute of Engineering and Technology

I give talks about design, as does Jake, to the undergraduates. I like to show them products where engineering and design are one and the same thing and all the better for it, like the Toio floor lamp Achille Castiglioni designed for Flos in 1962. It's designed around 'found' components: a fishing rod for the upright complete with fishing-line fair leads, a sealed car headlamp and an exposed transformer, better for cooling, that acts as ballast. It's a design that expresses its engineering, while at the same time being an elegant and clever work of art. It makes me smile. It's a very good lamp and it feels as fresh as ever, even though the design is now sixty years old. I don't ask undergraduates to go away and make a contemporary equivalent of the Toio lamp, but to understand that design is about expressing the engineering or technology inside the product, not styling or about trying to be contemporary for the sake of it.

Beyond our own nascent university, the Foundation had long been involved with the funding of research at established universities for purely altruistic reasons. The guiding principle of the Foundation is to inspire the next generation of engineers. We work closely with the University of Cambridge, Imperial College London and the Royal College of Art. We also have a wider relationship with research projects we fund at Southampton, Leeds and Newcastle universities, where many of our motors team studied, and, more recently, with universities in Singapore, Malaysia and the Philippines.

The James Dyson Building for Engineering at Cambridge, designed by Nicholas Hare Architects, provides space for 1,200 postgraduate engineers supporting world-leading research. The building itself is a test bed of building technology and performance. Fibre-optic sensors in its foundation piles, concrete columns and floor sections yield live data concerning temperatures and structural strain, providing a picture of how the building is behaving. 'The result,' says Nicholas Hare, 'is a building that's more of a living creature than a passive block of material. We can ask the building how it's feeling, and the building can reply.'

We have given £8 million to Cambridge to build a research laboratory and design-and-make workshop, and a further £10 million for a professorial chair in Fluid Dynamics. We work closely with the Whittle Laboratory at the university's Department of Energy specialising in thermo-fluids and turbomachinery. The laboratory is special for me not least because it was opened in 1973 by Sir Frank Whittle and it's where we've tested our tiny electric motor, an example of domestic turbomachinery, alongside mighty Rolls-Royce aero engines. Rolls-Royce is an industrial partner here alongside Siemens, Mitsubishi Heavy Industries and the Centre for Doctoral Training in Future Propulsion.

In May 2019 the Dyson School of Design Engineering at Imperial College, founded five years earlier, moved into a new home on the junction of Exhibition Road and Imperial College Road in South Kensington in a handsome four-storey Edwardian Baroque former Post Office building. The building dates from the time when Britain's imperial power reached its zenith. Britain gave up its empire after the Second World War, but its global reach in terms of advanced engineering and design remains considerable. The red brick and Portland stone building is right next door to the Science Museum. I'm really pleased about this. Many British children have been excited by science and technology here over the decades, some going on to study at Imperial College itself.

The Victoria and Albert Museum is just across the road, with its academic connection to the Royal College of Art. Actually, the RCA was founded here before it moved up the road to Kensington Gore opposite the Albert Memorial. The new school at Imperial College teaches a highly practical four-year, full-time MEng course in design engineering. It is already many times oversubscribed, with 50 per cent of its intake female, which, I hope, demonstrates to politicians both the importance and attractiveness of engineering education.

Having my name on these buildings and projects might seem like an act of personal vanity, but it's also, in the American tradition, partly a way, I hope, of encouraging others to put their money into the coffers of university colleges. The aim is not to privatise education, but to invest in areas of education where Britain is lacking and needs all the assistance it can get, whether in existing universities or in altogether new ventures like the Dyson Institute of Engineering and Technology.

Although I graduated in 1969 from the Royal College of Art – rather a misnomer as 80 per cent of its students study design and not fine art – I had retained links with it. In particular I was keen to support the RCA's pioneering teaching of both engineering with design through its Design Engineering course, a cause after my own heart and which I had proselytised during my career. I had recruited all of my early engineers from this course and, in the early 1990s, became an examiner of the final degrees. This led to my being voted onto the Council a few years later under the Provost Lord Snowdon. Sir Terence Conran succeeded Snowdon as Provost and I succeeded Terence in 2011.

Besides it being my alma mater, I was extremely fond and proud of the RCA. It was founded as the Government School of Design in 1837 by Prince Albert to improve the design of the manufactured objects at the height of the Industrial Revolution. The royal connection has continued through the special interest of Prince Philip, who was very astute on what is and what isn't good design and a champion of design engineering. The RCA's principal home is the classic 1960s Casson Conder building at Kensington Gore. The college is wholly postgraduate. Its annual degree show is generally considered to be the vision of future design and its stars, as well as being one of the most visited and popular attractions in London. The RCA has held the position of the world's leading art and design college for many years.

When the college needed to expand its campus at Battersea, the Foundation was happy to help. We gave £5 million for the new James Dyson Building, close to Battersea Bridge, part of which would house 'incubator units' where graduating designer–engineers could continue development of their degree projects, put them into production and then on sale. Here, they would have workshops as well as advice about patents and the commercial world. The RCA would link them with investors to provide early capital.

In its early years I chaired the Innovation RCA Board, which ran this area of the college. It has been hugely successful. Silicon Valley investors are lucky if 10 per cent of the start-ups they back succeed, whereas the RCA incubator unit has a 90 per cent success rate. In fact, the college as a whole has twenty times more start-ups emanating from their courses than the second highest university in the field, Cambridge. I believe this is because the product ideas come from imaginative engineers and designers with a burning passion for their products, rather than from people just trying their luck at being entrepreneurs.

On one occasion I showed George Osborne, who was Chancellor of the Exchequer at the time, around Imperial College and the student projects there. I was talking to him about the importance of engineering and technology, when he said, 'What about the importance of design?', so I explained. When the RCA was offered the chance to buy a whole block in Battersea by Chelsea Estates, and they needed £54 million to buy the block, I remembered what George had said at Imperial and wrote a very short letter, on the off-chance, mentioning only that the RCA spawned the most start-ups. I was most surprised and delighted when a letter came by return post offering the £54 million.

My interest in helping start the RCA incubator units matches my desire with the James Dyson Award to encourage undergraduates to develop solutions to problems using technology and design together. The success of both the RCA unit and our James Dyson Award winners is proof both that young people do create successful change in products and technology and that it comes about by combining the two disciplines of research with design. It also busts a myth that engineers and designers can't start and run successful businesses.

When I was trying, unsuccessfully, to raise capital to start my vacuum cleaner business, all the venture capitalists turned me down, with one even saying that they might consider the opportunity if I had someone heading up the company from the domestic appliance industry. This was at a time when that industry was vanishing from Britain because, taken as a whole, its products were uncompetitive.

By 2020, the James Dyson Foundation was reaching over 200,000 students a year and we had given £100 million for charitable causes. For me, one of the most satisfying ways of encouraging engineering students around the world has been the James Dyson Award. Established in 2002, this followed naturally from my plea that design should be integrated with engineering and technology, encompassing invention and problem-solving. The JDA was to challenge university students to solve a problem dear to their hearts through a product that could be put into commercial production.

What is astonishing is the sheer number of entries, the brilliant ideas students around the world have and their collective desire to make the world a better place. From the earliest days, students began solving problems encountered by the disabled, those suffering from debilitating disease and the elderly. In recent years there have been a growing number of ideas for improving sustainability. When selecting winners, it became invidious to choose between the wonderful projects in the wider brief and those focusing on sustainability, and we now offer a special new award for sustainability alongside the existing broader category.

We were running the award in twenty-seven countries by 2020 and had supported over 200 inventions financially. We award a national winner and two runners-up in each country and give two international awards, one for the best in the world and a second for the best sustainability-focused entry, both of which are worth £30,000 in prize money with an additional £5,000 going to their university. The prize money can help them start a business making the product they have worked on so diligently, as many have, or allow them more time to research. Incredibly, 60 per cent of the winners go on to successfully commercialise their ideas.

In 2020, for example, Judit Giró Benet from Tarragona in Spain developed The Blue Box, an at-home biomedical breast-cancer-testing device that uses a urine sample and an AI algorithm to detect early signs of the disease. Unfortunately, I have witnessed first-hand the harrowing effects of breast cancer in my family and, as scientists and engineers, we should do anything we can to overcome this terrible disease. Judit began this project when she read that dogs can detect cancer in humans. The device is linked to an app in the Cloud which collates data from millions of cancer patients and determines the cancer and type of breast cancer. This can enable more precise treatment while expanding our global knowledge of cancer. It also controls all communications to the user, putting them in touch immediately with a medical professional if the sample tests positive. The Blue Box strives to change the way society fights breast cancer, enabling women to test at home and giving all women the chance to avoid an advanced diagnosis, making screening a part of our daily lives.

This is really important because, according to the Centers for Disease Control and Prevention, 40 per cent of women skip their breast cancer screening mammogram, resulting in 1 in 3 cases being detected late, leading to a lower chance of survival. Forty-one per cent of those who skip mammograms say it is due to pain. Unfortunately, for a long spell Covid-19 prevented Judit from continuing her research at the University of California Irvine, but I am hopeful this will not have a long-term impact on her success.

The 2020 sustainability winner was Carvey Ehren Maigue, a student from Mapua University in Manila, who used a film of vegetable waste spread across windows to generate electricity – a solar panel on ordinary window glass. The thing that particularly impressed me about Carvey was his resolve and determination. Having failed to make the national stage of the award in 2018, he stuck at it and further developed his idea. This will be a vital character trait as he embarks on the long road to commercialisation.

The particles in Carvey's material absorb UV light, causing them to glow. As the particles rest, they remove excess energy. This excess energy bleeds out of the material as visible light, which can then be transformed into electricity. Because it works with UV light, electricity can be generated when the sun is not shining.

As mentioned, Carvey discovered that the film applied to the glass can be made of waste crops. This is resourceful because the Philippines suffers severe weather disruption, with farmers losing much of their produce as a result. Rather than leave crops to rot, Carvey sought to use them as a UV-absorbent compound for his substrate. After testing nearly eighty different types of local crops, he found nine that show high potential for long-term use.

These are two very bright breakthroughs. They are also just two of more award-winning students than I can possibly find room for in a chapter of a book. I have to mention a few more to show how invention courses through the minds of bright young people and can lead, at a very young age, to products that we can use in our everyday lives and that can create jobs and further livelihoods, too. These people are not just talking and grandstanding, they are actually driving change.

The UK national winners in 2020 were The Tyre Collective, four students who set out to solve the enormous yet little-discussed microplastic problem of tyre wear. Every time a vehicle brakes, accelerates or turns a corner, the tyres wear down and throw out tiny fragments, producing half a million tonnes of tyre particles annually in Europe alone. These particles are small enough to become airborne and can have an adverse effect on health. They can be swept into waterways and oceans, eventually entering the food chain.

The device The Tyre Collective designed is fitted to the wheel and uses electrostatics to collect particles as they are emitted from the tyres, by taking advantage of various airflows around a spinning wheel. Under a controlled environment on their test rig, the prototype can collect 60 per cent of all airborne particles from tyres. Once collected, the particles can be recycled or reused in new tyres or other materials.

In Australia, Edward Linacre from Melbourne's Swinburne University of Technology created his AirDrop irrigation system and was awarded for it in 2011. AirDrop pumps air through underground pipes in dry land, lowering the temperature to the condensation point. The water produced is moved to nourish the roots of the plants that would otherwise wither and die in conditions of extreme dry heat. Edward had studied the Namib beetle, an ingenious species that lives in one of the driest places on earth. With half an inch of rain per year, the beetle survives by consuming dew it collects on the hydrophilic skin of its back in the early mornings. Even the driest air contains water molecules, while biomimicry is clearly a powerful weapon in an engineer's armoury. Edward's research suggested that 11.5ml of water can be harvested from every cubic metre of air in the driest of deserts.

Other memorable projects have included Isis Shiffer's EcoHelmet, a radial honeycomb folding-paper helmet designed for use by bike rental companies; James Roberts's MOM, an inexpensive, electronically controlled, inflatable incubator constructed to decrease the number of premature child deaths within refugee camps; and Lucy Hughes's MarinaTex, a biodegradable alternative to single-use plastic, made of organic fish waste and locally sourced red algae. It is encouraging, through the award, to see a real gender balance among the winners.

Deirdre's and my interest in health has been reflected in the JDF's contribution to medical research. By 2020 we had donated £35 million to this truly vital endeavour through Breakthrough Breast Cancer, the children's cancer charity CLIC Sargent, the Meningitis Research Foundation, Cure EB (epidermolysis bullosa, a genetic skin-blistering condition that affects over 500,000 people worldwide) and the former Formula 1 champion Jackie Stewart's Race Against Dementia, with its groundbreaking work in keeping brain tissue alive.

We've also been able to contribute through both the JDF and our family charity, the James and Deirdre Dyson Trust, to the cost of the

Dyson Centre for Neonatal Care at the Royal United Hospital Bath, which opened in 2012. The hospital asked Deirdre to paint something. She gave them a large triptych that covers the wall facing the entrance. This gentle, sunlit building, designed by Feilden Clegg Bradley Studios, has allowed new forms of therapy for newborn, and often premature, babies in need of special medical care. It has a particularly satisfying plan by which children are moved around the building as they get progressively better and out, happily, into the wide world and home. Research we funded has found that, of babies studied, 90 per cent recuperating in the new unit went home breastfeeding, compared with 64 per cent in the former prenatal building. The study also showed that babies are better rested, sleeping on average for 22 per cent longer than in the old unit.

This research was the first in the world to adapt and use accelerometers to measure the respiratory and sleep patterns in babies in order to monitor their reaction to the surrounding environment, using an extremely low-power, self-contained wireless device. These are normally used in aircraft and smartphones, and increasingly in sports and athletics. Bath Rugby Club, for instance, use the technology to analyse player training techniques and fitness. It was far less invasive that the other intrusive methods, including ECG, and information from ventilator circuits which have been used to measure this.

Infrared tracking technology was used to pinpoint staff movements in the building and test the efficiency of the design. The study found that nurses in the new building spend 20 per cent more of their time in the clinical rooms, with the babies, meaning more time spent caring for them.

Lux meters were used to take light measurements according to specific times, dates and outside weather conditions. Up to 50 per cent more natural light was measured in the new building. This ensures a more natural circadian rhythm, allowing the babies, parents and staff to perceive the changing day, aiding the babies' sleeping and eating habits.

Sound pressure level meter readings were taken and an average level for each hour was documented in decibels. Noise levels in the special care unit have decreased by over 9dB on average from those in the old building. It is suggested that the increased sleep observed in the babies relates to this reduction in background noise.

If a prenatal clinic takes us all back to the very beginnings of our lives, Gresham's School takes me back to the earliest days I can remember. In 2019, Deirdre and I gave £18.75 million to Gresham's to pay for its new Dyson Building, known to pupils as the Steam Building, a rather old-fashioned name perhaps for a science, technology, engineering, art and maths department that will I hope supercharge my old school's performance in science. Science and the arts, that is. Here, the aim has been to reconcile C. P. Snow's 'Two Cultures' – science and art – and in doing so to weave the threads of invention, innovation, design, engineering and technology together in energetic young minds and, of course, in those of their hard-working and gifted teachers.

Deirdre and I enjoy supporting new ideas enthusiastically espoused by inspirational leaders. When I spoke with Douglas Robb, Gresham's headmaster, he said he wanted the school to be at the forefront of encouraging engineering and science among its pupils. I hope this building will do just that. I've observed that from the age of around six, children are very engaged. They are inventive, dreaming up ideas, and curious, wanting to know how things can be made. Yet these traits get stamped out of them, partly by the system and partly because the teaching of these subjects in schools has not kept up with the pace of technological change. By creating special new spaces in this new building, I hope that we can foster, inspire and educate more brilliant young minds.

Spaces will be equipped with the latest technology to ensure the highest levels of teaching, from robotics and programming, to artificial intelligence and machine learning. The building will also provide greatly improved opportunities for the outreach programme Gresham's runs with local schools. And the school will be nurturing links with the Dyson Institute of Engineering and Technology. The building itself, designed by Chris Wilkinson, balances the Arts & Crafts-style chapel at the very heart of the school, which it will face across a large grass quadrangle. Besides the ten years I spent singing in this wonderful chapel, I happen to have been christened there in 1947. There is also a charming remembrance plaque in Latin to my father beside the altar.

The new building is constructed of bold steel joists and huge glass panels. It is a building expressing engineering yet softened by the plants growing up it. Inside there will be unexpected spaces such as an auditorium based on a massive stair that doubles as a place to meet and sit on, space where budding young scientists and artists might share their education.

I hadn't been back to Gresham's all that much since I left to go to art college in London. I was invited by a delightful previous headmaster, John Arkell, to a speech day and to give the prizes. I was entertained by a group of teachers, one of whom had taught me, and pupils performing Malcolm Arnold's piece with three Dyson vacuum cleaners among the instruments. Normally speech day is held in their large and romantic open-air theatre, scene of the several Shakespeare plays that we put on for visiting parents.

I was delighted to see my wonderful old headmaster, Logie Bruce-Lockhart, shortly after we made the donation. Sadly, Logie died before I could take him around the Steam Building. He was a kind family friend and I would very much like to have done so, for this was my way of repaying him and the Norfolk school that had been so charitable to me after my father died.

Gresham's was my life. Not only did I begin my learning there, I lived there and took huge advantage of the grounds and facilities. It was very hard to leave all of that behind – that idyllic setting and my childhood friends – when I set off for London. My parents, of course, had lived that life even more than I had.

Gresham's is also a school that has produced more than its fair share of scientists and inventors, along with actors, musicians, poets, farmers, of course, and spies. Our family charity motto – Latin, of course and as my father would have expected – is *Numquam Tendere Cessa*, or Never Give Up Trying. This is my hope for future generations of young people at Gresham's and beyond through colleges, universities, laboratories and, yes, factories. Oh, and art schools both in Britain and around the world. I suppose we might have chosen *Custodiant Currit*, Keep on Running, but Never Give Up Trying serves, I think, perfectly well.

CHAPTER TWELVE

Making the Future

In many ways I have come to live what might seem to be a contradictory life – part in the future, part in the past. Most of my waking time is spent in our labs, surrounded by Dyson's engineers and scientists, exploring ideas that we hope might shape the future in five, ten, possibly many more years' time. Ours is a life of challenge and frustration, all of which is a fulfilling pastime. But I also have an interest, verging on obsession, with the past – with the stories, artefacts and spaces that have shaped our world. Repairing the old and adding the new has become as much an important part of my life as inventing has for the future. Renovation might sound an odd enthusiasm for a modern designer and engineer, yet there is much to learn from the past and from those who shaped the world before us. This is not about nostalgia, quite the opposite. It is about understanding and celebrating the progress that has been achieved, learning from it and building on it.

Perhaps it is my circumstances: losing my father early in life, being given responsibility beyond my years early in my career by Jeremy Fry, and being fortunate to have a supportive wife and family that have given me self-reliance and the faith to pursue ideas. I really do enjoy making things and, whenever possible, doing things myself, constantly learning in the process. This is as true at work as it is at home and is evident from the various 'projects' we have bought as houses. At various points I have had to be a plumber, a plasterer, an electrician and, best of all, a JCB driver. I've designed new furniture and fittings.

The extensive restoration Deirdre and I have undertaken at Dodington over the past nineteen years is the grandest example. Designed by James Wyatt, and built between 1798 and 1818, the house was in dire need of renovation when we took it on, but rather than outsource the work to one main contractor, we decided to do it ourselves, in a stepwise manner, learning as we went, and building our own team of people skilled in trades and crafts and with real passion and expertise. This has been a long, rewarding and as yet unfinished challenge, rather like the renovation of our 1930s yacht, *Nahlin*, which we rescued as an even greater – indeed quite literal – wreck.

Closer to work, the Whittle jet engine, which we have lovingly restored to the original specification and which we fire up from time to time in the car park in Malmesbury, is another good example of restoration of the past. Though it needed love when we took it on, the Whittle engine is not some old-world object. It is the embodiment of Frank Whittle's revolutionary concept, a way of solving the problem of how aircraft could fly at much greater height, speed and smoothness than they could possibly do in 1930. The idea that he formulated at the age of twenty-three, for a wholly new form of aero engine, was as extraordinary as it was fragile. And who wanted to believe Whittle was right? Certainly not government experts. He had to pursue his project alone, yet in doing so he revolutionised flight for everyone and changed the course of the Second World War.

While the Whittle engine is perhaps my biggest inspiration, there are a great many similar design and engineering icons scattered around our campuses, each with its own story. There is an English Electric Lightning jet hanging in one of our campus cafés, a Concorde engine in one of our office spaces, a dissected Mini parked in another and a Harrier Jump Jet in the car park. One of these, the Mini, is a good example of why you should not listen to market research. BMC (British Motor Corporation) cancelled one of the two proposed Mini production lines as a result of the feedback from market research. The corporation was told that people would refuse to buy a car with such small wheels. In the event, BMC was never able to catch up with the demand that followed the launch of the trend-setting Mini. There are many more examples I won't name here, yet each such artefact has its own story of against-the-odds progress and lessons on why having faith in your ideas and believing in progress is so very important.

What these pieces of our history demonstrate is that it is hard for other people to understand or get excited by an entirely new idea. This requires self-reliance and faith on the part of the inventor. I can also see that it is hard for an outsider to understand the challenge and thrill of inventing new technology, designing and manufacturing the product then selling it to the world. What is disgraceful is the portrayal, in television programmes like *The Apprentice* or *Dragon's Den*, of entrepreneurship and business as a brutish endeavour, which only serves to reinforce everyone's worst view of business. Some good products have come through *Dragon's Den*, though not always the ones chosen.

Being an entrepreneur is not the same as being the wide boy of the media's imagination. It is about creating new products and new opportunities, generating rewarding employment and opportunities in the process. At best, the entrepreneur is part of a cycle of renewal, driving progress. It is not easy, of course. The margin between profit and loss, between success and failure, is small. It may be necessary to change, to reinvent your business, as circumstances change around you. Just as you think you have understood a situation and how things work, it changes without warning. There are traps around every corner.

I have got to the point where I truly worry if everything is going smoothly. It's rather like being the US cavalry officer in an old western when he says, 'I don't like it, Sergeant – seems too quiet' and immediately gets an arrow in the chest. Every day is an adventure and a response to the unexpected. Even if things appear to be in some kind of stasis, a company must move on. It has to get better, evolve and improve in order to survive. There is no greater danger than satisfaction.

As soon as Dyson became successful, people in Britain asked when I was going to sell the company, as if I was only lowering myself temporarily in the dim and grubby world of uncreative manufacturing. Once I had made my first million, it was surely time for me to get away from the sweat, grime and grim routine of factory life and to become a reclusive landowner building duck houses, perhaps, and cleaning the moat around my house. In any case, why would a supposedly educated person be rootling about in factories when he could be doing something 'creative' and preferably from an office or studio?

Britain is about the only nation I know whose entrepreneurs, at the earliest opportunity, sell out or take their company public, often the first step towards that sell-out. This has always puzzled and saddened me, partly because I believe that a business loses something when it goes public but also because so many of these businesses end up in foreign hands, losing their way and ending up as vassal companies. Entrepreneurial family businesses, on the other hand, may be handed down through generations. Britain has remarkably few compared with other countries.

In Germany, for example, there are the famous *Mittlestand*, medium-size private businesses, often multi-generational, as well as giants like BMW and our competitor Bosch. France has big family-owned fashion houses; Italy and Spain the same. The US has the largest number of family businesses in the world, including Mars and Cargill, the agri-chemicals business. Yet in Britain it is hard to name a significant private business founded by an entrepreneur other than JCB. I believe Britain is weaker through not having a large number of successful family businesses. The advantages of family businesses are that they can think in the very long term, and invest in the long term, in ways public companies are unable to do. I also believe that family-owned enterprises have a spirit, conscience and philosophy often lacking in public companies.

It is as though British entrepreneurs want to cash out as fast as possible, either because they only ever did it for the money or because they are frightened of losing it all if they carry on. If it was for the money, this implies that they lacked a passion for what they were doing, for their businesses, their colleagues and even their customers. Theirs was a short-term way of becoming rich underpinned by a fear of losing the business before they have had time to cash in.

This lack of conviction, lack of belief in what they were doing, is worrying. Why do the British feel like this while other nations' entrepreneurs do not? Is it because, although we instigated the Industrial Revolution, we failed to continue it as well as other nations? Historically, British companies could supply their goods to the empire, providing easy sales, whereas other nations had to use their industrial might to compete at world trade and so made a greater long-term success. Is it because the British aristocracy never worked but, rather, inherited wealth, so the built-in psyche is to make enough money so that you never have to work again? Is it that the British do not see a company as an enterprise that makes wonderful products, employs and supports many others? Do they not see that as the owners they have a duty to nurture and sustain the enterprise for the benefit of those employed? Finally, is it because we admire brilliance and easy success over the longer stamina and grind of running a business for generations?

Many wise friends advised me to sell up in the early days when a few attractive offers came in. I suspect they feared that I might lose it all or they felt that I had achieved all that I needed to achieve. It is true that a family business has most of its wealth tied up in the business. Continuing to keep it as a family business is both a risk and a responsibility. But I like living on a knife edge, competing and building the business. I am passionate about developing new technology and working with a wonderful and creative team around me. If we fail, 'better drowned than duffers'.

Those kind people totally missed the point. I didn't work on those 5,127 vacuum cleaner prototypes or even set up Dyson to make money. I did it because I had a burning desire to do so. And, as do my thousands of colleagues, I find inventing, researching, testing, designing and manufacturing both highly creative and deeply satisfying. I also happen to think that future generations of children will be excited about careers and lives in invention, engineering and manufacturing, if only they can be encouraged to do so through education.

Early on in our story, the clear bin was another 'clear' example of going our own way regardless. Trusting our own instincts, we decided to ignore the research and the retailers. Pete and I had been developing the vacuum cleaner and we loved seeing the dust and dirt. We didn't want to hide away all the hard work the machine had done.

Going against established 'expert' thinking was a huge risk. No one could confirm that what we were doing was a good idea. Everyone, in fact, confirmed the reverse. The data were all against it. If, however, we had believed 'the science' and not trusted our instincts, we would have ended up following the path of dull conformity. The clear bin proved successful and, ever since, all the Johnny-come-latelies in the world of vacuum cleaner production have clear bins, too. They copied it because the clear bin was a symbol of successful new technology. Of course, they also clambered onto the clear bin bandwagon because it sold and continues to do so in ever-growing numbers.

In following a different path, obstacles will be put in the way of pioneering manufacturers, yet the process of creativity and of solving seemingly insoluble problems is rather wonderful. It is, though, hard to be pioneering because you don't know whether or not you are going to succeed. You will stumble and have to pick yourself back up believing that you will succeed. It is scary – I am scared all the time. Fear, though, can be a good thing as it pumps the adrenaline and motivates, as sportsmen will confirm.

A life of perpetual learning, pursuing science, engineering and technology, has certainly been a truly magical and fulfilling adventure – the quiet thrill of improving products through the application of technology and making them enjoyable and surprising to use. For an engineer, the creative impulse, the desire to improve things and the need to solve problems are a state of mind that cannot be switched off. It is there all the time, whether you're at work or at home. It is the intellectual challenge of seeing frustrations and problems and developing a product or system that solves them.

Scientists and engineers combined recognise today's problems and are capable of providing new solutions. History has proved this time and time again. The challenges of the future will be met and resolved by people who are young now, who can see the problems we face and have the urge – the impulse – to solve them. It is their world now and they don't need people ranting about the end of the world, nor Jeremiah-like politicians proclaiming, nor even old people like me telling them what to do. The most important thing is that we encourage them to do it and to make their own mistakes along the way, while providing a supportive framework within which they can operate successfully.

I have great faith that science and technology can solve problems, from more sustainable and efficient products to the production of more and better food, and a more sustainable world. It is technological and scientific breakthroughs, far more than messages of doom, that will lead to this world. We need to go forward optimistically into the future as if into the light, and with bright new ideas rather than the darkness and end to human ingenuity portrayed by doomsayers.

After the event, a revolutionary new idea can look so obvious – surely no one could possibly have doubted it? At their conception, though, new ideas are not blindingly obvious. They are fragile things in need of encouragement and nurturing against doubting Thomases, know-it-alls and so-called experts. Just as Frank Whittle discovered, it is easy for people to say 'no', to dismiss new ideas and to be stick-in-the-muds, pessimists, or even cynics. It is much harder to see how something unexpected might be a success.

The depressing thing is that harbingers of doom and gloom get far more attention than optimists and problem solvers. I feel very strongly that progress should be embraced and encouraged, and it is a duty of governments and companies to catalyse the ideas of the progressives and harness them to achieve good ends. There is no denying that the world faces some rather big problems, but there is no sense in trying to force people to go backwards as a result, or to rewind progress. I am certain that we will overcome our greatest challenges by moving forwards, confidently, and having faith in humanity's ingenuity. Problems are there to be solved, and this in itself is exciting and motivating.

For our part, Dyson has been lucky to recruit some wonderful young people who come bubbling with ideas and a desire to change things for the better. Many of these individuals have been fresh graduates, but now increasingly they are undergraduates, too, school leavers who are forging their education at the Dyson Institute while at the same time contributing to the future.

Meanwhile, the rest of us need to stay young at heart and never less than inquisitive. A case in point is one of our most inventive Dyson engineers, who is in his eighties. He has the curiosity of a young boy. He is an indefatigable optimist, challenging everything that needs to be challenged. He is enthusiastic and always willing to try new things. His is a truly creative mindset. There is hope, then, for us all as we look with clear and creative eyes and minds into a future where things can be so much better than they are today. We are at a truly fascinating juncture in human history and we must press on.

Dyson has certainly never stood still as it has 'grown up'. At times, this has been uncomfortable and not everyone has understood why we have had to change. From those first trips to Tokyo to negotiate a licence agreement for the G-Force, I have been excited about embracing the world. At its best, globalisation has the potential to raise everyone up. Success around the world provides the means to invest in research and better technology, driving our world forward. Today, we have people from fifty-four nations working at the Malmesbury campus, and people working for us in eighty-three countries around the world. We are stronger for it. They have a shared purpose and a global perspective but also a diversity of thought and culture. While age is no barrier at Dyson, most of the staff are young. They are open-minded and have the naïve intelligence to think differently.

These brilliant young people understand the problems facing the world on both a macro and micro scale. They are busy at work on a remarkable range of ideas and inventions that can and will make a difference to our lives. They are not academics publishing dry papers that set out simply to dismiss equally dry papers by rival academics. They are not would-be government scientists with all their smug expert advice which is often wrong. They are not driven by people shouting in streets and squares. They are working, hands-on, with fragile ideas fired by a creative impulse, and they need all the support I or any of us, running alongside to keep up, can give them. They are our future. Imagine if their spirit of optimism and invention was harnessed for the world.

These young people show the power that comes from being unafraid to pursue new ideas, to challenge orthodoxy and their own beliefs. While I am loath to give anyone advice, I do encourage others, and remind myself, too, to be adventurous. We must follow our own stars.

Deirdre and I are very pleased that our three children have done just this. As it happens, we are all designers of one kind or another, with a shared outlook. Jake studied design but veered towards technology as an engineering designer. His final-year project was a water turbine housed in a large waste pipe generating electricity from household waste water and rainwater. We installed it at Dyson and it was much admired by a visiting Prince Charles. He had taught himself to use a lathe and today has metalworking machinery at his home, often machining things he has dreamed up.

Sam is a natural engineer, designing an ironing board supported by gas struts at school, and was later an engineer at Dyson. He left to follow his real passion, music. He is a fearless performer, quite shocking us when he once played the lead role in Peter Schaffer's *Equus*. He started playing the flute aged seven, then taught himself the drums, the piano and finally his main instrument, the guitar. He is using his engineering to great effect, creating music and devising ways to do so online. He has his own band, Ramona Flowers, as well as Distillers, his record label. We love his music, and he is successfully navigating a path through the wreckage of the music business.

Emily is a talented fashion designer, who started her career at Paul Smith, where she met Ian, a gifted fellow designer and now her husband. Emily set up Couverture in 1999, originally in the King's Road, with her bedwear and bed-linen designs. Her designs were very original: one had the duvet cover as an envelope complete with a postage stamp of the Queen's head in profile with eyes closed, asleep. This incurred the wrath of the Post Office lawyers as they have a monopoly of the Queen on postage stamps! Now in London's Notting Hill in a three-storey shop, Emily and Ian have together created the successful Couverture & The Garbstore. Couverture has womenswear and home accessories, while The Garbstore has menswear. Ian's unusual designs are the main source for Jake's and my wardrobe. Deirdre and I are thrilled that Emily and Ian have pioneered their own field of design in such a competitive business, where they are recognised as ones to follow.

I have never pushed or even asked any of them to join Dyson, yet they have all said that they would like the business to remain in the family. Having done his own thing for twenty years, including establishing his own innovative and successful lighting company, Jake made his own decision to join the company in 2015, bringing his technology with him. We are lucky to have him, his inventiveness, and his ideas for the future. I love working with Jake and am very proud of what he is doing.

Indeed, it is hard for Deirdre and I to imagine how we could be more thrilled that all three of our children have chosen to be creative pioneers. They have also been loving and devoted parents to their delightful children, our grandchildren. They all join us on holidays, where we can see how they develop, their lust for life, their bravery and love for each other. While there have been family tragedies and difficult times, they have all maintained a spirit of optimism with a desire to create something new. I was overwhelmed to have all of them around us at my seventieth birthday party.

It is hard to know where the Dyson family stops and the Dyson family business begins. Having grown up with and in Dyson, Deirdre, Emily, Jake and Sam are passionate about the company and keen for it to retain the philosophy of adventure that has underpinned and spurred it along from the beginning. Of course, this will mutate and evolve. The future will be different, and it may even take the company off in a different direction. But the really important thing is that we maintain the spirit of adventure and don't become afraid of being original, pioneering and exciting.

What we do know is that companies always have to change to get better at what they do, plan to do and even dream of doing in the future. The adage that the only certainty is change is true, and this means not being afraid of change even if, for a company, it means dismantling what you have built in order to rebuild it stronger or killing your own successful product with a better one, as we did with our new-format battery vacuum cleaners.

People might not understand radically new concepts and products to begin with, yet if we persist and show their benefits, as we did with the cyclone and the new-format stick vac, they will understand and come on the journey with us along the path less travelled. One of the advantages of being a private family company is that we can do this – we can take the risk. This is because we are not subject to the vagaries of the stock market. Yes, a private company must have the same disciplines as a public company, yet in many ways we can take much bigger risks than a public company.

We certainly have taken big risks, with the digital electric motor, the washing machine, the electric car and our research into solid-state batteries. Not all have been commercially successful. That is the point. By its very nature, pioneering will not always be successful, otherwise it would be all too easy. We don't start these ventures with the inevitability of success – we are all too aware that we may well fail. Of course, I don't know the ultimate destination for Dyson – that is for someone else to determine – but the only thing I ask is that it is an exciting one.

Above all, how can we avoid being duffers? It comes from working away single-mindedly at solving a problem, however many setbacks befall us, even if that means 5,127 prototypes, and retaining an open mind. Remember that there is nothing wrong with being persistently dissatisfied or even afraid. We should follow our interests and instincts, mistrusting experts, knowing that life is one long journey of learning, often from mistakes. We must keep on running and we really can do better!

POSTSCRIPT

Jake Dyson

Dad was different from other fathers. He was always very individual, going against the grain and standing out. There was something that made us think he was special, whether this was the determination he showed in everything he did or the mismatched socks and the slightly odd clothes he wore when he picked us up from school. As children, we were never embarrassed by him being different. We were brought up around creativity. We also thought that he was going to succeed in something. We felt his drive, his passion for what he was doing, and believed in it.

Because Dad worked from home in the early days, we were always a part of what was going on. We were a family business from the start. I remember days when I was off school, ill with some bug, with the smell of soapy plastic wafting through the air and seeing Ballbarrows coming off the production line. This was on a very small scale, but I was proud and excited by what Dad was doing. I remember vividly, too, the day he came back from being booted out of the Ballbarrow company. That was a tough time for Mum and Dad.

For a child, though, it was all a great adventure. I remember having great fun zooming around on the wheelchair Dad developed with Jeremy Fry, with Lord Snowdon popping in and out to see how it was coming on. Or heading down to Poole Harbour to watch one of the prototype paddle boats being tested, or sitting in a Sea Truck going full speed across the open sea.

Dad made and experimented with things, seemingly oblivious to the dangers involved. I very clearly remember him making a vacuum-forming machine in the cellar of our house at Bathford. These are machines that shape plastic. They cost thousands of pounds to buy, so Dad decided to make one. They are highly dangerous with kilowatts of power going through big coils. I remember him down in the cellar for weeks. There was a lot of clattering and swearing. But he did it, and without destroying the house. It worked.

And then there was the vacuum cleaner. I remember him converting the coach house, which up to that point had been a wood store where I would hide with my friends, into a workshop. I spent a whole summer holiday making ten working vacuum cleaner prototypes with the model makers, while Dad was upstairs at a drawing board producing drawings to a level of accuracy that meant they could be handed over direct to the tool maker for production. It was hard work on everyone's part, but the atmosphere was a happy one. The young RCA graduates who came to work with Dad in the coach house were fun. We played tennis with them at lunchtime.

I remember Dad making and doing everything himself, at home as well as at work. I used to play a lot of cricket in the garden and broke many windows. He got really fed up with this, so he taught me to fix windows and cut glass myself. Over time, *I* got fed up with this, so I started replacing the glass with polycarbonate, which he only noticed when the panes started to go milky. I was learning from him how to do things myself.

Dad would not overly protect us or fight our corner, but he would debate things with our school, usually telling the teachers how old-fashioned they were. Particularly at dinner parties, he would be deliberately confrontational and come up with some alternative argument to make conversations more interesting. At one of our New Year parties, he abseiled down the wall. He was an attentive father and made things for us, like a very well-engineered zip wire. He plumbed and wired the entire house, in the evenings and at the weekends, as well as playing with us and looking after us. He was full of energy.

I do remember the champagne being popped when Dad said he had a licence agreement with a company in America. But things quickly went sour. The battle with Amway was long and painful. He had worked so hard for so long on his invention. The stress and hurt he felt was clear to see. There was a lot of tension in the house about money. Mum hosted life classes to help make ends meet. As an 11-year-old I could just about understand what was going on.

During the battle with Amway, I remember him reading every single word of every single legal document. That was intense. Many people would have left that up to a lawyer, but Dad wanted to understand every word to give him the best chance of winning. Working with a lawyer, but understanding every word, is probably why he won. Dad's attention to the smallest detail in anything he does is very much a part of his character. I don't remember any celebrations at all, though, when he finally won the case. All the focus was on how he could make the vacuum cleaner and start selling it.

He was away after this, travelling constantly, off to Japan for six weeks and to other countries, selling. As Dad mentioned earlier, when he was in Japan a squirrel ate through a water-tank pipe in the attic, the consequent flooding bringing down several ceilings. We all slept in the living room for two weeks. This gives a flavour of what Mum endured while Dad was on his mission. He came back from those six weeks quite numb, tired and overworked, but it was only another six weeks before we had a manufactured product, the shocking-pink G-Force, in our house.

The proudest moment for me was seeing what Dad had created on sale, and reading about it in magazines. Things started to change dramatically after that. When I first started college, at Central St Martins in London, everyone was talking about big-name stylist designers and how they were leading the way. By my third year, overhead projectors were showing pictures of Dad as an example of how you could combine manufacturing with design and business. It was overwhelming. I was incredibly proud and knew something interesting was happening.

There was clearly no respite as far as Dad was concerned. But it was amazing to see how he experimented with everything, including his own marketing. He launched the DC03 at The Sanctuary, a fashionable spa in Covent Garden, with vacuum cleaners emerging from the pool. It was bonkers, but people took notice. Dyson was different, as Dad had always been. He had taught himself how to invent, engineer and make a radical product while, at the same time, learning to be a lawyer, a salesman and his own advertising and marketing man. He was obsessed with both micro detail and the big picture.

Success came and it really hasn't changed him. He credits those who have brought and continue to bring success Dyson's way. He has constantly paid respect to Jeremy Fry, his mentor, for all that he taught him. As for what he taught me, my siblings and all those working for him, it can be summed up best as 'I can do that'. It doesn't matter if the challenge is making something entirely new, replumbing a house or taking on a powerful multinational – he just gets on and *does it*. He has no fear. This is the thing we need to uphold in the culture of Dyson: the no-fear culture, the need to be adventurous. The business is at a certain scale today and as a family we choose to plough the profits back into the business so that we can continue to take risks. I did the same in a small way, in Dad's image, with my lighting company, before I joined Dyson in 2015, taking on big lighting manufacturers with entirely new products.

Dad has a level of energy that he passes on to all the young Dyson engineers. He encourages them to turn things on their head and to avoid thinking in a conservative way. He wants them to understand that for new inventions and products to be successful, they have to be of an order of magnitude better than what went before. They must take risks. They must in their own ways be fearless.

'Learn as you go' is the other principle Dad has applied through life. Start with ideas, understand things, learn and improve. This is the principle guiding the Dyson Institute. Having employed lots of graduates at

Dyson, we had a pretty clear idea of how we can create a curriculum for these young people, allowing them to study academically while working on live research and manufacturing projects, so they learn as they work. They are given the freedom to experiment while working on really hard engineering principles.

As a business we understand that investing in technology with a long-term view is the vital ingredient. But we have also learned, in the process of manufacturing 25 million products a year, that the rest of the business has to be set up to support that. There are no reference points for this, since nobody makes the products we do, and certainly not on the same scale and quality. We have to continually think of new ways of supporting our customers through controlling the quality of our products and maintaining a direct relationship with them, and this requires a technology of its own. The technology to run the business is increasingly a sphere in which we are going to have to innovate to take Dyson forward.

We also want Dyson to remain a start-up in spirit, with the freedom this implies in terms of experimentation, learning and adventure. What we certainly don't want is layers of management. They may seem to be a safe way of doing things, but we are a collaborative business that changes and evolves weekly. We are always trying to work out how to improve systems and collaboration between teams. We need as little management as possible. Nor do we like being told what to do, as management tends to do. We want to discover this for ourselves, although we are totally open to people – creative individuals – coming to talk to us directly.

The business grows at a pace, and you have to redesign everything in it to keep up with that pace. Dad is totally in tune with this. He is unafraid to rebuild entirely everything he has created, in order to make a better version or to kill a successful product with an even better one. As a family-run business we believe in a long-term vision and think it is so important to invest and reinvest in future bets, in talented young people, investing in hunches, so that we can set teams on these technologies now, while being very patient. We are very open-minded in terms of going into new areas, and as a family business we are completely free to go wherever we think we can and should without shareholders pressuring us for a quick return. Our mentality is to make things we believe in and not what the market might expect us to make. As a family we were brought up to be creative, to experience and then to create new things, different things, things that go against the grain. And, as Dad has always encouraged us, to just *do it.*

POSTSCRIPT

Deirdre Dyson

James and I were on the same foundation course and used to bump into one another on the Tube sometimes. After a while we'd say, 'Second carriage down?' There were no romantic vibes, but we enjoyed chatting and joined up with two other students at lunchtimes. My art training had been hard-won and I had no intention of becoming involved with anyone at the time. I worked all hours.

It came as a complete shock when James started holding my hand during a drawing study day at London Zoo. The dilemma for me was, 'Oh! I can't take my hand away as I might upset my new friend.' Even travelling back to the studio, he still had my hand and by this time I thought I would just have to see how things played out. We were engaged by the end of the year!

CONFIDENT, ENGAGING, ENTHUSIASTIC, CHARMING

James and I are very different but also have many things in common. The attraction for me, apart from his beautiful blue eyes and his kindness, was James's extraordinary confidence. He was totally at ease with his plunge into art college and the uncertainty of the blank canvas before us. He was also totally at ease with difficult projects, which I would try to analyse to find the reason why we were asked to do them and what we were supposed to learn from them. He was just as serious as me, but keen to get on with whatever the project was and to see what developed.

James always had an opinion about anything and everything and was clever at persuading others to agree with him. He liked, and still likes, arguing, even if he secretly agreed with the opposing view. He was taught to do this at school. I found this rather confusing at first but admired his engaging enthusiasm that really convinced people. This proved to be a valuable gift in the years ahead.

NOTHING TO LOSE

In our second year, I was given a scholarship to stay at the Byam Shaw and James had been accepted at the RCA. We moved in together. One grant went on rent and we both lived off the other. Neither of us had cushy upbringings so the challenge of how to survive financially has always been with us. I believe this is the best grounding for any future success. We had nothing to lose. We both had enough skills to get work in the holidays, so we knew that we would be able to get through life with ordinary jobs if the art world didn't work out for us.

I moved on to Wimbledon College of Art for my three-year diploma course. We were married during our second year and we both finished our training at the same time. I had started to gain some private graphic design work as James was offered the Rotork position in Bath, but before finding new commissions in a new town we planned to squeeze in a family. We arrived in Bath with a six-week-old Emily. Looking back, this also was a huge risk.

RISK/COURAGE

During the Rotork boat period, James learned how to design, produce and sell but – and this is another character trait – he likes and needs to be in control and his Ballbarrow idea was the perfect project to allow him to break away on his own. With James running his own start-up, we felt quite comfortable for a short while and even moved house as, by then, we had two children and a baby. I was somehow managing to paint small canvases and selling some, plus doing some graphic design work for a hotel and illustrations for a book. However, we were let down badly by the shareholders (shafted is the truth!) of the Ballbarrow company and suddenly we were back to square one with a mortgage, two children and a tiny baby.

PRACTICAL

We are both practical people and would do everything ourselves if we could. For our new home, James hired a digger and drove it himself to create a swimming pool. We made a fish pond together and a vegetable garden in a half-acre patch. This took up most weekends. It was also, sadly, the first thing to go following the Ballbarrow crisis. We managed to sell our half-acre with planning permission.

I was terribly hurt by this whole disaster, especially as James had been working seven days a week, had built a production team of thirty people and had recently added other new designs to the product line, which the shareholders had decided should all be signed over to the company. This meant that James no longer owned his own designs.

I felt powerless to help, but soon started an evening class teaching art, which brought in a little money. I also had a monthly trip to London to illustrate a column for *Vogue*. We went, though, through a period of what felt like grief and chewing it all over day and night, until James said, 'Enough! I am going to develop the vacuum cleaner'. He had already dreamed of this cyclonic idea and presented it to the shareholders. They were not interested. 'I am never going to have shareholders and I am never signing my designs over to anyone again!' Lesson learned. We were helped enormously through this difficult time by close friends who remain our closest friends today.

MORALS

I was brought up to have a very strong moral compass, so the Ballbarrow event really shocked me. Over the years we have had various employees who have had their own little business running on the back of the main one or who have left the company and taken secrets with them. The bigger the business, the more of these incidents, and it is something I cannot grow immune to. I just believe in right and wrong.

When James first set up the Ballbarrow company, I remember saying how important it is to be squeaky clean in all dealings. This is why I have always supported James with his legal battles, in particular the US lawsuit over the vacuum cleaner, which lasted five years. I could not let this happen to him twice as it would have been truly soul-destroying, so I tried to encourage him to keep on trying to overcome this wrong in spite of the alarming leakage of funds for legal fees.

DETERMINATION

The following five years were ones of perpetual trial and error, yet stubborn determination carried James through the early development of the vacuum cleaner. Not only does he never give up, but he always does everything properly and pays attention to every last detail. Nothing less will do. I am stubborn and determined too, but I could never have persevered to the extent James did.

COMPETITIVE

James is extremely competitive with work and life in general. This is a really important characteristic in the world of trading and, competitively, his field is ruthless. I steer well away from possible conflict! I do the crossword secretly every day because if he sees me doing it, he immediately begins, too, and tries to complete it first. He never allowed the children to beat him at tennis, which induced tears and frustration. I am not sure how they feel about this now, but I think it was the result of being the youngest of three and having to fight his corner alone at school.

James is still working flat out, but is happy to be in control, although we now have the wonderful support of talented advisers – business, financial and legal. His need for control is reflected in his organisational abilities and his everyday tidiness. I keep the door of my untidy studio firmly shut!

ADVICE

James avoids giving advice because it may be the wrong advice. We both believe that people need encouragement, especially young people making their way in the world. They also need the opportunities to succeed. And this is why, like James, I'm so concerned with education. Education gave us both the chance to find out what we were good at and to do well. Thousands of children never discover their talents. The waste fills me with horror. Creativity really matters.

I'm really excited to see the Dyson Institute flourishing, with young people both studying and working creatively. Debt-free, too. They're prepared for a world of adventurous and creative work. I am particularly excited at the way in which Dyson undergraduates are being educated academically and practically at the same time to an exceptionally high standard and without incurring fees. I am passionate about the lack of respect for talent and skills in school education, which is the core of all creativity and problem-solving in products.

I could never have imagined that the journey would take us to where we are today. Although our children have their own careers and are creative people themselves, Dyson has become part of all our lives. A friendly monster, it's a family business so we all attend meetings and keep up with the progress in all areas. I am on the Foundation team, which has grown enormously over the years, helping schools across the world with design and technology education and sponsoring the James Dyson Award. Some major projects have been brought to me first as some people are afraid to approach James directly.

I have been able to follow my own path while riding this roller coaster. Right at the beginning of our journey together James said, 'It doesn't matter how hard it gets as long as we get through it together', words I've never forgotten.

Optimism has to be the most loving and supportive attribute that James has. He doesn't just hope for the best, he simply believes that everything will work out.

Afterword

I wrote the final chapter of the first, hardback edition of this book during the height of the Covid-19 pandemic. Inevitably, Covid had an impact on us at Dyson as it did at most other organisations or workplaces. To bring the paperback version of the book up to date, I wanted to add some of the big events and crosswinds that have attempted to blow us off course in the year since. If anything, these events show that though you may have a vision of where you want to take a business, getting there is seldom a linear path. It requires you to be nimble and reactive. From my classical learning, I might quote the Greek philosopher Heraclitus here. 'The only constant,' he wrote, 'is change,' as true a statement in 2022 as it was 2,500 years ago.

We used the downturn presented by the pandemic and the uncertainty that brought, to double-down on investments and research programmes, both inhouse and with universities. One of the main themes, therefore, of 2022 has been – thankfully – new products (though making them is the problem as I will explain), supercharging technologies like batteries, and the recruitment of new scientific and engineering disciplines. As I write, we have 900+ roles open in the UK and 2,000 globally. We are recruiting at all levels: undergraduate, graduate, through to world-expert. But there are many other events that have coloured the past twelve months which have threatened to undermine our efforts...

COVID CONTINUES...

You will recall that early in 2020 we had gathered members of the team together at Hullavington to discuss the next five years. Little did we realise that only days later we would be confronting what threatened to be a global human tragedy and the worst recession for a hundred years. We also had no idea that the space we were sitting in would be repurposed to develop an emergency ventilator. Within just a couple of weeks we had completely changed our business plan, borrowed £1.7 billion against contingencies and, of course, government orders meant that vast swathes of our organisation went home. We borrowed no government money, and took no Covid-related loans, anywhere in the world.

Hundreds of people working on the Dyson ventilator dutifully continued working on our campus, spending hours away from their families, and facing the personal risks that Covid brought. Meanwhile for an engineering, technology and product business like ours, having everyone else at home brought systemic challenges, slowing all momentum and progress. You can't engineer or set up a laboratory at home and it blocks collaboration. Future projects slowed down and, in parallel, production stopped as factories around the world closed – indeed, even well into 2022, they continue to struggle to meet demand due to ongoing labour and parts shortages.

A shortage of microchips, exacerbated by a worldwide rush to buy laptops for home working, led to a rise in their price from $2 to $75, and they were diverted to their biggest customers. These chips provide the brain to drive most of our products, which work on artificial intelligence; they are integral to the switching, control and power saving – without them the product doesn't work. In addition, it now took twelve rather than six weeks to ship products from our factories to the United States. I could add to this catalogue of Covid-related woes and yet, despite a very worrying January to March 2020, we shifted our focus, and our sales and performance accelerated throughout the rest of the year. We immediately accelerated the shift – which was already well underway pre-Covid – to direct online sales, which certainly stole the march on competitors.

But, even as I write in 2022, Covid looms large, threatens progress, and means we are taking actions we never dreamt necessary. An example is that, currently, our team in China remains under a strictly enforced lock-down – locked in their homes unable to receive food. Since we are an organisation of problem-solvers, the management in China developed creative solutions to ensure the day-to-day operations of the business continued while keeping people safe and helping to lift spirts in a very difficult period. This means we have just distributed the fourth round of food parcels, containing fresh vegetables, coffee and bread.

Change is also afoot on our UK campus. From the beginning we had wanted to create a healthy campus on which people chose to spend time. It is as much a social place as it is a workplace and we have not stopped adding to it – something we can do in our rural location, an advantage over city centre locations. There is no doubt in my mind that bringing people together in an inspiring environment is more important than ever – it fuels the inventive spirit.

On our UK campus, for example, you will now find: six cafes serving free food made with Dyson Farming produce (under the watchful gaze of Joe Croan – former Head Chef at l'Escargot), a free gym, meditation, yoga and fitness classes, a free professional hair salon, free employee transport to campus from nearby towns and cities, support for purchasing electric vehicles, multiple outdoor nature trails around Wiltshire countryside, a wellbeing centre – including a campus-based GP and physiotherapist, clubs and societies run by Dyson people (from gardening clubs, on campus allotments, beekeeping in our apiaries, to cycling and the annual Tour de Dyson event), extracurricular activities like our summer and Christmas parties – hotly anticipated dates in the Dyson diary where teams create inventive costumes. I want our campus to be a hub where people want to spend time with colleagues for work, for leisure and for exercise in the most inspiring of spaces.

In late February 2022, just as most of the world emerged from Covid which seemed much less threatening than it had been in those difficult first months of 2020, the Russian regime invaded Ukraine. The situation is tragic.

On 22 February, before the invasion, we proactively diverted products and stopped supplying into Russia. We closed our stores there and ended the very successful James Dyson Award – it was all very upsetting and hugely disappointing for the hundreds of Russian employees who had loyally built that business over many years. Dyson has no employees in Ukraine but we are supporting our staff who have family there.

Just as the Russian regime set out on its destructive path, we opened our new global headquarters building, St James Power Station in Singapore, launched a revolutionary new wearable product (the Dyson Zone™ air-purifying headphones), are building a pilot battery plant in Singapore and entered negotiations to build elsewhere a new gigafactory the size of 35 football pitches – both factories producing the new-technology batteries we have been working on for years.

Where best to start with what we have been up to in 2022? The Dyson Zone™ air-purifying headphones, I think. Jake and the engineering team launched this worldwide from our Hullavington Campus in March. It's our first wearable, personal product consisting of what appears to be a rather smart set of headphones attached to a contactless mask. While providing high-quality stereo sound and cancelling out noise pollution – from building sites, planes passing overhead, cars, buses, lorries, tube stations – the headphones incorporate highly effective pollutant filters that, working with our miniature electric turbines (one in each ear), provide a flow of clean air to the wearer's nose and mouth through an adjustable and washable mouth visor. This means you can walk through a noisy and heavily polluted part of a city, or ride its underground metro system, while breathing clean, cooling air and listening in peace to your favourite music, podcasts and books. The device is rechargeable with the batteries concealed in the adjustable headband.

A product like the Dyson Zone™ headphones has never existed before. This is very much a pure invention, representing a new product and the latest evidence of our willingness to take risks.

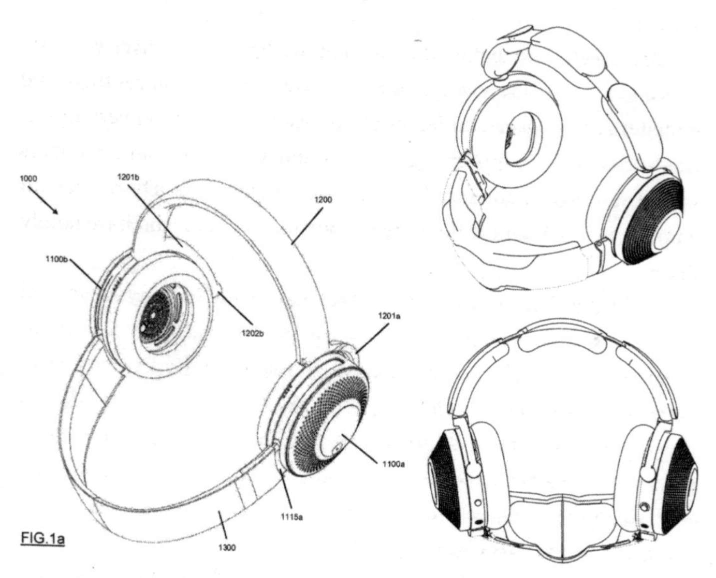

The World Health Organisation (WHO) estimates that nine in ten people globally breathe air with high levels of pollutants and in Europe is it estimated that around 20 per cent of the population are exposed to long-term noise exposure above the guidelines. It's most likely to appeal to a younger, city-based demographic, especially in the cities in Asia where pollution is very well understood and where we are already well known for our air filters. To date these have been for use in the home or the workplace. The product takes our filters out onto the streets, into the sulphurous cores and particulate-infested depths of crowded cities.

This, though, is not some fashionable product dreamed up during the Covid pandemic. Far from it. It's the product of six years' research, some 500 prototypes and, I should say, thirty years of filtration experience. The Dyson Zone™ headphones might look nothing like the diesel particulate filter I worked on with Pete Gammack all those years ago, but it is an extension of the thinking behind it. My concern with the dangerous filth we breathe simply walking along a city street whether from vehicle exhausts, brake dust and tyre particles has never gone away during the time we've developed several new technologies, including our digital electric motor and a variety of filters and compressors. The Dyson Zone™ is a new product built on and combining successive layers of Dyson technology. Although we do a huge amount of research into sound, it is usually about reducing sound and noise cancellation, in this case we learned something entirely new – how to create high fidelity audio.

As for air quality research, we have our own scientific advisory board comprised of world experts, work with universities, and pursue our own inhouse research. This has included developing pollution sensing backpacks for 250 young students to measure the different types of pollution that they are exposed to on their journeys to and from school. At the more unusual end of the spectrum is 'Frank', a manikin with medical grade mechanical lungs and sensing equipment, which was developed in one of our Hullavington laboratories.

Our DIET undergraduates engineered a sensitive and accurate air-quality monitor, small and reliable enough to be carried by children. It had to have a battery life of at least ten hours, and it needed to be sufficiently strong to cope with the robust treatment some, if not all, young children dish out to anything they carry around with them during busy, rough and tumble days. In the event, the undergraduates succeeded in creating a successful monitor fitted into school rucksacks. The information gathered was invaluable, not least for the 250 school students, 31 per cent of whom changed the route they took to school after the experiment to ensure they breathed in cleaner air than they had before.

The Dyson Zone™ air-purifying headphones is a highly visible example of our obsession with polluted air and the kinds and level of filtration needed to combat its dangers to our health and wellbeing. It's an uphill battle, especially now we know so much more about particulate health hazards. It's not only those produced, for example, by cars, lorries and buses, but also those like formaldehyde, released from such seemingly innocuous products as carpets, sofas, candles, even cooking and polyester-nylon clothes. Even when systems have been designed and engineered to produce artificial fresh air to isolate us from the fumes and heat of city streets, we need to think about these carefully: it's often better, for example, to open a window, if you're able to, in a city centre hotel to breathe in the polluted air outside than it is to breathe air from uncleaned air-conditioning systems of which there are all too many.

New Dyson spaces as St James Power Station lights up again...

Incidentally, Singapore is one of the least polluted and certainly the cleanest of Asian cities. The city state sets high standards for itself in many ways and those spelt out by Singapore's long-serving prime minister, Lee Hsien Loong, in his speech at the opening of our new global headquarters, St James Power Station, are some of the key reasons we find this industrious island such an attractive place to work.

'Singapore,' said the prime minister, who studied mathematics and computer sciences at Cambridge, 'is open and connected to the world both physically and in the ethos of its society. It welcomes new ideas and is never resistant to change or complacent about the need to stay ahead. Free trade is a mainstay of its economy as is a culture of innovation built on research and development, while the government encourages STEM [Science, Technology, Engineering and Mathematics] students, recently merging the NUS [National University of Singapore] Faculty of Engineering with the School of Design and Environment to form the College of Design and Engineering.'

The prime minister cited a group of young inventors – Kelu Yu, Si Li and David Lee – from NUS Engineering who won the international category of the 2021 James Dyson Award. Their invention overcomes the problem that patients are unwilling to go for a glaucoma test, which involves firing compressed air onto the eyeball, as it is painful and off-putting. The team aims to bring their invention to the market as soon as possible to help solve the problem, and their ambition is very much that of Singapore's.

You can feel the energy in this extraordinary island city. Not surprisingly, we're investing heavily in Singapore in the coming years, hiring more than 250 engineers and scientists with roles spanning robotics, machine learning, AI, high-speed digital electric motors, sensing and vision systems, connectivity, software, power electronics and energy storage. Today over 1,400 Dyson people work in Singapore, 560 of whom are engineers and scientists. It seems surreal that I was once accused of moving elements of our operation to Singapore in the search for cheap labour. We wanted its dynamism, openness and intelligence as well as its geographic position at the heart of global trade.

EDUCATION

The prime minister also spoke of the importance of STEM. Speech Day, four months later in July 2022 at my old school, Gresham's in north Norfolk, coincided with the opening of the Dyson Building, also known as the STEAM Building, where art meets science, technology, engineering and maths, a complete circle of studies. For me, the occasion was moving not least because Chris Wilkinson, the building's architect, had died at the in December 2021. Chris was a long-standing collaborator and friend of mine. He designed our wavy-roofed factory at Malmesbury and developed our campus there over the years, including the Dyson Institute of Engineering and Technology. We've renamed the Institute's Roundhouse Building, the Wilkinson Building in his honour and memory.

Chris also oversaw the restoration of our campus at Hullavington Airfield, designed the Dyson flagship stores in Paris, London, San Francisco and New York with its giant window fronting Fifth Avenue. We worked together until Chris's sudden death, by which time he had become a major international architect celebrated for projects such as the 440-metre Guangzhou Tower and the futuristic Gardens of the Bay in Singapore. Despite his great success, Chris was never less than a modest and genial man. We will miss him enormously.

One of his last projects for Dyson was the Dyson Village – the pods where our undergraduates live on campus – along with the Roundhouse. On that subject, one of the reasons that I wrote this book in the first place was as a tribute to the pioneering first cohort of undergraduates who joined us at the Dyson Institute. Since the publication of the first edition of the book the first cohort has graduated. In September 2021, we celebrated all that they had achieved – personally and professionally – in a ceremony on Hullavington Airfield and a day on campus. They all achieved excellent results, but more than that, they contributed to real products too, all while driving the culture and spirit of Dyson forward. One of the most moving things that the graduation ceremony revealed was how the group formed extremely close bonds and helped each other through what is inevitably a very challenging course – it is a powerful combination!

Their reflections over that day showed that, in coming to study at Dyson, they made a serious career choice, a proactive decision to break from the expected norms of education today. In choosing to come to Dyson, rather than going to a traditional university, they took a risk, but one which excited them and benefited them – perhaps proved by the fact that they decided to remain with us post-graduation. As they are the first cohort, I was keen to understand why their parents had let them come and took the pre-dinner reception as my opportunity to ask directly: what on earth were they doing sending them as pioneers to Dyson, when we were totally unproven? I was taken aback by the response... It wasn't the fact they would be paid a salary – as I thought it might be – rather it was the fact that they liked the idea of their children working with a company working on real products – I was delighted.

On the day of the graduation, Bart Jennings, one of our first cohort, gave the most moving speech. I think it worth extracting here as he so neatly sums up what DIET means for us all at Dyson, not least the undergraduates themselves: 'No one knew how it would work with thirty-odd teenagers playing at engineering on live Dyson projects with absolutely no qualifications, knowledge or experience, while also trying to squeeze in one of the most time-consuming degree courses, an apprenticeship, and build out of nothing some form of social life in the countryside. But thanks to our fantastic colleagues for giving us so much of their time to mentor us in our workplace rotations, our lecturers for teaching us wide-ranging technical subjects and the Dyson Institute team who worked tirelessly behind the scenes, pulling the strings like puppet masters to pull everything together – we didn't just survive – we thrived. We were proud of who we were and what we were part of. We cared about our degrees, our work and how we were perceived... as all eyes were on us. [...] But throughout the ups and downs, the highs and the lows, we had one very special thing that helped us find a way through and come out stronger than before. Each other. Thirty-three of us joined the Dyson Institute, and thirty-three of us are graduating today.'

A good education and exposure to teachers who allow us to think independently and be creative, rather than learn simply by rote, is crucial. And so, I was delighted to hear from a former junior colleague of my mother's, at Fakenham Grammar School, last year who wrote so kindly that she '*was an inspirational leader who gave me the most incredible amount of freedom to teach. She did so much to put me on the right lines at the very beginning of my career, giving me liberty but always showing me how best to use it, completing my teacher in the most exemplary way... After her death the school wrote to her family to ask what to do about all Mary's personal things (mainly books) in her desk and cubby-hole in the Common Room. We were told to keep whatever we wanted and so the selection was displayed. My trophies were a very good French dictionary and a tiny little volume of a dictionary of mythology. This pocket-book proved invaluable throughout my career – whenever I needed to know who, or what, say, Entellus, was – I just looked it up and there it was. It was in my right-hand school desk drawer in all of my classrooms, even in Monaco. The volume is inscribed 'J Dyson, Gresham's'.* This tiny volume has been returned to me. How remarkable! After all these years, I wondered where it had gone.

INTELLIGENT MACHINES

Hullavington, meanwhile, is now en route to becoming Britain's largest robotics centre. While we'll be using robots increasingly in Dyson factories and farms, we have, of course, also been developing robots for domestic use over many years, starting with the vacuum cleaner. Jake is at the robotics helm today. He has a huge and fascinating task ahead of him. What do we want robots to do about the house? What are their capabilities now and into the future? We are well advanced with floor cleaning robots, but it won't be too long before they do other things as well while you are at work or otherwise away from home.

And this is just the tip of the domestic robot revolution, a revolution rooted as much in mechanical dexterity as it is in software, vision systems and communications. I find it all endlessly fascinating. There is so much new to learn, to play and experiment with, so very much yet to invent, and so many talented and free-thinking young people around the world excited by the future and what design, engineering and invention can do for it. Whatever fears and threats appear from the wings of the world stage, we must never give in to them, much less give up.

BECOMING DELIGHTFULLY SUPERFLUOUS

I started Dyson 30 years ago in my Coach House. I was lucky to have a space adjacent my house where I could have a workshop and later an office. Some wonderful engineers and other disciplines took a risk by joining a small start-up enterprise. Although I tried to interest investors to back our business, they all thought that taking on established household names was folly, and that Dyson should be run by a professional manager from the industry who could help sell our technology.

I had some money coming in as royalties from licensees. Lloyds Bank was kind enough to lend me the £600,000, for which I am eternally grateful. This was just enough, with close control of finances, to get us to the point of start of manufacture. We had no money for a factory, no money for stock and no money for advertising. How we managed to have a production line, to supply customers and start advertising is described earlier in the book. We were what might be described as 'under financed'. To investors we were a pariah. The flip side of being a company with no outside investors was that, by default, I owned the whole company. Although at the time that was more of a liability than an asset.

As we grew, I became aware that we had become what you would call a family business: one owned by a single family rather than outside shareholders or the stock market. As our success was noticed, approaches and offers were made to purchase the company from me. This was flattering and would have meant a quick and substantial profit from the sale. Friends advised me to take up the offers of unimaginable riches, removing risk and guaranteeing wealth for my family.

It was tempting, I must say. However, it was easy to decline their kind offers. Though I couldn't have dreamt what the company would grow to become, I felt we had only just started. We had a business that could develop technology, manufacture new products and launch them into markets. Why would I give up a successful enterprise that I had worked to create, with many false starts along the way? Besides, we had more ideas in the pipeline. I also realised we were a rare animal in the corporate world: without shareholders we were free to make our own decisions.

During my career, I had set up the marine division of Rotork, a successful quoted company, and I had worked for the Ballbarrow company that I had started where I was a minority shareholder with outside investors. I had felt the vicissitudes of both models. Both the quoted company and the one with outside shareholders held little attraction for me. I wanted every decision to be our own, not one influenced by others. Those of us running the company understand every nuance, we see the opportunities and the pitfalls. Any mistakes we make are our mistakes. It is our money that we are risking, not other people's. This makes us cautious with our money on the one hand, but free to take calculated risks should we see opportunities. It is our responsibility.

You might argue that we as executives are so involved in the business that we might not avoid pitfalls. For this reason, quite early on, I decided we should have some experienced outside directors on our board. They have provided much-needed external views and political insights. When faced with serious problems, they have been the wisest of contributors to the discussions at board meetings. I decided we should run Dyson with the discipline and corporate responsibility of the best quoted companies, but with the agility and care of private family ownership. We could never have grown as we have, without the wisdom and help brought to us by our non-executive directors.

Suggestions have been made over time from several sources that we should go public. In other words, sell part of the company to outside shareholders. This is, after all, the route that most successful start-ups follow. It allows the shareholders to pocket a large sum of money as well as continue to see their remaining shares grow alongside with outside investors. In many cases the founders often sell up after a discreet period and cease being part of the company.

I discussed the idea of going public with Deirdre, Emily, Jake and Sam. Their reaction was one of horror. 'You mustn't do that, Dyson is a private family-run-business.' They were right. I did not start the business to make money by selling out. After all, I had started Dyson so we could develop technology and engineer exciting products. This would change if we went public. My family confirmed their commitment to Dyson being a private family business. It was their decision not to sell shares in Dyson. I couldn't have been more relieved, proud and delighted. At the same time, it was the beginning of the possibility that they were willing to take on the responsibility of owning and running Dyson. However, that's rushing things. They were assuming I would continue to run Dyson myself. My children all had their own start-up businesses to which they were fully committed. No time or space for Dyson.

A few years later, Jake, who had been designing new technology lights for his own manufacturing business, admitted to me that what Dyson was doing was far more exciting and that he envied the engineers in Dyson for the opportunities they had and said he'd like to join us. We merged with his business, Jake Dyson Lighting, and Jake became head of engineering at Dyson. Both Jake and Sam, my second son, had worked as engineers at Dyson before starting their own separate endeavours so they were very familiar with how Dyson worked. However, there is a caveat: it's well documented that children of founders have difficulty working for their parents. Jake and I were well aware of this danger. Yet Jake had spent fifteen years founding and running his own complex manufacturing business. He had learnt all the lessons the hard way and could join as a successful engineer and entrepreneur in his own right. Everyone at Dyson was aware that there was so much more to Jake than being the founder's son – he had 'made it' on his own.

As I passed the 'normal' retirement age of 65, I was urged to take action on what is called 'succession planning': the running of the company without me. Jake and Sam came onto the board. Jake as an executive director and Sam as a non-executive. Sam worked at Dyson for two years after leaving college, as an engineer. He left to make a career in music, his passion. He formed his rock band, Ramona Flowers, and started his independent record label, Distiller Records. Not a propitious time to start either venture, but he has persevered and is making good headway in that toughest of businesses, creating music and new ways of composing as well as learning much about starting businesses and running them successfully.

Sam joined the Dyson board having learnt the hard way how to run businesses and manage a team. He also chairs the family controlling entity. Although Sam is titled as a non-executive, he is more than that as he takes part in management meetings every week and has been a key advisor on our headphones project. I'm aware I haven't spoken much about my daughter, Emily, who runs her own successful business, and also devotes considerable time to Dyson. It is always a delight to see how well she works with Jake and Sam. They are a very strong family unit together. How lucky I am that all three of them have chosen to run their own businesses and worked out how to make them successful.

Every business reports to its shareholder(s) or partners. Dyson has three family members, shareholders, engaged with the company and sitting on its board. However, the company is run by a CEO and the top team of managers. Dyson is perhaps unusual because Jake and I sit in many meetings within the business. These are chiefly in the research, engineering and design areas. This works because technology and our products are so key to our future, and also because we love doing it. It is our profession. We also sit in meetings about how to launch and describe products. We were part of their creation, so we ought to know how they should be described. Incidentally, Dyson uses no external advertising agency – we do everything ourselves. Although we have tried in the past, we decided that however brilliant an external agency, we should be able to say in simple terms what is good about our products. This may make for less entertaining copy, but I have a feeling that people see through blatant marketing and prefer the simple truth anyway. Snake oil salesmen and all that!

The CEO and the top management team run the business. They are professionals, many of whom have joined us as graduates (hopefully in the future as undergraduates) and have risen to top positions. They are dedicated and run the business through all the turmoil and headwinds, as well as growing our enterprise exponentially. They are the best in the business, and they too believe in the values of a family business, the driving force of technology development and building a creative team.

You might wonder, if that is the case, why the family needs to be so involved in what goes on? Are we not superfluous and apt to meddle? I believe families can influence the management of family businesses in a positive way. The priority of most commercial enterprises is to make progress and to create wealth. This safeguards the future of its employees as well as creating more jobs and ever more people being supported by the enterprise, not to mention paying an increasing amount of tax to, among other things, fund education and social care. Companies perform this vital function for society. Quoted companies' first duty is to the shareholders, to maximise profits and the company's value, usually in the short term. Family businesses can take a long-term view and be less driven by short-term profits. If the family is so inclined, they can take greater risks with capital and investment. The management of a quoted company might shy away from risk, as their jobs and security might be threatened if the risk turned to failure. When we cancelled our electric car project, the family had supported the high-risk car project and the management were not to blame. Far from it, they had done an amazing job on the car. The stock market might not react in the same way over the loss of investors' funds.

The family can take a long-term view. After all, the family is a constant, while management might not be. Other aspects can have a greater significance than the pursuit of profit for its own sake. Take the example of architecture for the workplace. It is more expensive to create good architecture than to put up purely pragmatic buildings. A family cares about what it stands for and the long-term reputation of the company. A family also establishes the ethos and values of its company. That is not to say the executives cannot influence it for good or develop in a different direction, as they often do, but the family is ultimately responsible.

Which brings me back to continuity and the importance and relevance of our children, Jake, Sam and Emily, in shaping the future of Dyson. Dyson will only thrive if it can continue to develop new and relevant technology and bring out products engineered to perform better and better, in every way. This is Jake's forte and one he has taken up with enormous enthusiasm. Sam provides insights and support for Jake, as does Emily. I can see myself becoming delightfully superfluous.

Acknowledgements

On a visit to George Osborne in 2009, just before he was Chancellor of the Exchequer, I met Oli Blair, who was an intern there. I was instantly impressed, and he came to join Dyson. Eleven years later, this book was his idea. Thank you, Oli, for driving through its production with your usual intelligence and elegance. A huge thank you to a long-standing friend and brilliant writer, Jonathan Glancey, for the most enjoyable of collaborations and with whom one had to exercise great discipline to stick to the subject, so fascinating was the conversation; to Ian Marshall, our editor at Simon & Schuster, with his incisive questions and excellent suggestions modestly hidden behind his encyclopedic knowledge; to Deirdre, for her invaluable input and insight during the drafts. My eldest son Jake, who, although he had invented several designs of his groundbreaking LED lights and built up a successful business on his own, took the step of merging with Dyson: a brave step from a loving son whom I admire so much and who has brought his extraordinary inventiveness and design talent to Dyson. To our other son, Sam, and daughter, Emily, who are creative talents with their own wonderful businesses yet spend much of their time helping and advising, nurturing our family business. A chap is utterly blessed to have Emily, Jake and Sam as loving children, putting up with a roller-coaster life and then bringing up our adorable grandchildren. To our loving friends who gave unfailing support and encouragement, whatever their doubts, throughout our married life. My PA, Helen Williams, who has unfailingly and cheerfully supported me at work whatever time of day or night for the past eighteen years, keeping me on track and without whom my professional life would not be possible. Peter Gammack, the greatest of engineers, who has brilliantly masterminded every product and design venture for the past thirty-three years, with thoughtfulness and enthusiasm – he is the only tennis player I have ever seen who can return a tennis ball while horizontally airborne. Alex Knox, who joined in 1992, designing almost every product and who heroically developed our ventilator for production within just six weeks during the pandemic. To Sir James Bucknall, who, with great charm and brilliance, leads our other family businesses. My immense thanks to Ian Robertson, Tony Hobson, John Clare, Allan Leighton, Bob Ayling, Andy Garnett, Tian Chong Ng, Boon Hwee Koh, Kishin RK, Warren East, Rodney O'Neil, Sebastian Prichard-Jones, David Fursdon, Mike Brown, Mark Slater, Sir Richard Needham, Jo Keddie, John Chadwick, Chris Wilkinson, Tony Muranka, Mark Taylor, Martin Bowen, Jim Turner, Lennard Hoornik, Roland Krueger, Jorn Jensen, Scott Maguire, John Churchill, Nicholas Barker, Guy Lambert, John Shipsey, all of whom have most skilfully helped Dyson and the family along the way. Sarah Chalfant, at the Wylie Agency, for her encouragement and for guiding us so helpfully. Pippa Burgess, who took on the task of sourcing illustrations and photographs, even braving my dusty archive room. Finally, the greatest of thanks to all my delightful colleagues at Dyson and Weybourne, who have forged a global technology company and who make this adventure so exciting.

Plate section image credits

Roofscape of Malmesbury. Photograph by Mike Cooper.

The Dyson Symphony. Photograph by Martin Allen Photography.

Contrarotator. Photograph by Mike Cooper.

DC01, DC02, DC03, DC04, DC05, DC06, DC07, DC11, DC12, DC14, DC15, DC16, DC35. Photographs by Mike Cooper.

Computer rendition of the Dyson EV, car interior seating and cutaway of Electric Drive Unit. Rendered by Daniel Chindris.

Sunday Times magazine page. Image courtesy of *The Sunday Times* / News Licensing.

Potato planting, pea-vine harvesters at work and strawberries at Carrington. Images courtesy of Dyson Farms.

DIET desk pod. Image courtesy of WilkinsonEyre.

Student pods on Malmesbury campus. Images courtesy of WilkinsonEyre.

James Dyson and DIET students. Photograph by Mike Cooper.

The Roundhouse and the Dyson STEAM Building, interior and exterior. Images courtesy of WilkinsonEyre.

Carvey Maigue. Photo courtesy of The James Dyson Foundation.

James and Jake Dyson in cut-through mini. Photograph by Laura Pannack, Camera Press London.

All other images from the author's personal collection or property of Dyson.

Index

Page references in *italics* indicate illustrations.
JD indicates James Dyson.

A

Aalto, Alvar 129
academy schools 282–3
accelerometers 303
advertising 54, 81–2, 83, *83*, *87*, *124*, 125, 127, 138–9, 196, 201–2, 322
Aedas architects 214
Age of Austerity 17
AI (artificial intelligence) 23, 186, 193, 204, 214, 257, 288, 300
Airblade, Dyson 173–6, *174*, *176*
Airblade Tap, Dyson 174, *174*
AirDrop irrigation system 302
Air Ministry 67, 68, 244
Air Multiplier, Dyson 176, 177, 178, *179*, 180, 181–2, 207–8, 217
Airwrap, Dyson 185
Albert, Prince 297
Alexandra Technopark, Singapore 210–11
Amway 91, 105–6, 109, 111–14, *112*, 119, 321
Apax Partners 117–18
Apex 106–7, 110, 187, 198, 200
Architectural Association 71
Argos 129, 130
Arkell, John 305
Arnold, Malcolm 155, 205
ASEAN free-trade area 239
Aston Martin 237–8
Austin Allegro 77
Austin-Healey 37, 85
Austin Morris 56, 64, 110, 195
Australia, Dyson in 110, 195–6
Aveling, Thomas 263

B

Bache, David 237
Baker Electric Cars 228
Ballbarrow 73, 74–90, *79*, *81*, 92, 93, 94, 96, 115, 117, 158, 263, 319, 326–7, 328
  builders market, redesigned for 86
  cyclonic separators, JD's first use of and 83–5
  design 78, *79*, 80
  grass box extension 83, *83*
  JD ousted from company 89–90
  origins of 73, 74–5
  patent 79, 88–9
  plagiarism of 88–9
  puncture-proof ball/wheel 86–7
  sales/selling process 80–90
  Tomorrow's World, appearance on 82
ballistics 193
Balls, Ed 283–4
Bangladesh Liberation War (1971) 59–60
Banham, Reyner 63
Bannenburg, Jon 46
Barker, Nicholas 201, 202

Bashford, Gordon 237
Bath Rugby Club 303
Bath Spa University 283
battery technology 104–5
  Corrale straightener and 186
  DC16 and 165–7
  DC35 Digital Slim and 168
  EV, Dyson (electric vehicle) (N526) and 225, 226–30, 235, 238, 241
  R&D, Dyson 172–3, 209, 210, 225, 226–30
  360 Eye robot vacuum cleaner and 188, 190
Baxter, Dick 113, 114
Baxter, Raymond 82
Baynes, Leslie 14
BBC 14, 82–3, 97, 100, 143, 252–4, 289
Bedwell, Bob 121
Beecroft, Adrian 117–18
Beharrel, Clive 109
Bell 47 helicopter 60, 137
Benet, Judit Giró 300
Bernoulli's principle 176
Bertoni, Flaminio 60, 66, 98
Best Buy 201
Beuttel, Alfred 131
Bex Bissell 112
Bhattacharyya, Professor Lord 289
Bhola cyclone (1970) 59–60
Bissell 105, 112, 114
Black & Decker 105, 111
Black, Misha 38
Blair, Tony 52–3, 198, 225, 277, 282, 283
Blomfield, Reginald 244
Blue Box, The 300
Blue Peter 143, 224
BMC (British Motor Corporation) 64, 308–9
BMW 52, 310
boardroom table design 151–2, *152*
Bonham-Carter, Maurice 68
Bosch 100, 310
Boulanger, Pierre-Jules 65, 98
Bourdon, Charles 260
Brazier, Alan 111
Breakthrough Breast Cancer 140, 302
breast cancer 140, 300, 302
Brexit 217–22, 253, 264
Bridgestone 231
Briggs, Alan 196
Briggs, Rebecca 196–7
Bristol University 95
British Aerospace 86–7
British Design Council kite mark 81
British Leyland 64, 75–6
British Rail 77
Britten, Benjamin 14, 20
Bronowski, Jacob 97
Brooke, Charles Vyner 62
Bruce-Lockhart, Logie 13, 14, 20, 22, 305
Brunel, Isambard Kingdom 95, 142, 282
brush bar 120, 144–5, 147, 156, 193, 208
Bulloch, Archibald 244
Burke, Peter 133
Byam Shaw School of Drawing and Painting 28, 29–31, 33–4, 326

C

Cabinet Office 249, 250–1
camera, 360-degree 189–90, *189*, *191*

Cameron, David 242, 284, 285, 287
Cameron, Ross 110, 195–6
Carden-Baynes Auxiliary 14
Carrington estate 261, 268, 271
Carsen, Robert 56
Casson, Sir Hugh 38, 39, 44, 47
Castiglioni, Achille 295
Çelik, Tuncay 168
Centre for Doctoral Training in Future Propulsion 296
Chamberlin, Powell and Bon 33
Chelsea Flower Show 149–51, *150*
Children's Theatre design 40–1, *41*
China 203
  Covid-19 and 215
  Dyson enters 207, 208
  Dyson Pure Hot + Cool purifier and 180
  Dyson recruitment in 206
  Dyson vacuum and 120
  electric vehicle market in 233, 239, 240, 241
  EU and 218
Chippenham, Dyson vacuum cleaner factory in 50, 62, 69, 128, 129, 131, 132
Chrysler 281
Churchill, Winston 17, 68
circular pedalo 43
Citroën 48, 49, 51–2, 66
  DS 51, 62, 65, 66, 98
  HY corrugated delivery van 98
  SM 51, 237
  Traction Avant 60, 65, 98
  2CV 60, 61, 65, 98, 237
Citroën, André 48, 49
civil service, British 250–1
Claas 269–70
Clark, Greg 238
Clark, Ossie 36
CLIC Sargent 302
Clothier, Andy 168
CNBC 177
Coandă effect 185
Cockerell, Sir Christopher 14, 44, 102
Codrington family 82
Collis, Charles 154
Colombe, Paul 97
Colombo, Jo 37
Comet 129–30
Conair 111
Concert for Bangladesh 60
Concorde 77, 83–4, 278, 308
Confederation of British Industry (CBI) 287
Conran Shop 128–9, 130, 199, 297
Conran, Terence 36–7, 128–9
Conservative Party 252–3
Contrarotator, Dyson 140, *141*, 142
Cooper, John 61
core technologies, Dyson 144–93. *See also individual technology and product name*
Cornish, Evan 115
Cornish, Michael 263
Corrale, Dyson 185–6, *187*, *188*, 207
Cossens, Sir Neil 283
Couverture 316
Covid-19 pandemic 186, 215, 216, 247–55, 294, 300
Creative Industries 52–3
Crichel Down Rules 243
Critchley Light 60
crop rotation cycle 258, 262

Cryptomic technology 178, 180
Cu-Beam Duo, Dyson 138
Cure EB 302
Currys 129, 130
cyclonic separators
  Cinetic vacuum cleaner and 158
  DC07 upright cleaner and/first vacuum to have multi cyclones 147
  Diesel Trap and 143
  Dyson core technology 144, 168
  JD first uses 84, 85–6, 93
  prototypes of Dyson vacuum cleaner and 9, 11, 94–9, 107, 201, 274, 311, 317–18
  technology evolution *148*

D

Daily Telegraph 149
Daimler 60
DA001, Dyson (vacuum cleaner) 122
Davidson, Andrew 190
DC01, Dyson (vacuum cleaner) 117–43, *123*, *124*, 144, 145–6, 168
DC01 Antarctic Solo, Dyson (vacuum cleaner) 140
DC02, Dyson (vacuum cleaner) 139, *139*, 140, 145, 147
DC02 Recyclone, Dyson (vacuum cleaner) 140
DC03, Dyson (vacuum cleaner) 145, *146*, 322
DC04, Dyson (vacuum cleaner) 145, *146*
DC05, Dyson (vacuum cleaner) *146*, 147, 156
DC06, Dyson (vacuum cleaner) 147, *147*, 147, 186, 188–9
DC07, Dyson (vacuum cleaner) 147, *148*, 201
DC08, Dyson (vacuum cleaner) 156
DC11, Dyson (vacuum cleaner) 156, *157*, 158
DC12, Dyson (vacuum cleaner) 158, 164–5, *165*, 199–200
DC14, Dyson (vacuum cleaner) 158, *159*
DC15, Dyson (vacuum cleaner) 158, *159*, *160*
DC16, Dyson (first handheld vacuum cleaner) 165–7, *166*
DC35 Digital Slim, Dyson (vacuum cleaner) 167–8, *169*, *170*, 171–2
Dean, Roger 37
Decauville, Paul 260
de Gaulle, General 221
degree-awarding powers (DAPs) 288, 292
Design Centre, Haymarket 30
Design Museum 129
De Stijl 140
Devlin, Polly 56
DeVos, Richard 105
'Dieselgate' 240–1
diesel pollution 142–3, *143*, 223–5, 226, 227, 240–1, 242
Diesel Trap, Dyson 142–3, *143*
Digital Motor, Dyson 162–8, *162*, *170*, 171, 173, 180, 182, 190, 192, 199, 210–11, 225, 229
Dillon, Charles 30
Dodington Park 82, 94, 155, 263, 308
Dorman, R. G. 97, 98
Dowson, George 61
Doyle Dane Bernbach 54

drypowder carpet shampoo machine 108–9
Dry Stone Walling Association 267
Dymaxion 247
Dysolve 109
Dyson (Ltd)
core technologies 144–93
see also individual technology and product name
education and 274–306
see also education
farming and 256–73
see also Dyson Farming
future of 307–18
global reach of 194–222 *see also individual nation name*
management structure 251–2
middle-class buyers and 130
origins of 90, 91–142
philanthropy *see* James Dyson Foundation
premises *see individual area name*
private company 89–90, 307
products *see individual product name*
recruitment *see* recruitment, Dyson
website 215–17
Dyson, Alec (JD's father) 14, 15, 16, 17, 18–20, 21, 22, 24, 25, 26, 88, 204, 223, 260–1, 305, 307
Dyson, Alexander 'Shanie' (JD's sister) 15–16, 17, 19, 25
Dyson Centre for Neonatal Care, Royal United Hospital Bath 303–4
Dyson, Deirdre (JD's wife) 11, 21, 37, 43, 70, 75, 77, 122, 128, 201, 259, 268, 315, 316–17
Amway lawsuit and 114
Ballbarrow and 80, 82, 94
Byam Shaw School of Drawing and Painting 30, 31, 33–4
Dodington restoration and 308
Dyson family finances/financial risk and 11, 32, 73, 80–1, 90, 91–2, 94, 96, 108, 114, 118, 131
evening of opera arias, hosts 155
family background 32–3
James and Deirdre Dyson Trust and 302–3, 304
JD first meets 30–1
JD, on life with 325–30
marries JD 31
rug designer 32
Tube Boat and 73
Dyson Demo stores 198, 215, 216
Dyson, Emily (JD's daughter) 24, 31–2, 70, 96, 316–17, 326
Dyson Farming 256–73
anaerobic-digester power plants 271
drystone wall rebuilding 267–8
Dyson Ltd and 257–8, 271–2
electricity, making and selling of 263, 265
food sales direct from farms 263–5
greenhouses 271
'greenwash' and 257
growth of 270
landlord, Dyson Farms as 270–1
machinery, agricultural 262–3, 266, 268–70
media reaction to 270
Nocton estate 259–67, 268
origins of 258–64

polylactic acid (PLA) extraction from corn starch 272
size of farms 256
subsidies and 264–5
underinvestment and lack of maintenance in farms acquired by 258
Dyson Institute of Engineering and Technology/Dyson University 10, 134, 212, 287–95, *290*, 291, *293*, 297, 298, 304
Dyson, Jake (JD's son) 24, 31–2, 70, 96, 138, 185, 216, 295, 315, 316–17, 319–24
Dyson, James
advice, avoids giving 329
Ballbarrow and *see* Ballbarrow
birth 17
bloody-mindedness 10, 125
Brexit/European Union, opinions on 142, 143, 206, 217–22, 224, 264, 265
Byam Shaw School of Drawing and Painting 28, 29–31, 33–4, 326
childhood 13–27, 259
companies *see individual company name*
competitive nature, Dierdre Dyson on 328–9
confidence, Dierdre Dyson on 325–6
determination, Dierdre Dyson on 328
Dyson Ltd and *see individual area of Dyson Ltd*
Dyson products and *see individual product name*
Eagle painting competition (1957), wins 26
education, opinions on *see* education
everyday use, become interested in designing products for 70
family and *see individual family member name*
finances 31, 32, 47, 73, 80–1, 83, 88–9, 90, 91–2, 94, 106, 108, 110, 111, 113, 115, 118, 119, 131, 327
gap year 28
Gresham's School 13–16, 17, 18, 19, 27, 155, 259, 304–6
'Ingenious Britain: Making the UK the leading high tech exporter in Europe' (report) 284
Kirk-Dyson and *see* Kirk-Dyson
manufacturing, on British attitudes towards 49–54, 68–9, 74–6, 203–4, 217–22, 250–5, 273, 274–306
marries 31–2
media attacks upon 205, 212–13, 252–4
morality, Dierdre Dyson on 328
optimism, Dierdre Dyson on 330
philanthropy *see individual philanthropic venture name*
practicality, Dierdre Dyson on 327
Rotork and *see* Rotork
Royal College of Art, attends 22, 30, 34–46
Royal College of Art, Provost of 37, 297
Sea Truck and *see* Sea Truck
Dyson Japan 198–9
Dyson, Mary (JD's mother) 13–14, 17, 18, 19, 20–2, 24, 27, 31, 33, 61

Dyson, Sam (JD's son) 24, 31–2, 87, 91, 96, 315, 316–17
Dyson Symphony performance, Cadogan Hall, London (2018) 155
Dyson, Tom (JD's brother) 15, 17, 19, 20, 25

E

Eagle 25–6
Eames Soft Pad chair 233
Eames, Charles 37, 105
Eastern Electricity Board 126–7
ECKO 132
EcoHelmet 302
editorial coverage, value of 82–3
education 274–306
- academy schools 282–3
- Bath Dyson School project 282–5
- Bath schools, Dyson beings engineering into classrooms of 286
- Concorde engineers 278–9
- Dyson Centre for Neonatal Care, Royal United Hospital Bath 303–4
- Dyson Institute of Engineering and Technology/Dyson University 10, 134, 212, 287–95, *290*, 291, *293*, 297, 298, 304
- Dyson School of Design Engineering, Imperial College 190, 296–7
- exam setting 278–9
- Great Exhibition and 276–7
- Gresham's School and *see* Gresham's School
- 'Ingenious Britain: Making the UK the leading high tech exporter in Europe' report 284
- Innovation RCA Board 298
- James Dyson Award 280–1, 298–302
- James Dyson Building for Engineering, Cambridge University 295–6
- James Dyson Foundation and 281–2, 284, 286, 287, 289, 295, 298, 299
- learning by doing/trial and error/failing 274–5
- manufacturing, British attitudes towards and 274–306
- 'master classes', Dyson engineer 281–2
- Mechanics' Institutes and 275–6
- Millennium Experience and 277
- 'Roadie' boxes, Dyson 282
- Royal College of Art James Dyson Building 298
- Royal College of Art, JD becomes Provost 37, 297
- science and technology, British attitude towards teaching 35
- technical schools 277–8

Egyptian Special Boat Brigade 58
Eilmar, monk of Malmesbury Abbey 132
Electricity Board 126–7
Elizabeth II, Queen 38, 150, 316
EMC (electro-magnetic compatibility) chamber 209
energy-label performance data 218–20
English Electric F1.A Lightning Mach 2 RAF interceptor 137
English Heritage 283
entasis 237
Environment Agency 283

epoxy powder 83
ERM (EEC Exchange Rate Mechanism) 119
Essity AB 175
European Commission 219, 220
European Court of Human Rights (ECHR) 113
European Court of Justice (CJEU) 220
European Economic Community (EEC) 38
European General Court (EGC) 220
European Tissue Symposium (ETS) 175
European Union (EU) 142, 143, 206, 217–22, 224, 264, 265
EV, Dyson (electric vehicle) (N526) 66, 223–41
  acceleration 238
  aerodynamics 235
  air-filtration technology 234
  batteries 227, 228–9
  chassis 226
  cost 239
  dashboard 234
  digital micromirror 234
  driving position 231
  electric drive unit (EDU) 229–30
  electric motors 238
  head-up display, HUD 234
  heating and air conditioning 234–5
  length 230
  lights 235
  production halted 240
  quarter-scale models in clay 236
  range 226–7, 235
  sales, direct 239
  seats 231, *232*, 233
  site for construction of 238
  space, feeling of 231
  start-up acquisition and 229
  suppliers 239
  testing 235–6
  top speed 238
  torque 238
  turning circle 231
  tyres 231
  weight 238
  wheels 230–1
  windscreen 233–4
EVA (ethylene vinyl acetate) 78, 80, 87
experience, value of 10–11, 24, 38, 55, 197

F

failure, importance of 9–10, 11, 24, 89, 90, 95, 96, 144, 274, 280, 309
  failed hand-made prototypes of cyclonic vacuum, 5,126 9, 96
Fairey Battle bomber 132
family businesses 310–11
Fantom 108, 110–11, 113, 114
Faraday, Michael 168, 228
farming. *See* Dyson Farming
Faurecia 231
Feilden Clegg Bradley Studios 303
Fenlands 261
Fiennes, Ranulph 140
Finch, John M. 84
First World War (1914–18) 62, 71, 193, 259, 266
Flos 295
focus-group-led designing 55–6
Ford, Henry 54, 130–1, 228, 269
Foster, Norman 35, 37, 39, 131

Foxconn 70
France, Dyson in 186, 194, 197–8, 246
Friends 202
Fry, Jeremy 115, 282
  Amway lawsuit and 106
  Ballbarrow, offers to back 73, 94
  influence upon JD 54–6, 77, 88, 97, 131, 160–1, 307, 322
  JD begins to work for 41–3
  Le Grand Banc and 70
  motorised wheelchairs and 103, 104, 319
  Prototypes Ltd 101
  Sea Truck and 43–5, 47–9, 56, 60, 61–2, 70, 73, 74, 131
Fuller Brush company 101
Fuller, Buckminster 35, 39–40, 41, 62, 136, 155, 247, 283
Fuller, Thomas 283
Furst, Anton 39–40

G

Gammack, Peter 115, 119, 192, 312
  Chelsea Flower Show garden and 149
  DC01 and 125, 139
  DC12 and 199
  DC35 Digital Slim and 167, 168, 171
  Digital Motor and 161
  EV, Dyson (electric vehicle) (N526) and 235, 236, 238, 312
  Pure Hot + Cool purifier and 178
  recruited by JD 109–10, 115
  Supersonic hairdryer and 180
Garbstore, The 306
Garnett, Andy 56, 197
gas-filled shock absorbers 51
General Electric (GE) 68
Genius of Britain 135
geodesic domes 39, 40, 155, 282
Germany, Dyson in 196–7
G-Force 107, *107*, 108, 199, 200, 314, 321
Gloster Meteor 68
Good Morning America 177
Goodwood Revival 67
Google Glass 154
Gordon, Alexander 213–14
Gove, Michael 284, 286, 287
Great Exhibition (1851) 276–7
Great Smog (1952) 223–4
Great Western Railway 276
'greenwash' 134–5, 240, 257–8
Gresham's School 13–16, 17, 18, 19, 27, 155, 259, 304–6
  Dyson Building 304–5
Gresley, Sir Nigel 51
GUS (General Universal Stores) 120–1
Gustin, Daniel 83

H

Habitat (exhibit), Montreal Expo 291
Hall, Jerry 151
Hardie, George 36
Harrison, George 60
Harvey Norman 196
Hastings, Sir Max 270
Heath, Edward 74–5
Heinkel 178 68
Henry IV, King of England 100, 172
HEPA filters 145, 174–6, 178, 249
Heyerdahl, Thor 71
Hill, Jane 36

Hillman Imp 281
Hitler, Adolf 54, 67
Hockney, David 26, 36, 37
Honda
  Accord 64
  Civic 77
  50 Super Cub 28–9, 31, 199
Honey, Jim 149, 151
Hoover 68, 92–3, 100, 118, 121, 127
  Junior 92–3, 94, 95
hovercraft 14, 44, 102
Howe, Elspeth 129
Howe, Geoffrey 129
Hughes, Lucy 302
Hullavington Airfield campus, Dyson 192, 203, 212, 213, 238, 239, 242–8, *245*, 249, 255, 294
Hunt, David 120, 122
Hunt, Tony 39, 40, 101, 131, 132
hydrolastic suspension system 62

I

ICI 78
Ideal Home Exhibitions 100
Imperial College London 35, 109–10, 139, 190, 276, 288, 295, 298
  Dyson School of Design Engineering 190, 296–7
India, Dyson in 216–17
Indonesia 206, 208, 217
Industrial Revolution 50, 262–3, 275, 280, 297, 311
infrared tracking technology 303
'Ingenious Britain: Making the UK the leading high tech exporter in Europe' (JD report) 284
injection moulding 80, 145
Institute of Mechanical Engineers 53
Institution of Civil Engineers 53
International Electrotechnical Commission (IEC) 218, 220
Iona Appliances 108–9, 110, 111, 113, 116
Isle of Wight Pop Festival (1970) 72
Issigonis, Alec 26, 35, 47, 49, 60, 61, 62, 64, 66, 88, 98, 230, 231, 281

J

Jacob, Lord Justice Robin 171, 172
Jaguar 16, 61, 64
  D-Type 25
  Mark 2 61
  XJS 237
  XK8 237
James and Deirdre Dyson Trust 302–3
James Dyson Building for Engineering, Cambridge University 205–6
James Dyson Foundation 253, 281–2, 285, 286, 287, 289, 295–6, 298, 299, 302–3, 329
Japan
  DC12 designed for 164
  Dyson cyclone vacuum motor sourced from 135, 158, 160
  Dyson Japan 198–200, 202, 208, 229
  G-Force and 106, 107, 108, 110
Jaray, Paul 247
JCB 133, 308, 310
jet engine 66–9, 134, 161, 164, 279, 280, 308
JLR (Jaguar Land Rover) 238
John Lewis 124, 126, 127–9, 130
Johnson, Boris 247–8

Johnson, Jo 287–8, 289
Johnson, Scott 110
Johnson Wax 110, 195
Jones, Allen 37
Junkers, Hugo 246
Jupp, Simeon 109, 110, 115, 125

K

Kalms, Stanley 130
Karajan, Herbert von 65
Kettering, Charles 228
Kevlar 102, 282
Kier & Co. 246
Kimberly-Clark 175
King, Sir David 143, 225
King, Spen 237
King's Road, London 32, 36, 316
Kirby 101
Kirk-Dyson 73, 74–90
  Ballbarrow and *see* Ballbarrow
  cyclonic separators, first use of 84
  JD ousted from 88–90
  origins of 73, 74
  Trolleyball and 87–8
  Waterolla and 87–8, *87*
Kirkwood, Stuart 80
Kite Light 30
Kleeneze Rotork Cyclon 100–1
Klimov, Vladimir Yakovlevich 68
Knickerbocker Corporation 84
Kon-Tiki raft 71
Kuenssberg, Laura 252

L

Lagerfeld, Karl 185
Lamella hangars 246
Lamont, Brian 120–1
Land Rover 57, 60, 105, 231, 237, 238
lean engineering 23, 134–5, 258
Le Corbusier 33, 37, 129
Ledwinka, Hans 247
Lee Hsien Loong 239
Lee Kuan Yew 204
Lefèbvre, André 60, 65–6, 98
Le Grand Banc 71, 73, 104–5
Lightcycle task lamp, Dyson 216
Linacre, Edward 302
Linolite 131, 132
Linpac 115, 263
lithium-ion batteries 165, 172–3, 188, 226–7, 228, 229, 238
Littlewood, Joan 40–1
Littlewoods 120, 121, 126
Ljungström, Gunnar 60
London Fire Brigade 59
London Olympics (2012) 52
Lutyens, Edwin 244
Lux meters 303

M

MacDonald, Ramsay 244
Magès, Paul 51, 66
Maigue, Carvey Ehren 300–1
Malaysia 49, 136, 137, 202–3, 204, 205, 206, 208, 209, 210, 212, 295
Malmesbury campus, Dyson 88, 142, 149, 163, 182, 195, 202, 203, 206, 209, 255, 259, 288, 289, 291, *293*, 308, 314
  D4 building 225–6
  DIET Village *290*, 291, *293*
  D9 building 137–8
  The Hangar 137
  Lightning Café 137
  Morris Mini Minor at 64

origins of 88, 131–8, *132–3*
renewal of offices and upgrading of laboratories 212
Rolls-Royce RB.23 Welland at 66–7
management structure, Dyson 251–2
manufacturing, British attitudes towards 49–54, 68–9, 74–6, 203–4, 217–22, 250–5, 273, 274–306
Maple Tree 214
MarinaTex 302
Masura Ibuka 65
May, Theresa 217, 239
McCullin, Don 139
Mechanics' Institute 276
Meikle, Andrew 262
Meningitis Research Foundation 302
Michelin 65, 66, 231, 291
middle-class buyers, Dyson products and 130
Mig-15 fighter 68
Millennium Experience 277
Minards, Ian 237–8
Mini 26, 35, 37, 38, 49, 56, 60, 61, 63, 64, 66, 98, 137, 230, 231, 281, 308–9
Ministry of Defence 102, 242
Minoru Mori 199
MIRA (Motor Industry Research Association) wind tunnel 235
Mitsubishi Heavy Industries 296
Mittlestand (medium-size private German businesses) 306
Miyake, Issey 152–4, 200
Model T, Ford 228
MOM 302
monocoque 61, 238
Moore, Charles 149
Morris
Marina 64, 77
Mini Minor 64
Minor 20, 26, 37, 54, 64
1100 47, 62
Oxford 31
Traveller 26
Morris, Estelle 283
Morita, Akio 48
Morton, Lord 132, 134
Moulton, Alex 61–3
Moulton bike 62–3, 282
Moulton, John Coney 62
Mr Freedom shop, Kensington 36
Muranka, Tony 125
Musk, Elon 240
Myers, Bernard 35

N

Nahlin (1930s yacht) 308
NASA 77
Nasmyth, James 276
National Grid 263, 265, 271, 272
National Health Service 276
National Union of Students 288
Needham, Richard 129
Newcomen engine 135
Nicholas Hare Architects 296
Nocton estate 259–67, 268
Nori Ohga 65
Norman, Torquil 42
Notre-Dame du Raincy 247

O

O'Connor, Jim 36
Ofgem 265
oil crisis 75

oleo struts 51
Omni-glide, Dyson 208
Op Art 29
Orion Orchestra 155
Osborne, George 242, 285, 298

P

paddle wheel 44
Page, Katie 196
Page, Mike 118–19
PAM (Philippines Advance Manufacturing) 211
paper towels 173, 175, 176
Paris Motor Show (1955) 66
Paris October fashion event (2002), JD designs show for Issey Mayake 152–3, *153*
Paris store, Dyson 186, 198
Patent Office 102, 113
patent system
  Ballbarrow, JD losses patent 88–9
  cyclonic vacuum and 99–100, 102, 110–11, *112*, 113–14, 116
  Digital Slim vacuum cleaners and 171–2
  Dyson Digital Motor and 168
  inventors keeping hold of, JD on importance of 63, 67, 88, 89
  renewal fees 113, 172
  system overhaul 171–2
particle counter, aerodynamic 224
Patten, Chris 106, 288
Paxton, Joseph 271, 276
Péchot, Prosper 260
PEEK (polyester ether ketone) 163
Perret, Auguste and Gustave 246–7
Peter Jones 128, 130
Peugeot 66
Philippines 136, 137, 162, 204, 206, 208, 209, 210, 211, 295
Philip, Prince 297
Phillips, Andrew 80, 89
Phillips, Derek 149, 151, 199–200
Phillips Plastics 120, 121–2
Pike, Jeff 108
Pink Floyd 36
plagiarism 89, 171–2, 207
planning permission 41, 69, 132, 202, 239, 283, 327
Planté, Gaston 228
pollution 143, 180, 192, 207–8, 216–17, 224–5, 240
Porsche, Ferdinand 54, 269
Power Jets 68
Preece, Cardew and Rider 213
PricewaterhouseCoopers 120
Prime Minister's Business Advisory Group 287
Prior Art 172
private company, Dyson as 89–90, 307
Prototypes Ltd 101
Pure Hot + Cool, Dyson *179*, 180
Purser, Toby 155

R

Race Against Dementia 302
Raleigh 63
Range Rover 77, 230, 231, 237
Ransome, Arthur: *Swallows and Amazons* 25
Raspberry Pi microcomputer 155
recruitment, Dyson
  Brexit and 216–17
  Dyson Institute and 314–15

Malaysia and Singapore 202–3, 205, 210, 239
Malmesbury factory and 134, 136, 195
Modern Languages graduates 196–7
motor and motor drive experts 161, 163
Peter Gammack and Simeon Jupp 109–10
problems with 195, 239
RCA Design Engineering course and 297
recruitment agencies and 198
Reliance Controls 39
remote working 254
Renault 5 77, 104
Rennie, John 261
Richardson, Tony 42
Rickaby, Caroline 27
Riley, Bridget 29
Rizzuto, Lee 111
'Roadie' boxes 282
Robb, Douglas 304
Roberts, James 302
Roberts, Tommy 36
Robinson, Derek 75
robotics 23, 192, 214, 242, 255, 257, 271, 273, 275, 279, 288, 304
DC06 vacuum cleaner 147, 186, 188–9
factory production lines and 210–11
360 Eye robot vacuum cleaner 189–90, *190*, *191*
360 Heurist robot vacuum 190
Roche, David 156
Rogers, Richard 35, 39, 131, 277
Rolls-Royce 38, 39, 66–9, 133, 164, 247, 296
Aero Engines 134
Merlin piston engine 67
Nenes 68
RB.23 Welland 66–7
Ronald Ward and Partners 41
Rootes Group 281
rotational, or blow-moulding 80
Rotork 74, 77, 197, 282
JD takes job at 41–4, 326
licenses Dyson vacuum cleaner 100–1, 105
Sea Truck 43–6, 47–73, *57*, *58*, *59*, *72*, 82, 101–2, 131, 135, 137, 216, 320
Wheel Boat and 44, 102–3
Roundhouse, Chalk Farm 42
Rover 68
Royal Academy 37
Royal Air Force (RAF) 15, 17, 67–8, 69, 82, 132, 133, 137, 242, 243, 244, 246, 247
Royal College of Art (RCA) 10, 22, 26, 109–10, 131, 139, 276, 280–1, 295, 296–7, 320
James Dyson Building 298
JD as Provost 37, 297
JD attends 30, 34–46, 47, 48, 56, 74, 85, 297, 326
Innovation RCA Board 298
Royal Fine Art Commission 244
Royal Institute of British Architects 53
Royal Navy 58
Royal Society 64
Royal Yacht *Britannia* 58
rubber suspension 61–3, 64

Rubinstein, Leopold 33
Rumbelows 124, 126
Rutter, Mike 100

S

Saab 60, 247
Safdie, Moshe 291
Sakti3 229
Sason, Sixten 60, 247
Saturday Night Live 201
Sausmarez, Maurice de 29–30, 34
Sayer, Malcolm 237
Scanning Electron Microscope and Hair Mapping Analysis 206–7
Scarfe, Gerald 26
Schrader valve 78, 80
Science Museum 276, 278, 296
Scott, Ridley 36
Scottish Hydro 127
Scottish Power 127
Sea Truck, Rotork *57*, *58*, *59*, 61, 70, 71, 73, 137, 320
  JD and sales of 57–60, 80, 82, 101, 131, 216, 320
  JD asked to engineer 43–6, 47–8, 56–7, 135
  origins of 43–6
  Wheelboat and 101–2
Sears 109
Sebo 127–9
Second Law of Thermodynamics 35
Second World War (1939–45) 14, 16, 17, 26, 32, 36, 50, 51, 52, 62, 67, 68, 75, 83, 132, 204, 213, 243, 244, 266, 276, 277, 296, 308
Sedgeley, Peter 29
seed drill 262
semi-anechoic chambers 183, 209
Shiffer, Isis 302
Siemens 100, 296
Silver Seiko Ltd 107
Singapore 49, 62, 136, 137, 162, 163, 182, 190, 192, 203–6, 221, 229, 238–9, 248, 249–50, 294, 295
  Alexandra Technopark, Dyson premises at 210–11
  Dyson global headquarters move to 211–15, *214–15*
  SAM (Singapore Advance Manufacturing) 210–11
  Singapore Technology Centre 211
  St James Power Station campus 137, 213–15, *214–15*
  vacuum cleaner production, Dyson moves to 202–6, 208–9
Slater, Jim: *The Zulu Principle* 98–9
Slater, Mark 98
Smith, Paul 108, 316
Smith's Crisps 260–1
Snow, C. P.: 'The Two Cultures' 34–5
Snowdon, Lord 101, 103, 105, 297, 319
Soichiro Honda 28, 29, 48
solar panels 242, 272–3, 300
Sony Pressman 65
Sony Walkman 48, 63–5, 199
Souhami, Mark 130
sourcing 69–70, 120–2, 194–5, 249–50
South Bank, London 38
South Korea 206, 208
Spitfire 67, 82, 173, 247
Squirrel four-wheel drive power-steered wheelchair 101, 103–5, *104*
Stalin, Joseph 68
steam traction engine 260, 263

Stephenson, Robert 276
St James Power Station campus, Singapore, Dyson 137, 213–15, *214–15*
Stokes, Donald 64
structural engineering 40, 131
Studio Himmelb(l)au 52
Sunday Times 212
Supersonic hairdryer, Dyson 180–2, *183*, *184*, 185, 206, 211
Swiss Army penknife 60
switch selling 127

T

Taiwan 206
Takeo Fujisawa 28, 29
Tatra cars 247
tax 114, 211–12, 252, 270, 285
technical schools 277–8
Tesla 226, 240, 241, 282
Thatcher, Margaret 91
Theatre Workshop 40
thermoplastic 71, 127, 163
Thoburn, William 294–5
Thomson, Alice 270
Thomson, Andrew 139
Three-Day Week 74–5
360 Eye robot vacuum cleaner, Dyson 189–90, *190*, *191*
360 Heurist robot vacuum, Dyson 190
Times, The 270
Toio floor lamp 295
Tokyo Design Museum 154
Tomorrow's World 82–3
Townshend, Charles 'Turnip' 262
tractor 263, 268
trade unions 74–5, 278
transverse car engine 60
Triodetic structural system 41
Trolleyball 87–8
Tube Boat 71, *72*, 73, 101
Tull, Viscount Jethro 262
Type E hangars 244, *245*, 246
Tyre Collective, The 301–2

U

United States, Dyson in 200–2
United States Environmental Protection Agency 240
University of Bath 196
University of Bradford 175
University of California Irvine 300
University of Cambridge 62, 120, 133, 155–6, 204, 279, 288, 289, 298
  Fluid Dynamics, Dyson sponsor professorial chair in 296
  James Dyson Building for Engineering 295–6
  JD's family and 15, 16, 17, 20, 21, 33
  Whittle Laboratory 164, 296
University of Minnesota 224
University of Southern Florida 59
US Bureau of Mines 224

V

vacuum cleaners, Dyson cyclonic 23–4, 47, 89
  advertising 125, 127
  Amway and 91, 105–6, 109, 111–14, *112*, 119, 321
  'bagless' vacuum name 125–6
  brush bar 120, 144–5, 147, 156, 192–3, 208
  Chippenham factory 50, 62, 69, 128, 129, 131, 132

cost of 125, 130
cyclonic separators, JD first uses 84
DA001 122
DC01 117–43, *123*, *124*, 144, 145–6, 168
DC01 Antarctic Solo 140
DC02 139, *139*, 140, 145, 147
DC02 Recyclone 140
DC03 145, *146*, 322
DC04 145, *146*
DC05 *146*, 147, 156
DC06 147, *147*, 186, 188–9
DC07 147, *148*, 201
DC08 156
DC11 156, *157*, 158
DC12 158, 164–5, *165*, 199–200
DC14 158, *159*
DC15 158, *159*, *160*
DC16 (first handheld vacuum cleaner) 165–7, *166*
DC35 Digital Slim 167–8, *169*, *170*, 171–2
Dual Cyclone technology 148, *148*
Dysolve and 109
Dyson Digital Motor and 162–8, *162*, 199
electric motor, Japanese 1400W 135, 160–1
Fantom model 108, 110–11, 113, 114
G-Force model 107, *107*, 108, 199, 200, 314, 321
Kleeneze Rotork Cyclon 100–1
licensing of design 100–1, 105–15
mathematics, calculation of cyclone 97
origins of 91–4
patents 99–100, 102, 110–11, *112*, 113–14, 116
production line 126
prototypes 9, 11, 94–9, 201, 274, 311, 317–18
range of, first 130–1
retailers, first 124–9
robotic vacuum cleaner 188–9
RootCyclone technology 148, *148*
Rotork, licensing to 100–1
sales, first 122–5
south-east Asia, production moved from England to 202–3
360 Eye robot vacuum cleaner 189–90, *190*, *191*
360 Heurist robot vacuum 190
torture course 136
2 Tier Radial cyclones 149, *149*
UK vacuum cleaner market share 128
V11 cordless hand-held vacuum cleaner 216
weight of, reducing 135–6
Zorb-It-Up and 109, 121
Van Andel, Jay 105
Vax 111, 115–16
V11 cordless hand-held vacuum cleaner, Dyson 216
velodrome, see-through 199–200, *200*
ventilators 247–55, *251*
venture capitalists 73, 93, 117–18, 299
Vickers/Vickers Tower 41
Victoria and Albert Museum 296
Victoria Line 38–9
Volkswagen 64, 240–1, 269
Beetle 26, 54
Golf 77
Vorwerk 101

W

Waterolla 87–8, *87*
Watt, James 135
Wealleans, Jon 36, 37, 109, 121
Webber, Major 'Jock' 259–60
website, Dyson 215–17
Welsh Development Agency 119–20
Wentworth, Richard 30
Westonbirt School 282
Weybourne (holding company of Dyson operations) 98
Wheel Boat 44, 101–2
wheelbarrows 70, 73, 77, 78, 81, 88. *See also* Ballbarrow
Wheeler, Schuyler 177
Whittle, Sir Frank 66–9, 133–4, 164, 279, 296, 308, 313
Whitworth, Joseph 276
Whitworth Scholarships 276
Wilkinson, Chris 131, 132–3, 137–8, 197, 203, 243, 283, 291–2, 304–5
Wilkinson, Diana 133
WilkinsonEyre *132–3*, 225–6, 244, *290*, *293*
Williams Racing 235
Williams, David 115
Wilson, Harold 75
Wilson, Jim 18
Wiltshire Engineering Festival 253
Wimbledon School of Art 34
wind turbines 272
Winter of Discontent (1979) 91
Witherow, John 270
WMG (Warwick Manufacturing Group) department at the University of Warwick 289
Wolfson, Isaac 120–1
Wood, Leslie Ashwell 26
Wood, Sir Martin 14
Worboys, Nick 263
Wrong Garden 149–50, *150*, 151
Wyatt, James 82, 308

Y

Yes (band) 37
Yom Kippur War (1973) 58, 75
Young, Arthur M. 137–8

Z

Zanussi 100, 119
Zorb-It-Up 109, 121